T0302198

A Transition to Proof
An Introduction to Advanced Mathematics

Textbooks in Mathematics
Series editors:
Al Boggess and Ken Rosen

AN INTRODUCTION TO NUMBER THEORY WITH CRYPTOGRAPHY,
SECOND EDITION
James R. Kraft and Lawrence Washington

MATHEMATICAL MODELING: BRANCHING BEYOND CALCULUS
Crista Arangala, Nicolas S. Luke and Karen A. Yokley

ELEMENTARY DIFFERENTIAL EQUATIONS, SECOND EDITION
Charles Roberts

ELEMENTARY INTRODUCTION TO THE LEBESGUE INTEGRAL
Steven G. Krantz

LINEAR METHODS FOR THE LIBERAL ARTS
David Hecker and Stephen Andrilli

CRYPTOGRAPHY: THEORY AND PRACTICE, FOURTH EDITION
Douglas R. Stinson and Maura B. Paterson

DISCRETE MATHEMATICS WITH DUCKS, SECOND EDITION
Sarah-Marie Belcastro

BUSINESS PROCESS MODELING, SIMULATION AND DESIGN,
THIRD EDITION
Manual Laguna and Johan Marklund

GRAPH THEORY AND ITS APPLICATIONS, THIRD EDITION
Jonathan L. Gross, Jay Yellen and Mark Anderson

A FIRST COURSE IN FUZZY LOGIC, FOURTH EDITION
Hung T. Nguyen, Carol L. Walker, and Elbert A. Walker

EXPLORING LINEAR ALGEBRA
Crista Arangala

A TRANSITION TO PROOF: AN INTRODUCTION TO ADVANCED
MATHEMATICS
Neil R. Nicholson

A Transition to Proof
An Introduction to
Advanced Mathematics

Neil R. Nicholson

CRC Press
Taylor & Francis Group
Boca Raton London New York

CRC Press is an imprint of the
Taylor & Francis Group, an **informa** business

CRC Press
Taylor & Francis Group
6000 Broken Sound Parkway NW, Suite 300
Boca Raton, FL 33487-2742

Printed on acid-free paper
Version Date: 20190208

International Standard Book Number-13: 978-0-3672-0157-9 (Hardback)

Library of Congress Cataloging-in-Publication Data

Names: Nicholson, Neil R., author.
Title: A transition to proof : an introduction to advanced mathematics / Neil R. Nicholson.
Description: Boca Raton : CRC Press, Taylor & Francis Group, 2018. | Includes bibliographical references and index.
Identifiers: LCCN 2018061558 | ISBN 9780367201579 (alk. paper)
Subjects: LCSH: Proof theory.
Classification: LCC QA9.54 .N53 2018 | DDC 511.3/6--dc23
LC record available at https://lccn.loc.gov/2018061558

Visit the Taylor & Francis Web site at
http://www.taylorandfrancis.com

and the CRC Press Web site at
http://www.crcpress.com

To my mom and dad,
for believing in me since day one.

To Elizabeth and Zeke,
for love and support every day.

And to every cat I've ever known.

Contents

Preface

Why write another textbook aimed at the student "beginning" the study of theoretical mathematics? I imagine my answer is not too different from many other math textbook authors when asked why they are writing another book about *fill in the blank*. Having taught this course many times over, I have developed my approach for the material, tested, tweaked, tested again, adjusted and fine-tuned semester-after-semester. Certain topics in one textbook fully support part of my approach, while others in another text fulfill a few of my other goals. Collections of handouts fill in the remaining gaps, rounding out what I feel is *my course* taught in *my style* emphasizing what I feel are the important aspects of a bridge-to-higher-math course.

All of that work has led me to this place and this text. Is it better than others? Hardly. There are fantastic textbooks on the market introducing the up-and-coming mathematician to theoretical math. Each comes with its own flavor and style, and those texts may speak to a certain instructor's approach. By all means, those instructors can effectively teach this material in their own manner and if that approach is more systematic, axiomatic, inquiry-based or through different subject areas, then this text is not for them.

So who then is this textbook for? Rather than describe the intended instructor who would use this book, because this text is written for students, perhaps it is better to describe the outcomes that you, a *student*, could expect to obtain and why those outcomes are important in your development as a mathematician.[1] If you are the instructor and these outcomes align with your goals for your course, then this book may be the right book for you.

First and foremost, you will *understand mathematical proof* in all its contexts: why proofs are necessary, when to use them and how to write them. Learning mathematics is a journey and you began the trip years ago. As you come to different intersections along the way, it is not just important to know where roads take you, but it is critical to know where those roads came from and why they have led to the point at which you currently stand. You have to understand the entire landscape.

As you learn to write proofs, you will learn to *develop your mathematical voice*. There is not a single correct way to prove a mathematical result. Think of it like creating a persuasive paper. How you piece it together, employ mechanical techniques or choose language make the writing *yours*. You will learn

---

[1] The "why they are important" is debatable; the views expressed here were chosen by the author for emphasis.

the required fundamentals of proofs, but from there, the approaches vary as to how you create final drafts.

However. . . there is a lot that goes into creating a mathematical proof before you actually get around to writing it. So, in addition to actually writing correct and personalized proofs, ample discussion of how to figure out the "nuts and bolts" of the proof takes place: *thought processes, scratch work* and *ways to attack problems.* You will learn not just how to *write* mathematics but also how to *do* mathematics. Putting these two components together will allow you to *communicate mathematics effectively.*

There must be some vehicle for which to learn these aspects of proof. Here, different concepts from abstract and discrete mathematics play that role; you will be exposed to *fundamental definitions and results* from multiple areas of mathematics. The choice of topics to include in this text was purposeful; symbolic logic, sets, elementary number theory, relations, functions and cardinality are included because they appear throughout mathematics. In learning the theory of these areas, you create a firm foundation for later studies. Additionally, the final chapter of the text introduces point-set topology. It is included to show you that "high level" mathematics is often grounded in basic ideas (in this case, sets and their properties).

The material is developed and presented in a systematic fashion: easier to more challenging. You may find early concepts in chapters "simple." This is intentional. By the end of the chapter, proofs are presented with little scratch work, and at times, are quite challenging to work through. However, if you work through the material as it is presented, you will develop *confidence* in your abilities and build *rigor* into your approach. A strong mathematician does not simply muddle through the "easy" material; a mathematician attacks the "hard" questions *for the love of the puzzle.* It is my intent for this puzzle-solving love to grow as you proceed through this book.

To the student

The pure mathematician is often asked, "What is that used for?" To that I respond bluntly: "I do not know nor do I care." In no way is this meant to belittle the questioner. The question is rightfully fair; why do something if there is not a specific end-goal in sight?

In 1843, William Hamilton wanted to generalize the complex numbers to three dimensions. Three dimensions turned out to be one dimension too few; Hamilton had to generalize the complex numbers into *four* dimensions, creating what became known as the *quaternions* [18]. While the quaternions were mathematically "nice," what could the use of something in four dimensions possibly be? Perhaps Hamilton was a visionary, because it turns out that quaternions are fundamental in making every animated movie or video game. To look realistic, rotating objects and dynamic lighting effects rely on the four-dimensional mathematical objects. Hamilton did math *for math's sake*; the direct use of his work is merely a side effect.

In short, do not approach this material by asking, "What will this be used for?" It is all part of the larger mathematical picture. Someday, if you choose to directly apply mathematics to an outside field, the tools you learn here will surely be necessary. But if you travel the pure mathematical route, then you are learning the machinery necessary to *do* higher mathematics. This is ultimately the reason for learning the ideas you are about to learn.

Knowing where this material will take you, it is perhaps necessary to lay out some basic assumptions, the *prerequisites* for using this book. It is assumed that you will have seen at least two semesters of calculus. It is not the material of those courses that is necessary, though, for understanding this book. It is the development of a certain "mathematical sophistication" that is needed. You ought to be able to read mathematics, follow mathematical arguments and connect-the-dots across multi-step problems. If you have not seen this calculus material but are mathematically astute, you will be able to follow nearly all of this text (with the exception of a few calculus-based exercises).

Supposing you have read this far and are prepared for the course, you may be wondering how to be successful in working through the material. Part of that success comes from how you have been successful in previous mathematics courses: *working problems.* You cannot learn mathematics without doing mathematics; the more problems you work, the better you will become at attacking similar sorts of questions. However, if this is your first taste of theoretical mathematics, then you may not be aware that there is more to learning this material than simply doing problems. You have to *understand* the theory in order to *apply* the theory. What does this mean? You need to spend time *thinking* about the concepts. This is more than just memorization, however. It involves understanding the concepts *deep down inside* and developing an intuitive sense for definitions and results.

This understanding of the material forms the foundation for your success in the course. Once obtained, you can move on to *implementing* the material through the aforementioned problem-working. The problem sets provided are intended to be at times routine and at other times a serious challenge. Work the problems, assess your solutions, tweak, adjust and ultimately craft your final draft. Spend *time* on the problem solving process; a final draft of a proof should not appear quickly. The final draft is the "cremè de la cremè" of your efforts; think of it as the capstone of the learning pyramid. The final draft allows you to put all the pieces together and show mastery of the material.

It is in this capstone-step of the process that you can incorporate your style and direct your solution for *particular audiences.* The audience for whom you write guides how you write. This awareness, and how your style is dependent upon the audience, is one of the last steps to polishing your technique.

To the instructor

You are aware that your students will learn to *construct* and *write* proofs.

The topics covered are likely of particular interest to you. Consider the following road map to the text.

- Chapters 1 − 2: Logic and sets are the "building blocks" for the proof techniques presented here. An understanding of the terminology, notation and language is essential for the later chapters. Logical deductions (Section 1.4) are intended to be a "taste" of mathematical proof. Bookend these with statements and arguments with multiple quantifiers (Section 2.4) and the necessary preliminaries for proof techniques are formed.

- Chapters 3 − 4: Chapter 3 is *the* introduction to a multitude of proof methods: direct, set element, contrapositive, contradiction and cases. Every section of this chapter is critical for the remainder of the text, as are the first two sections of Chapter 4 on induction. Not only are the techniques presented in these sections important, but basic number theoretic concepts (definitions and results) appear here. These are used throughout the rest of the text as well. Section 4.3 includes material used in later sections (such as the Quotient-Remainder Theorem and results on primes). For the inclined instructor, this section could be supplemented with additional materials from number theory.

- Chapter 5 − 8: The second half of the text is the "putting proof techniques to work" part of the book. Sections 5.1 and 5.2, along with all of Chapter 6, should be covered. All remaining sections are optional and can be covered completely, or if time is limited, can be covered by picking-and-choosing the major ideas from each section. Section 5.4 and Chapter 8 open the door to further exploration beyond this textbook, if so desired.

Whatever sections you choose to cover in the text, proceed linearly, following the order presented. Various results and exercises throughout each section call upon items from previous sections. The homework exercises are designed to vary, from routine (building a basic understanding of concepts) through challenging (introducing new concepts and applying them to ideas from the section). The solutions and hints provided at the end of the text should be considered "bare bones." In particular, proofs given in the solutions section are *not* well-written. They are designed to simply show "how the proof works."

Acknowledgments

A multi-years long process cannot occur without the support of friends, family and colleagues. The love and support (and *patience*) of Elizabeth, Zeke, and my parents is unrivaled. Hours tucked away staring at a computer screen, checking for typos (of which there are some still lingering, and for that I apologize) or simply hearing me talk about "where I'm at" with my book ... I am grateful and I am lucky.

One feeling I have hoped to convey throughout this book is that *math is fun*. Becoming a mathematician is not simply a career choice. It is who you become: how you think, how you act and what types of jokes you tell. I decided I wanted to be "this way" in large part due to the mathematicians at Lake Forest College. Many, many thanks for their inspiration and guidance.

It is tough to classify my North Central College colleagues simply as such. They are friends and mentors who push me to be a better version of myself, both inside and outside of the classroom. Katherine, Karl, Marco, Mary, Matthew, David, Rich: teachers, scholars and wonderful human beings, in the finest sense of the words. And to the large number of students at North Central College who have helped me hone my skills in teaching this material: you will never know how much you have actually taught me. One can only improve at something through making mistakes. I would be remiss without thanking them for pointing out such flaws and suggesting ways to improve.

Lastly, an utmost sincere and heartfelt appreciation goes out my contacts at CRC Press: Bob Ross, Jose Soto, Karen Simon and her editing team and the multitude of reviewers who provided wonderful suggestions. One could not ask for more supportive and knowledgeable guides through the book-writing process.

Dr. Neil R. Nicholson
Department of Mathematics and
Actuarial Science
North Central College
Naperville, IL 60540
nrnicholson@noctrl.edu

1

Symbolic Logic

Logical reasoning is foundational to every field of study. New knowledge is created by drawing conclusions about certain phenomena. In the social sciences, for example, the behavior of a small, specific part of a population may be observed, and then, these observations are generalized to form a hypothesis about the entire population. To test that theory, other specific groups within the larger population are observed to see if the hypothesis is accurate. Should any of these additional test cases yield results that do not agree with the original generalization, the social scientist must change or reject the hypothesis.

This approach of taking the *specific to the general* is what is called **inductive reasoning**. It is fundamental to every scholar, even the mathematician. Much mathematical research is accomplished by looking at specific cases and generalizing those to some basic theory about *all* cases. But the mathematician knows that if something is true for *some* things, that same thing need not be true for *all* things. An example of this comes from one of history's more famous mathematicians: Pierre de Fermat (1601-1665).

Though he made notable contributions in numerous areas of mathematics, the Frenchman Pierre de Fermat is perhaps most highly regarded as the father of modern number theory (the study of properties of the integers). We will see some of his results in later chapters, but it is his conjecture regarding prime numbers that serves us well for this introduction to logic. Fermat claimed to have found a generating function for primes: for any positive integer n, the number $2^{2^n} + 1$ is prime [32].

A quick check for the first few values of n yields easy-to-verify integers: 5 ($n = 1$), 17 ($n = 2$), 257 ($n = 3$), and 65,537 ($n = 4$). When we let $n = 5$, we obtain the integer 4,294,967,297. Today, a computer can quickly check if this number is prime, but in the 17th century, this would be quite the task (for fun, check, by hand, to see if any of the primes between 2000 and 3000 divide the number; when you've finished, you'll only have about 6400 more primes to check!). Yet, even if we were to see that it *is* prime, would that make *every* number of the form $2^{2^n} + 1$ prime? Certainly not; if we randomly chose a few larger values of n and found that they made $2^{2^n} + 1$ prime, we still could not claim that $2^{2^n} + 1$ is prime for all choices of n.[1]

How then does the mathematician validate such claims? Rather than reasoning from the specific to the general, she reasons from the *general to the*

---

[1] It turns out that Fermat's conjecture was wrong. It took nearly a century until Leonhard Euler (1707-1783) showed that $4,294,967,297 = 641 \times 6,700,417$ [17].

specific, a process known as **deductive reasoning**. It is fundamental to all of mathematics and is the basis for mathematical proof. No mathematical theory is deemed as true unless it is proven. In order to construct these proofs, we must understand the rules of logic. But where did this concept of treating logical thinking as its own study begin?

It is the mathematician and philosopher Aristotle (384-322 B.C.), a student of Plato's Academy in ancient Greece, who is regarded as the founder of logic [25]. He investigated the laws of reasoning in everyday language. Nearly two millennia later, Gottfried Leibniz (1646-1716, of calculus fame) sought to formalize this reasoning into a symbolic language [3]. It was this idea of his that spurred deeper investigations by Augustus De Morgan (1806-1871) in his work *Formal Logic* [35] and George Boole (1815-1864) in *The Mathematical Analysis of Logic* [5]. De Morgan and Boole took the symbolism of Leibniz and introduced a system of algebra on it to form what has become known as *symbolic logic*. It is the goal of this chapter to present these tools of symbolic logic that will support our discussions into mathematical proofs.

1.1   Statements and Statement Forms

Charles Lutwidge Dodgson, known by his pen name Lewis Carroll, is perhaps best known for his fictional pieces *Alice's Adventures in Wonderland* [10] and *Through the Looking-Glass* [9], yet he was a noted mathematician and logician [13]. Consider the following puzzle from his *Symbolic Logic* [8].

Example 1.1. If we assume the following four sentences as fact, then a conclusion can be made. What is that conclusion?

(1) None of the unnoticed things, met with at sea, are mermaids.

(2) Things entered in the log, as met with at sea, are sure to be worth remembering.

(3) I have never met with anything worth remembering, when on a voyage.

(4) Things met with at sea, that are noticed, are sure to be recorded in the log.

If we try to piece together all of the statements, it quickly becomes quite perplexing. It turns out that the conclusion to Carroll's puzzle that can be made is, "I have never met with a mermaid at sea."

It is easy to see that everyday language is riddled with confusion. Placement of phrases within sentences suddenly matters a great deal and word choice is of the utmost importance. For example, consider the very simple sentence, "At any given moment, there are two points on opposite sides of Earth that have the same temperature." Does this sentence have the same meaning as, "There are two points on opposite sides of Earth that, at any given moment, have the same temperature," or, "There is a moment for which two points on opposite sides of Earth have the same temperature?" It turns out that all three of these sentences have very different meanings.[2]

In the real world, we often have to consider numerous facts simultaneously in order to draw a single conclusion, such as in Example 1.1 (though you may never find yourself meeting mermaids and keeping a log while on a voyage at sea). How can we wade through the complexities of the English language? Is there a way to simplify the process, eliminating the variety of ways to phrase a single thought? Symbolic logic does just this. It uses symbols to aid in reasoning, and the building block of all symbolic logic structures is a *statement*.

Definition 1.1. A *statement* is a declarative sentence that is either true or false but not both.

Notice the key words in the previous definition. A statement must be a *declarative* sentence. Interrogative sentences, exclamations or commands are not statements. Moreover, a statement must *always* be true or *always* be false. It cannot change over time, and it must be one or the other, not both.

Example 1.2. The following are all statements.

(1) Theodore Roosevelt was born on October 27, 1858.
(2) LaTeXis a typesetting system for creating high-quality scientific documents.
(3) The moon is made of cheese.
(4) $12 \times (5 + 6) = 132$
(5) The real-valued function $f(x)$ is continuous at the real number a if

$$\lim_{x \to a} f(x) = f(a).$$

(6) The function $f(x) = \dfrac{1}{x}$ is continuous at $x = 0$.

Each of these is a declarative sentence that is always true or always false.

---

[2]The original statement is actually true; it is a consequence of a famous topology result named the Borsuk-Ulam theorem. The basics of topology serve as a capstone for this text and appear in Chapter 8.

Statements (1) and (2) of Example 1.2 are true. Theodore Roosevelt was indeed born on October 27, 1858, and LaTeXis a common tool for typesetting scientific (and in particular, mathematical) content. Upon reading (1.2) of Example 1.2, you may have thought, "I know the moon is not made of cheese!" The fact that it is a statement does not depend upon whether or not it is actually true. In this case, (3) just so happens to be a false statement.

Even though (4) of Example 1.2 does not appear as a "properly written sentence in grammatically correct English," it is a statement. It just as easily could have been written expositorily: "Twelve times the quantity five plus six equals one hundred thirty two." The choice to use a shorthand method does not prohibit it from being a statement.

The last two statements of Example 1.2 come from calculus. You should notice that (5) is the definition of a function $f(x)$ being continuous at a particular real number a. There is no dependence upon "if $f(x)$ is this function and a is this value" for (5) to be or not to be a statement. Simply put, this always will be the definition of continuity. Thus, it is a true statement. Statement (6) is false. The function $f(x) = \frac{1}{x}$ has an asymptotic discontinuity at $x = 0$. As with (3), the fact that this sentence is false does not prohibit it from actually being a statement, however.

Example 1.3. The following are not statements.

(1) Make your bed.
(2) Did you see the rainbow?
(3) $x^2 + 3x + 2 = 0$.
(4) The function $f(x)$ is not continuous at $x = 1$.

While it is obvious why (1) and (2) of Example 1.3 are not statements (they are not declarative sentences), it may not be as obvious why (3) or (4) are not. In (3), depending on what value the variable x takes in the expression, the equation may (such as $x = -1$) or may not ($x = 1$) be true. A statement must always be true or always be false. This same reasoning holds for (4) and the choice of function $f(x)$. In Section 2.3 we will develop a method for creating statements from these sorts of mathematical expressions.

Similar to this use of variables in mathematical expressions is the use of open-ended language in everyday English. The following example lists two sentences that are also not statements because they do not have a fixed truth value.

Example 1.4. The two sentences below are not statements.

> (1) This year is a leap year.
> (2) She aced her exam!

Sentence (1) of Example 1.4 would be true if it were said in 2012 or 2016. In 2017, it would be false. Because it does not have one fixed truth value, it is not a statement. Similarly, (2) is open-ended in that we do not know who "she" is. Generally, this is not a statement. However, if in context (such as a specific class), "she" is a fixed person, then this is indeed a statement. To avoid this confusion, we will assume throughout the text that such context is assumed and sentences like, "I am a cat owner," and "They live down this street" are statements.

At this point, you should be questioning the English language. Is this really a statement? Is it not a statement? For our purposes, we brush aside these concerns and introduce a mathematical notation to solidify our approach to logic. We will denote particular statements with variables. For example, let P be the statement, "The chemical symbol for water is H_2O." We know P is true. However, if one were to vaguely introduce a statement by saying, "Let Q be a statement," then we do not know if Q is true or if Q is false. We simply know it represents *some* statement but we do not know which one. Expressions like "Q" are called *statement variables*, though we will refer to them simply as statements.

Then, suppose P and Q are statements. Linguistically we can combine them, with the use of certain conjunctions to create new sentences that themselves are statements. The expressions "P and Q" or "P or Q" are part of our everyday vocabulary and we understand their meaning. For example, "The chemical symbol for water is H_2O and the chemical symbol for nitrogen is N," is itself a statement, but it can also be thought of as two separate statements joined via the conjunction "and." This is an example of forming new statements from old by using *logical connectives*. The terms "and" and "or" are our first examples of these. Expressions that are formed by combining statement variables with logical connectives are called *statement forms*, though as with statement variables, to avoid confusion, we refer to them simply as statements.

Definition 1.2. Given statements P and Q, the *conjunction of P and Q* is the statement denoted $P \wedge Q$, pronounced "P and Q," that is true precisely when both P and Q are true and is false when at least one of P or Q is false.

Because every variable present in a statement form can take one of two truth values, a statement form with n different statement variables in it will have 2^n possible truth assignments. To list these out in a systematic and visual manner, we introduce the notion of a *truth table*. They provide an equivalent method for defining logical connectives.

Definition 1.3. A *truth table* for a statement form P with n statement variables $P_1, P_2, \ldots, P_n$ is a table with 2^n rows, consisting of columns for, at the minimum, each of the statement variables P_i and for P. Each row of the table consists of a unique assignment of a specific truth value (true or false) to the statement variables, and the resulting truth value of P.

Example 1.5. For statements P and Q, $P \wedge Q$ is the statement defined via the following truth table.

P	Q	$P \wedge Q$
T	T	T
T	F	F
F	T	F
F	F	F

Definition 1.4. Given statements P and Q, the *disjunction of P and Q* is the statement $P \vee Q$, pronounced "P or Q", that is true precisely when at least one of P or Q is true and is false when both P and Q are false. It is exhibited via the following truth table.

P	Q	$P \vee Q$
T	T	T
T	F	T
F	T	T
F	F	F

In the English language, the conjunction "or" can considered one of two ways, *inclusively* or *exclusively*. A sports coach may be told, "You will keep your job if your team wins their division or wins 100 games." Would the coach be fired if his team wins 104 games and then proceeds to win the division championship? Of course not! This is the idea of "or" being used in the *inclusive* sense. Alternatively, at a restaurant, upon ordering an entrée, the server may present you with the option of a side: soup or salad. You are allowed to choose one single side but will be charged extra if you choose both. This is "or" being used *exclusively*.

Note that Definition 1.4 is defining "or" in the inclusive sense. Any mathematical use of the word "or" or the symbol "∨" is meant to be inclusive.[3] When exclusivity is required, it will be explicitly stated, as in the following definition.

Definition 1.5. Given statements P and Q, the *exclusive disjunction of P and Q* is the statement $P \veebar Q$, pronounced "P exclusive or Q", that is true when exactly one of P or Q is true and is false when both P and Q are false or both P and Q are true. It is exhibited via the following truth table.

P	Q	$P \veebar Q$
T	T	F
T	F	T
F	T	T
F	F	F

The last basic logical connective is perhaps the simplest: the negation of a single statement.

Definition 1.6. Given a statement P, the *negation of P* is the statement $\sim P$, pronounced "not P", that is true when P is false and is false when P is true. It is exhibited via the following truth table.

P	$\sim P$
T	F
F	T

These four logical connectives allow for the creation of more complicated statement forms from previously defined statement forms or variables. To clarify this language, consider Example 1.6.

Example 1.6. Consider the following.

(1) P is a statement variable in the statement form $\sim P$.

(2) P and Q are statement variables in the statement form $P \wedge Q$ and $P \vee Q$.

(3) The statement $P \wedge \sim(\sim Q \vee \sim R)$ is a statement form with variables P, Q and R.

---

[3] A classic mathematical joke is based on this idea: A mother has just given birth to a baby when she turns to the baby's father, a mathematician, and asks, "Is it a boy or a girl?" He responds, "Yes!"

As previously mentioned, however, our use of the term "statement form" will be rare. Simply calling $P \wedge \sim(\sim Q \vee \sim R)$ a statement serves our purposes well.

Armed with just these four logical connectives, we can construct truth tables for countless statements. Grouping symbols such as parentheses may be used, but as with arithmetic operations, there is an "order of precedence" for logical connectives. Of highest precedence is $\sim$ while $\wedge$, $\vee$, and $\veebar$ have the same precedence, taken left-to-right (similar to $+$ and $-$ in the order of arithmetic operations). Thus,

$$\sim P \wedge Q \vee \sim R$$

is interpreted as

$$((\sim P) \wedge Q) \vee (\sim R).$$

The following two examples do more than just provide the truth table for the given statements. They exhibit the *process* of creating a truth table. Note that all of the variables of the statement appear in the first columns. From there, the variables are combined, one logical connective at a time, until the desired statement form is constructed. This is a systematic way to not just form the columns of the truth table but also a very simple method for filling in the truth values of the columns. When a column is formed by logically connecting at most two previous columns, filling in the truth values is as simple as pointing to the two (or one, in the case of $\sim$) "columns" the logical connective is "connecting." If the new column is of the form, "Column 1 $\wedge$ Column 2," then proceed down Columns 1 and 2, pointing at truth values in the same row. If you are pointing at two T values, then the result is to place a T in the new column. Otherwise, place an F in the new column (this is how a conjunction "works").

Example 1.7. Construct the truth table for $(P \vee Q) \wedge \sim(P \wedge Q)$.

We begin by listing the columns. First, the variables of the statement are P and Q. From there, we build up one connective at a time. Thus, we need a column for $P \vee Q$ and a column for $P \wedge Q$. Note that we would not jump straight to a column for $\sim(P \wedge Q)$; this involves two connectives. Then, before we place a column for our desired statement, we do need a column for $\sim(P \wedge Q)$. Then, the final statement's truth values will be obtained by looking only at the truth values in the columns for $P \vee Q$ and $\sim(P \wedge Q)$.

P	Q	$P \vee Q$	$P \wedge Q$	$\sim(P \wedge Q)$	$(P \vee Q) \wedge \sim(P \wedge Q)$
T	T	T	T	F	F
T	F	T	F	T	T
F	T	T	F	T	T
F	F	F	F	T	F

Regardless of the variables present or the logical connectives used, truth tables are easily constructed in the fashion described above.

Example 1.8. Construct the truth table for $\sim(P \veebar \sim R) \wedge Q$.

P	Q	R	$\sim R$	$P \veebar \sim R$	$\sim(P \veebar \sim R)$	$\sim(P \veebar \sim R) \wedge Q$.
T	T	T	F	T	F	F
T	T	F	T	F	T	T
T	F	T	F	T	F	F
T	F	F	T	F	T	F
F	T	T	F	F	T	T
F	T	F	T	T	F	F
F	F	T	F	F	T	F
F	F	F	T	T	F	F

In the previous two examples, the statement for which the truth table was constructed was sometimes true and sometimes false, depending on the truth assignments to the specific variables. Statement forms that always takes the same truth value, regardless of the truth assignment to its variables, have one of two special names.

Definition 1.7. A statement form that is true for every truth assignment to its variables is called a *tautology*. One that is false for all truth assignments to its variables is called a *contradiction*.

What is the intuitive sense of a tautology? It is a statement that is *always true*. It is possible to construct rather complicated statements that are tautologies, but one in particular is simple to construct. It relies on the fact that a statement is either true or false. Because of this, either the statement or its negation must be true.

Example 1.9. The following truth table shows that, for a statement P, $P \vee \sim P$ is a tautology.

P	$\sim P$	$P \vee \sim P$
T	F	T
F	T	T

Using a similar thought process, we have the following contradiction.

Example 1.10. Let P be a statement. Show that $P \wedge \sim P$ is a contradiction.

P	$\sim P$	$P \wedge \sim P$
T	F	F
F	T	F

The third column of the truth table shows $P \wedge \sim P$ is a contradiction.

Consider the possible truth values for $(P \vee Q) \wedge \sim (P \wedge Q)$ in Example 1.7. They are identical to the truth values of $P \veebar Q$, for all possible truth assignments to P and Q. Such statements are called *logically equivalent*.

Definition 1.8. Two statement forms with the same variables are called *logically equivalent*, with equivalence denoted via the symbol $\equiv$, if the statement forms have the same truth value for every possible truth assignment to the statement variables of the statement form.

Example 1.11. The statements $P \veebar Q$ and $(P \vee Q) \wedge \sim (P \wedge Q)$ are equivalent, as verified by the truth tables in Definition 1.5 and in Example 1.7. We write

$$P \veebar Q \equiv (P \vee Q) \wedge \sim (P \wedge Q).$$

In general, to determine if statements are or are not logically equivalent, construct a single truth table with a column for each of the statements. If the truth values are identical in every single row, then the statements are logically equivalent. If they differ in at least one row, the statements are not logically equivalent.

Example 1.12. For statements P and Q, the truth table below shows that $P \vee Q \equiv Q \vee P$.

P	Q	$P \vee Q$	$Q \vee P$
T	T	T	T
T	F	T	T
F	T	T	T
F	F	F	F

Example 1.12 should feel natural. Saying "red or blue" has the same meaning as saying "blue or red." This is the intuition behind logical equivalence. It is a mathematical way to communicate that two statements "mean the same thing."

It is important to also note that $\equiv$ is not a logical connective. Rather, it is *relational* in nature, much like the symbol $=$ is relational on real numbers. It is not a tool for creating new statements from old ones, like $\wedge$ and $\sim$ on statements or $+$ or $\sqrt[3]{}$ on real numbers. That is, $P \equiv Q$ is not a statement; it is not possible to construct a truth table for it.

The concept of logical equivalence leads to our first theorem. Though we have not discussed mathematical proof, we prove parts of this theorem here. For now, think of a proof as "irrefutable justification" of the results.

Theorem 1.9. *Let P, Q and R be statements, t a tautology and c a contradiction. The following logical equivalences, called logical equivalence laws, hold.*

1. Commutative	$P \wedge Q \equiv Q \wedge P$
	$P \vee Q \equiv Q \vee P$
2. Associative	$P \wedge (Q \wedge R) \equiv (P \wedge Q) \wedge R$
	$P \vee (Q \vee R) \equiv (P \vee Q) \vee R$
3. Distributive	$P \wedge (Q \vee R) \equiv (P \wedge Q) \vee (P \wedge R)$
	$P \vee (Q \wedge R) \equiv (P \vee Q) \wedge (P \vee R)$
4. Identity	$P \wedge t \equiv P$
	$P \vee c \equiv P$
5. Negation	$P \vee \sim P \equiv t$
	$P \wedge \sim P \equiv c$
6. Double Negative	$\sim(\sim P) \equiv P$
7. Idempotent	$P \wedge P \equiv P$
	$P \vee P \equiv P$
8. Universal Complement	$\sim t \equiv c$
	$\sim c \equiv t$

Proof Proofs of the negation equivalence (5) and the commutative equivalence (1) appear in Examples 1.10 and 1.12, respectively. We prove here only one of the distributive laws; the remaining proofs are left as exercises.

To prove the first of the two distributive laws, let P, Q and R be statements. For brevity, only the columns for the statement variables and desired statement forms are presented.

P	Q	R	$P \wedge (Q \vee R)$	$(P \wedge Q) \vee (P \wedge R)$
T	T	T	T	T
T	T	F	T	T
T	F	T	T	T
T	F	F	T	T
F	T	T	T	T
F	T	F	T	T
F	F	T	T	T
F	F	F	F	F

Regardless of the truth assignment for every variable of the statements, the statements $P \wedge (Q \vee R)$ and $(P \wedge Q) \vee (P \wedge R)$ have the same truth value, as exhibited above. Thus,

$$P \wedge (Q \vee R) \equiv (P \wedge Q) \vee (P \wedge R),$$

the desired result. $\square$

There is one extremely important logical equivalence law that could have been included in Theorem 1.9. Because of its importance, however, we state it as its own theorem. It is attributed to one of the founding fathers of symbolic logic, the aforementioned Augustus De Morgan.

Theorem 1.10. (De Morgan's Laws) For statements P and Q, we have

$$\sim(P \wedge Q) \equiv \sim P \vee \sim Q$$

and

$$\sim(P \vee Q) \equiv \sim P \wedge \sim Q.$$

Proof We prove the first of De Morgan's Laws, leaving the second as an exercise. For statements P and Q, we have

P	Q	$\sim P$	$\sim Q$	$P \wedge Q$	$\sim(P \wedge Q)$	$\sim P \vee \sim Q$
T	T	F	F	T	F	F
T	F	F	T	F	T	T
F	T	T	F	F	T	T
F	F	T	T	F	T	T

Because $\sim(P \wedge Q)$ and $\sim P \vee \sim Q$ have the same truth value for every truth assignment to P and Q, the result holds. $\square$

The basic logical connectives of *not*, *and* and *or* allow us to create meaningful logically equivalent statements. In the next section, we introduce a pair of logical connectives that open the door to mathematical conjecture, theory and proof.

Exercises

1. Determine which of the following are statements.

   (a) There are 30 days in April.
   (b) What happened on April 23, 1980?
   (c) There are 31 days in April.
   (d) Look out!
   (e) Brian wants to know what brand of bike that is.
   (f) Gasoline is not a liquid.
   (g) That tiger is ferocious!

2. Write the negation of each of the following statements.

   (a) Blue is Caroline's favorite color.
   (b) The smartphone has no more than 32 gigabytes of memory.
   (c) I was not born yesterday,
   (d) It was a home game and they lost.

3. Let statements P, Q and R be defined as follows.

   P: I stayed up past midnight.

   Q: I overslept.

   R: I passed my exam.

   (a) Write each of the following symbolically, using logical connectives and P, Q, and R.

      i. I stayed up past midnight and overslept.
      ii. Even though I stayed up past midnight, I did not oversleep.
      iii. I stayed up past midnight and overslept, but I still passed my exam.
      iv. I either overslept or I passed my exam, but not both.

   (b) Write the following symbolic statements as sentences in English.

      i. $P \wedge Q \wedge R$
      ii. $\sim P \vee \sim R$
      iii. $(P \veebar R) \vee \sim Q$

4. Let P, Q, and R be the following statements.

   P: This pen is out of ink.

   Q: I must complete this assignment by tomorrow morning.

   R: This assignment is worth 10 points.

   In everyday English, exhibit the logical equivalences of Theorem 1.9 for the particular statements.

   (a) Commutative: Q, R

   (b) Associative: P, Q, R

   (c) Distributive: P, Q, R

   (d) Double negative: Q

5. Exhibit, in everyday language, De Morgan's Laws for the statements P and Q of the previous problem.

6. For each of the following, (a) write it in symbolic form (you will need to define the statement variables), (b) write the symbolic negation of each and (c) write the negation of each in everyday English.

   (a) The integer 2 is a prime number but is also an even number.

   (b) Either 7 is even or 7 is prime.

   (c) The number 10 is neither prime nor odd.

7. For each of the following sentences (some of which are not statements), determine if the intention is for the *or* to be considered inclusively or exclusively.

   (a) Would you like cream or sugar in your coffee?

   (b) Flip the switch up or down.

   (c) You get soup or salad with your dinner.

   (d) Was the baby a boy or a girl?[4]

8. Give an example of a statement in the English language that is a tautology.

9. Give an example of a statement in the English language that is a contradiction.

10. Construct a truth table for each of the following statements. Assume that P is a statement.

    (a) P

    (b) $P \wedge \sim P$

    (c) $P \vee \sim(P \vee \sim P)$

    (d) $P \veebar \sim P$

---

[4]This question is an integral part of a classic mathematical joke. *A couple is in the delivery room, and upon giving birth, the wife asks her husband, a mathematician, "Is it a boy or a girl?" He responds, "Yes."*

11. Construct a truth table for each of the following statements. Assume that P and Q are statements.

   (a) $Q \wedge (P \vee Q)$
   (b) $\sim\!P \wedge Q$
   (c) $\sim\!P \vee \sim\!Q \vee (P \wedge Q)$
   (d) $\sim\!(P \vee \sim\!Q) \vee \sim\!(P \wedge Q)$

12. Construct a truth table for each of the following. Assume that P, Q, and R are statements.

   (a) $P \wedge (Q \vee R)$
   (b) $(\sim\!P \wedge Q) \wedge R$
   (c) $\sim\!((P \vee Q) \vee R)$
   (d) $[(P \wedge \sim\!Q) \underline{\vee} R] \vee (\sim\!R \vee Q)$

13. Assuming that P, Q and R are statements, determine if any of the following statements is a tautology or a contradiction. Justify your answer using a truth table.

   (a) $\sim\!(P \vee (\sim\!P \wedge Q))$
   (b) $P \underline{\vee} \sim\!P$
   (c) $\sim\!(P \vee Q) \underline{\vee} R$

14. Suppose a new logical connective $\times$ is defined in the following truth table. Using this definition, construct a truth table for the following statements, assuming P, Q, and R are statements.

P	Q	$P \times Q$
T	T	F
T	F	T
F	T	F
F	F	T

   (a) $P \vee (Q \times P)$
   (b) $P \times (Q \times R)$
   (c) $(P \times Q) \underline{\vee} (P \times R)$
   (d) $(P \underline{\vee} Q) \times (P \underline{\vee} Q)$

15. Determine if the logical connective $\times$ defined in the previous problem is:

   (a) Associative
   (b) Commutative

16. Show the following logical equivalences hold. Assume that P, Q, and R are statements.

   (a) $P \wedge Q \equiv Q \wedge P$

(b) $P \veebar Q \equiv (P \wedge \sim Q) \vee (\sim P \wedge Q)$

(c) $P \vee Q \equiv \sim(\sim P \wedge \sim Q)$

(d) $P \vee (Q \wedge R) \equiv (P \vee Q) \wedge (P \vee R)$

(e) $\sim(P \wedge Q) \equiv \sim P \vee \sim Q$

17. Show the Associative laws of Theorem 1.9 hold.

18. Show the Identity laws of Theorem 1.9 hold.

19. Show the Negation laws of Theorem 1.9 hold.

20. Show the Double Negative law of Theorem 1.9 holds.

21. Show the Idempotent laws of Theorem 1.9 hold.

22. Show the Universal Complement laws of Theorem 1.9 hold.

23. Determine if each of the following pairs of statements are logically equivalent. Assume all variables represent statements.

(a) P, Q

(b) $P \veebar Q, (P \vee Q) \vee (\sim P \wedge \sim Q)$

(c) $\sim(\sim P) \vee \sim P, Q \vee \sim Q$

(d) $P \vee (P \wedge \sim Q), P \wedge \sim Q$

24. Are (a) through (c) below statements? Explain.

(a) This statement is false.

(b) The barber shaves everyone who does not shave himself.

(c) A being with unlimited physical powers can create a wall taller than he is to scale.

1.2   Conditional and Biconditional Connective

Think about decisions you make as you go throughout your day. Maybe you did not know what to wear, so you turned on the weather forecast and thought to yourself, "If it is supposed to be chilly, then I'll wear long sleeves. If the forecast is for warmer weather, then I'll choose a t-shirt to wear." These types of if-then sentences, part of our everyday language, are the basic tool for inference. Mathematicians call these *conditional statements*.

Definition 1.11. A *conditional statement* (or *implication*) is one of the form "if P, then Q," where P and Q are statements, and is denoted $P \Rightarrow Q$. The statement P is called the *hypothesis* (or *assumption*, *premise*, or *antecedent*) of the conditional statement. The statement Q is called the *conclusion* (or *consequence*) of the implication. The conditional statement $P \Rightarrow Q$ is false only when P is true and Q is false. It is exhibited via the following truth table.

P	Q	$P \Rightarrow Q$
T	T	T
T	F	F
F	T	T
F	F	T

If we consider a real-world example and ask the question, "When is this implication true," we may find ourselves somewhat confused. For example, consider the conditional statement, "If it is raining, then I will carry an umbrella." It could either be raining or not raining and I could either be carrying or not carrying an umbrella. This means there are four possible cases to consider. If it is raining and I have my umbrella open, then it is clear that our original statement is true.

But what about if it is not raining? Is the statement, "If it is raining, then I will carry an umbrella," true if it is not raining yet I do have an umbrella open? Is it true if it is not raining and I do not have my umbrella open? According to Definition 1.11, the original statement is true in both these cases. But *why*? It is in no way intuitive. Looking at it from a different perspective, the reasoning for defining the conditional in this way makes complete sense.

What point-of-view clarifies Definition 1.11? Rather than asking when you would say a conditional statement is *true*, what if we asked when it is clearly *false*? Even four-year old children know the answer to this. Suppose a parent tells her child, "If you eat your broccoli, then you can have dessert." What happens if the child cleans his plate of broccoli but is told that he won't be getting any dessert? Perhaps the child screams, "That's not fair! You lied!" No matter the child's response, he realizes that the original statement was clearly *false*.[5]

Because one can explicitly say a conditional statement is false only when the hypothesis is true and the conclusion is false, mathematicians define implications in this way.

Once a truth table for a statement is established, under the guise of logical equivalence we can ask, "Does this statement *say the same thing* as any other statement?" Analyze the truth table for $P \Rightarrow Q$. It is true in the bottom two rows (when P is false), or in the first and third rows (when Q is true). This inspires the following logical equivalence.

Theorem 1.12. *Let P and Q be statements. Then,*

---

[5]You can partake in applied symbolic logic by attempting this with your favorite toddler. The author assumes no responsibility for the reactions of the youngster.

$$P \Rightarrow Q \equiv \sim P \vee Q.$$

Proof Let P and Q be statements. Then,

P	Q	$P \Rightarrow Q$	$\sim P$	$\sim P \vee Q$
T	T	T	F	T
T	F	F	F	F
F	T	T	T	T
F	F	T	T	T

Because $P \Rightarrow Q$ and $\sim P \vee Q$ have the same truth value for all truth assignments for P and Q, they are logically equivalent.   □

Theorem 1.12 provides an equivalent way to verbalize conditional statements. Saying "if P, then Q" is logically equivalent to saying, "not P or Q." For example, saying, "It is not raining or I'm carrying an umbrella" has the same logical interpretation as, "If it is raining, then I carry an umbrella." But this alternative pronunciation is just one way to present conditional statements in the English language. Example 1.12 provides some of these equivalent sayings, while others are left as exercises.

Example 1.13. The following three sentences are equivalent to saying, "if P, then Q."

(1) Q if P.
(2) P implies Q.
(3) P is a sufficient condition for Q.

The third equivalence to saying "if P, then Q" of the previous example, "P is a sufficient condition for Q", is a common mathematical expression. Why is its interpretation the same as "if P, then Q?" The latter statement is interpreted as, "Whenever P occurs, Q must happen." View it as a domino-effect; the domino representing P falling over must tip over the domino representing Q. Then, knowing that the P domino falls is sufficient for knowing the Q domino falls over.

A similar mathematical expression involves the word *necessary*. Saying, "P is a necessary condition for Q" is equivalent to saying "if Q, then P." Notice the order of the conditional statement; Q is the antecedent. Why? In this scenario, the Q domino cannot fall over without the P domino falling. Thus, for Q to fall, it is necessary for P to fall.   •

Example 1.14. Rewrite each of the statements in the standard if-then form of a conditional statement.

---

(1) Being at least 18 years old is a necessary condition to vote in Iowa.

Equivalent statement: If a person votes in Iowa, then he or she is at least 18 years old.

(2) It is necessary for Joey to have a helmet to compete in the bike race.

Equivalent statement: If Joey competes in the bike race, then he must have a helmet.

(3) Knowing that an integer is divisible by 8 is sufficient to know it is even.

Equivalent statement: If an integer is divisible by 8, then it is even.

It is important to be aware that being necessary for and being sufficient for something to happen are two different things. For example, the third statement of Example 1.14 states that being divisible by 8 is sufficient in knowing an integer is even. Is it necessary? Of course not. There are plenty of even integers that are not divisible by 8, such as 2, 4 and 6.

In Theorem 1.9 we saw that the logical connectives $\vee$ and $\wedge$ are associative. We show below that $\Rightarrow$ is *not* associative. Because of this, the use of grouping symbols is necessary to create properly defined compound statements.

Example 1.15. The statements $P \Rightarrow (Q \Rightarrow R)$ and $(P \Rightarrow Q) \Rightarrow R$ are not logically equivalent, written

$$P \Rightarrow (Q \Rightarrow R) \not\equiv (P \Rightarrow Q) \Rightarrow R.$$

The two statements will have identical truth tables if they are logically equivalent. Consider the table for each.

P	Q	R	$Q \Rightarrow R$	$P \Rightarrow (Q \Rightarrow R)$	$P \Rightarrow Q$	$(P \Rightarrow Q) \Rightarrow R$
T	T	T	T	T	T	T
T	T	F	F	F	T	F
T	F	T	T	T	F	T
T	F	F	T	T	F	T
F	T	T	T	T	T	T
F	T	F	F	T	T	F
F	F	T	T	T	T	T
F	F	F	T	T	T	F

Because $\Rightarrow$ is not associative, statements like $P \Rightarrow Q \Rightarrow R$ require, as mentioned above, either grouping symbols or, as discussed in the previous section, an "order of operations." The connective $\Rightarrow$ has its precedence directly after $\wedge$ and $\vee$. For example, $P \wedge Q \Rightarrow R \vee S$ is interpreted as $(P \wedge Q) \Rightarrow (R \vee S)$. To avoid confusion, we will consistently use grouping symbols. Before proceeding in our development of symbolic logic, we use this as an opportunity for our first discussion on *style*.

To this point in your mathematical upbringing, you have been mostly concerned with simply *getting the right answer*. You may have learned to develop scratch work and then write up your final solution "neatly," but have you ever stopped and pondered what is meant by "neatly presented mathematics?" In calculus, this might be showing the appropriate steps of a lengthy integration, clearly noting substitutions you have used and how those substitutions impact the bounds on the integral. Your page is organized, your handwriting clear and your page contains no "messy stuff." This organization and in-page display is the foundation of proper mathematical presentation, but as we will learn throughout this text, there is much more to writing *good* mathematics.

For now, there are two points about writing a proper solution worth discussing. Both involve keeping your reader informed. When you write, you are writing *for* somebody. You do not want your reader to be confused. Starting every problem with a proper introduction and finishing with a summarizing conclusion are simple ways to bookend your work and keep your reader focused. "This is what we are going to do ... We have shown what we intended to show." For example, if you are showing two statements are logically equivalent, do not simply present a truth table. Even if the truth table is perfectly correct, your reader does not know why it is there. A short, concise introduction, such as, "We will show A and B are logically equivalent by considering the following truth table," tells your reader exactly what you are about to do and why you are doing it.

Along this same line, if you do not tell your readers why a certain conclusion holds, they are left confused. "She told me she was going to show A and B were logically equivalent, and I see the truth table, but is there *more* to the solution?" A quick sentence squashes all potential issues: "Because A and B have identical truth tables, the desired result holds." You have *justified* your reasoning, something mathematicians demand.

Lastly, as you present your work, keep your reader in mind. This idea circles back to the choice to use grouping symbols when presenting statements such as $P \Rightarrow Q \Rightarrow R$. While it is not required, because of the order of precedence with logical connectives, it allows your reader to more easily follow your work. There is no need for a reader to pause and think, "Does $P \Rightarrow Q \Rightarrow R$ mean $(P \Rightarrow Q) \Rightarrow R$ or $P \Rightarrow (Q \Rightarrow R)$."

Example 1.15 showed that the logical connective $\Rightarrow$ is not associative. Is it commutative? In terms of everyday language, asking if the conditional connective is commutative equates to asking if saying, "If I eat my broccoli, then I get dessert," is the same as saying, "If I had dessert, then I must have eaten my broccoli?" It is not, which we prove after the following definition.

Definition 1.13. The *contrapositive* to the statement $P \Rightarrow Q$ is the statement $\sim Q \Rightarrow \sim P$. The *converse* to $P \Rightarrow Q$ is the statement $Q \Rightarrow P$, and the *inverse* statement to $P \Rightarrow Q$ is $\sim P \Rightarrow \sim Q$.

Example 1.16. Write the contrapositive, converse, and inverse to the statement below.

Statement: If I water the plants daily, then they will bloom.

Contrapositive: If the plants do not bloom, then I did not water them daily.

Converse: If the plants bloom, then I watered them daily.

Inverse: If I do not water the plants daily, then they will not bloom.

Which of the statements in Example 1.16 have the same logical meaning? All of the statements are conditional statements, so it makes sense to determine the sole case when each is false. The original statement is false only when the plants are watered daily and they do not bloom. The contrapositive is not true precisely when the plants do not bloom and they are not *not* watered daily (that is, they are watered daily). Thus, the statement and its contrapositive have the same meaning.

The converse, however, is false only when the plants bloom and they are not watered daily. This is not the same as the original statement. Likewise, the inverse is not true only when the plants are not watered daily and they do not *not* bloom. So, the converse and the inverse have the same meaning. Theorem 1.14 summarizes these ideas.

Theorem 1.14. *A conditional statement is logically equivalent to its contrapositive but not to either its converse or inverse.*

Proof The truth table below proves that $P \Rightarrow Q$ and $\sim Q \Rightarrow \sim P$ always have the same truth value while $P \Rightarrow Q$ and its converse or inverse do not.

P	Q	$\sim P$	$\sim Q$	$P \Rightarrow Q$	$\sim Q \Rightarrow \sim P$	$\sim P \Rightarrow \sim Q$	$Q \Rightarrow P$
T	T	F	F	T	T	T	T
T	F	F	T	F	F	T	T
F	T	T	F	T	T	F	F
F	F	T	T	T	T	T	T

□

Note that proof to Theorem 1.14 shows that the converse and the inverse of an implication are themselves logically equivalent. This can also be seen either in the fact that they are contrapositives of one another, and by Theorem 1.14, they must be logically equivalent.

One last thing to note about conditional statements is that $P \Rightarrow Q$ and $Q \Rightarrow P$ are *not* negations of one another. That is, knowing that one is not true does not automatically imply that the other is. Mathematically, we are claiming that $\sim(P \Rightarrow Q)$ and $Q \Rightarrow P$ are not logically equivalent. The truth table below verifies this.

P	Q	$P \Rightarrow Q$	$\sim(P \Rightarrow Q)$	$Q \Rightarrow P$
T	T	T	F	T
T	F	F	T	T
F	T	T	F	F
F	F	T	F	T

What then do we mean if we mention the *negation* of a conditional statement $P \Rightarrow Q$? The negation of a statement is itself a statement, and it is true precisely when the original statement is false. In the case of $P \Rightarrow Q$, this is when P is true and Q is false.

Definition 1.15. The *negation* of the conditional statement $P \Rightarrow Q$ is the statement $\sim P \wedge Q$.

Sometimes in our everyday conversations we want to state that knowing one piece of information is identical to knowing another. That is, knowing statement P means we know statement Q, and, knowing statement Q means we know statement P. This would require saying two separate conditional statements: $P \Rightarrow Q$ and $Q \Rightarrow P$, respectively. In everyday language this can be a bit cumbersome: "if you eat your broccoli, then you will get dessert, and, if you have had dessert, then you must have eaten your broccoli." Luckily, there is a shorthanded way of saying this, and it uses the logical connective known as the *biconditional*.

Definition 1.16. A *biconditional statement* is one of the form "P if and only if Q" where P and Q are statements, and is denoted $P \Leftrightarrow Q$. It is true when P and Q have the same truth value and false when their truth values are different, exhibited via the following truth table.

P	Q	$P \Leftrightarrow Q$
T	T	T
T	F	F
F	T	F
F	F	T

Note the process for creating the truth table for a biconditional. In the spirit of "building" a column by pointing at previous columns, $P \Leftrightarrow Q$ is true if you are pointing at the same truth value (regardless of what the actual truth value is) and false if you are pointing at different truth values.

Example 1.17. Build the truth table for $P \Leftrightarrow (Q \Rightarrow \sim R)$.

Even though this statement involves a new logical connective, we construct the truth table just as before. Begin by creating columns for the variables P, Q, and R. Before creating a column for $P \Leftrightarrow (Q \Rightarrow \sim R)$, we must have a column for $Q \Rightarrow \sim R$, necessitating a column for $\sim R$.

P	Q	R	$\sim R$	$Q \Rightarrow \sim R$	$P \Leftrightarrow (Q \Rightarrow \sim R)$
T	T	T	F	F	F
T	T	F	T	T	T
T	F	T	F	T	T
T	F	F	T	T	T
F	T	T	F	F	T
F	T	F	T	T	F
F	F	T	F	T	F
F	F	F	T	T	F

The statement "P if and only if Q" is the conjunction (in plain English) of two other statements: "P if Q", and "P only if Q." This is not by accident. Both of these latter statements are themselves conditional statements: $Q \Rightarrow P$ and $P \Rightarrow Q$, respectively. The discussion leading into Definition 1.16 introduced the biconditional as such. Theorem 1.17 formalizes this concept. Its proof is left as an exercise. While Theorem 1.17 has repercussions in symbolic logic, its real value will appear in our discussions on proof techniques.

Theorem 1.17. *For statements P and Q, $P \Leftrightarrow Q \equiv (P \Rightarrow Q) \wedge (Q \Rightarrow P)$.*

Pairing Theorem 1.17 with the multitude of ways to verbalize conditional statements creates even more ways to verbalize biconditional statements. This is just one more suggestion as to the power of symbolic logic. While there are dozens of ways to say two statements P and Q have the same meaning, symbolically, any such way will be logically equivalent to $P \Leftrightarrow Q$.

Example 1.18. Rewrite the statement, "An integer is even if and only if it is divisible by 2," two different ways.

While there are many options, we first explicitly use the result of Theorem 1.17:

(1) If an integer is even, then it is divisible by 2, and, if an integer is divisible by 2, then it is even.

Next, refer to the discussion on necessary and sufficient conditions. The given statement is therefore equivalent to:

(2) Being divisible by 2 is a necessary and sufficient condition for an integer to be even.

With this understanding of symbolic logic and the five logical connectives of $\wedge$, $\vee$, $\sim$, $\Rightarrow$ and $\Leftrightarrow$, we are prepared to understand what it means to mathematically *argue*. Having a firm grasp on the notion of *argument* will not only help you better understand mathematical reasoning, it will prepare you to substantiate arguments in the public realm (or point out that certain arguments carry no merit).

Exercises

1. Write the following statements in standard "if, then" form.

   (a) We will go on vacation if your mother does not have to work.
   (b) Your injury will heal as long as you rest it.
   (c) Classes are canceled if it snows more than 12 inches.
   (d) Vertical angles are congruent.
   (e) Linda's cat Rudy hides whenever he hears thunders.

2. Write the following statements in standard "if, then" form.

   (a) It is necessary for you to do your chores in order to earn your allowance.
   (b) Giving the plants water is necessary for their survival.
   (c) The commuter train being on schedule is sufficient for you arriving on time.
   (d) A sufficient condition for traveling from Chicago to St. Louis is taking a direct flight.

3. Write the following statements in standard "if, then" form.

   (a) I always remember my first kiss when I hear this song.
   (b) A failing grade is a consequence of plagiarizing.

    (c) Winning the game follows from outscoring our opponents.

    (d) The video game will reset only if you hit those buttons simultaneously.

    (e) I begin to sweat whenever I think about snakes.

4. Write the following statements in standard "if and only if" form.

    (a) The exam is curved precisely when the average falls below 70%.

    (b) If I go to the pool, then I will sleep well tonight, and if I sleep well tonight, then I went to the pool.

    (c) Passing the final exam is both necessary and sufficient for you to pass this class.

    (d) That dashboard light being on is a consequence of a low oil level, and vice versa.

5. Which of the following are true statements?

    (a) If the earth is flat, then 15 is a negative number.

    (b) A whole number is a perfect square if its square root is also a whole number.

    (c) $x + 7 = 9$ is necessary for knowing $x = 2$.

    (d) $x + 7 = 9$ is sufficient for knowing $x = 2$.

    (e) You sweat only if you run a marathon.

    (f) You sweat if you run a marathon.

6. For each of the following statements, write its converse, inverse and contrapositive statements.

    (a) If the restaurant is closed, we will eat at home.

    (b) The question should be answered if the material is covered in class.

    (c) The tree falling down would be a consequence of a bad storm.

    (d) I want to go only if he does too.

7. Write each of the following statements in standard "if, then" form, and then write their converses and contrapositives.

    (a) Harper apologizing implies that she caused the accident.

    (b) She will win the tournament if she birdies the hole.

    (c) Brad cries whenever he watches this movie.

    (d) To get into the club, it is necessary that you know the secret handshake.

    (e) The cake will set only if you use the correct ratio of ingredients.

8. Create a truth table for each of the following. Assume P, Q, and R are statements.

    (a) $P \Rightarrow {\sim}Q$

   (b) $P \Rightarrow (Q \Rightarrow R)$

   (c) $(P \Rightarrow Q) \Rightarrow R$.

   (d) $(\sim P \vee Q) \Rightarrow (\sim Q \Rightarrow P)$

   (e) $(P \vee Q) \Rightarrow (R \wedge Q)$

9. Create a truth table for each of the following. Assume P, Q, and R are statements.

   (a) $(P \Leftrightarrow Q) \Leftrightarrow R$

   (b) $R \Leftrightarrow (Q \Leftrightarrow R)$

   (c) $(P \Leftrightarrow R) \wedge (P \Leftrightarrow Q)$

   (d) $P \Leftrightarrow (Q \veebar (R \Rightarrow P))$

10. Create a tautology using symbolic statements (P, Q, R, etc.) and the logical connectives $\sim$ and $\Rightarrow$, and show that it is indeed a tautology.

11. Create a contradiction using symbolic statements (P, Q, R, etc.) and the logical connectives $\sim$ and $\Rightarrow$, and show that it is indeed a contradiction.

12. Create a tautology using symbolic statements (P, Q, R, etc.) and the logical connectives $\sim$ and $\Leftrightarrow$, and show that it is indeed a tautology.

13. Create a contradiction using symbolic statements (P, Q, R, etc.) and the logical connectives $\sim$ and $\Leftrightarrow$, and show that it is indeed a contradiction.

14. Show that Theorem 1.17 is true: $(P \Leftrightarrow Q) \equiv (P \Rightarrow Q) \wedge (Q \Rightarrow P)$.

15. State the negation of each of the following.

   (a) If the answer is negative, then I was incorrect.

   (b) The carpet will be replaced only if the basement floods.

   (c) Campus officials will attend if and only if they are invited.

   (d) If I order the salad bar, then I will get a free bowl of soup or a free dessert.

16. State the negation of each of the following using only the logical connectives $\wedge$ and $\sim$. Assume P, Q, and R are statements.

   (a) $\sim P \Rightarrow Q$

   (b) $(P \wedge Q) \Rightarrow (Q \vee R)$

   (c) $\sim(R \vee \sim P) \Rightarrow (P \Rightarrow Q)$

   (d) $P \Leftrightarrow \sim Q$

   (e) $(P \Rightarrow Q) \Leftrightarrow (R \Rightarrow P)$

17. Suppose P and Q are statements and that the statement $P \Rightarrow Q$ is false. Can you conclude anything about whether the following are true or false? Justify.

(a) $P \wedge Q$

(b) $P \vee Q$

(c) $P \veebar Q$

(d) $P \Leftrightarrow Q$

(e) $Q \Rightarrow P$

18. Suppose P and Q are statements and that the statement $P \Leftrightarrow Q$ is false. Can you conclude anything about whether the following are true or false? Justify.

(a) $P \wedge Q$

(b) $P \vee Q$

(c) $P \veebar Q$

(d) $P \Rightarrow Q$

(e) $Q \Rightarrow P$

19. What can you conclude about the truth of $P \Rightarrow Q$ if you know the statement in each of the following is false? (Note that these are separate problems; for each, you know that only the given statement is false.) Assume all variables represent statements.

(a) $(P \Rightarrow Q) \Rightarrow \sim(P \Rightarrow Q)$

(b) $\sim(P \Leftrightarrow Q)$

(c) $R \Rightarrow (P \Rightarrow Q)$

20. Determine, with justification, if each of the following is a tautology, contradiction, or neither. Assume all variables represent statements.

(a) $P \Rightarrow (P \Rightarrow Q)$

(b) $(Q \Leftrightarrow P) \Leftrightarrow (P \Leftrightarrow R)$

(c) $\sim(Q \Rightarrow P) \Rightarrow P$

(d) $[P \wedge (Q \Leftrightarrow R)] \Rightarrow \sim R$

(e) $[P \Rightarrow (Q \Rightarrow P)] \veebar (Q \Leftrightarrow P)$

21. For each of the following, show that the first statement is logically equivalent to the second. Assume all variables represent statements.

(a) $(P \Rightarrow Q) \wedge P$, Q

(b) $(P \Rightarrow Q) \wedge \sim Q$, $\sim P$

(c) $P \Rightarrow Q) \wedge (P \Rightarrow R)$, $P \Rightarrow (Q \wedge R)$

(d) $(P \Rightarrow R) \wedge (Q \Rightarrow R)$, $(P \vee Q) \Rightarrow R$

(e) $P \Leftrightarrow Q$, $\sim P \Leftrightarrow \sim Q$

22. Suppose P and Q are logically equivalent statements, and, Q and R are logically equivalent statements. Must P and R be logically equivalent? Explain.

23. Suppose P, Q, R, S and T are statements. Show that

$$P \Rightarrow (Q \Rightarrow (R \Rightarrow (S \Rightarrow T)))$$

and

$$P \Rightarrow (Q \Rightarrow (R \Rightarrow (S \Rightarrow \sim T)))$$

are not logically equivalent.

1.3   Arguments

Symbolic logic not only is a necessary tool for deductive mathematics but it also introduces us to the world of proof via *arguments*. For example, suppose Katie and Luke are discussing the upcoming weekend's events.

Katie: I heard that if you go bowling tonight, you will get a coupon for free entrance to the zoo on Sunday.
Luke: Really? If I get a coupon like that, I always go.
Katie: You promised that if you ever go to the zoo you would buy me a souvenir.
Luke: I did, didn't I? Well, you know what, after giving it some though, I'm going to go bowling tonight.

If we assume all of these statements are true, should Katie expect a souvenir from the zoo? Of course. That's a logical conclusion to make from the given information. Luke said he was going to go to bowling, which in turn means he will receive the zoo coupon. His first statement, if assumed true, means he will use the coupon and consequently will result in his keeping of a promise a souvenir.

Symbolically, we could have represented this as follows.

B: Luke goes bowling tonight.
C: Luke receives a coupon for free entrance to the zoo on Sunday.
A: Luke uses his coupon for free entrance.
S: Luke buys Katie a souvenir.

Translating the conversation to symbols, we assume the following statements to be true: $B \Rightarrow C$, $C \Rightarrow A$, $A \Rightarrow S$, and B and the conclusion we draw is S.

Symbolically, that this is a "good" argument should seem somewhat natural. We assume the statement B to be true. Then, since B implies C is true, we can conclude that C must be true. By the same reasoning, since we have

$C \Rightarrow A$, the statement A must hold. Consequently, S must be true (since $A \Rightarrow S$ is assumed true). We formalize this idea by defining an *argument*.

Definition 1.18. An *argument* is a finite list of statements P_1, P_2, ..., P_{n-1}, called the *assumptions* (or *premises* or *hypotheses*) of the argument, and a single statement P_n, called the *conclusion* of the argument. We write such an argument in the following way. The symbol $\therefore$ is read as "therefore."

$$
\begin{array}{l}
P_1 \\
P_2 \\
\vdots \\
\underline{P_{n-1}} \\
\therefore P_n
\end{array}
$$

The argument in the example above "makes sense;" the conclusion seems to logically follow from the assumptions. But this does not need to be the case for every argument. An argument is simply defined as a finite list of assumptions and a single conclusion. There is no requirement of that conclusion to "follow" from the assumptions. For example, consider the following basic argument between two parents.

Parent A: I said that if Cory did the laundry, then he would get to use the car.
Parent B: Well, Cory has the car, so the laundry must be done.

Symbolically, if L represents Cory doing the laundry and C Cory getting to use the car, the argument is

$$
\begin{array}{l}
L \Rightarrow C \\
\underline{C} \\
\therefore L
\end{array}
$$

Being comfortable with implications, we know that this conclusion somehow "isn't right." Assuming both $P \Rightarrow Q$ and Q to be true does not mean P must be true. In fact, if P is false, then the statement $P \Rightarrow Q$ is true regardless of the truth value of Q.

This idea about "good" arguments and "poor" arguments is formalized by classifying arguments as either *valid* or *invalid*.

Definition 1.19. An argument is called *valid* if whenever every assumption of the argument is true, then the conclusion is also true. If there is a case when the assumptions are true but the conclusion is false, then the argument is called *invalid*.

This definition does more than just define valid and invalid arguments. It instructs us how to check an argument's validity. To do so, build a single truth table with a column containing every single assumption and the conclusion of the argument. Then, consider every row of the truth table where *all* assumptions are true. Is the conclusion true in every single case? If so, the argument is valid. If there is even just one row where all the assumptions are true yet the conclusion is false, then the argument is invalid. Let us exhibit this process by showing that the laundry/car example is an invalid argument.

Example 1.19. Show that the following argument is invalid.

$$L \Rightarrow C$$
$$\underline{C}$$
$$\therefore L$$

We build a truth table with a column for every assumption and the conclusion.

L	C	$L \Rightarrow C$	
T	T	T	
T	F	F	
F	T	T	$\leftarrow$
F	F	T	

The third row, as indicated by the arrow, shows that this argument is invalid. The assumptions are both true but the conclusion is false.

Using a truth table to show an argument is invalid requires finding a *single* row that satisfies certain conditions (every assumption is true and the conclusion is false). Showing an argument is valid often takes a bit more work.

Example 1.20. Show the argument below is valid.

$$A \Rightarrow C$$
$$B \Leftrightarrow C$$
$$\underline{\sim A}$$
$$\therefore A \Rightarrow B$$

We build a truth table with a column for each assumption and a column for the conclusion.

A	B	C	$A \Rightarrow C$	$B \Leftrightarrow C$	$\sim A$	$A \Rightarrow B$	
T	T	T	T	T	F	T	
T	T	F	F	F	F	T	
T	F	T	T	F	F	F	
T	F	F	F	T	F	F	
F	T	T	T	T	T	T	$\leftarrow$
F	T	F	T	F	T	T	
F	F	T	T	F	T	T	
F	F	F	T	T	T	T	$\leftarrow$

We must consider *every* row where *all* assumptions are true. For this argument there are two such rows to consider, as indicated by the arrows. Once we have identified these rows, we ask: is the conclusion also true in every one of these rows? Here, the conclusion, $A \Rightarrow B$, is true. Thus, the argument is valid.

It is important to note that by claiming an argument is valid, we are making a claim only about the *argument* rather than the individual statements within the argument. This is highlighted by the following two examples, both of which contain valid arguments.

Example 1.21. The following argument is valid. Showing it is valid is left as an exercise.

> The moon is made of cheese.
> If the moon is made of cheese, then the world is flat.
> Therefore, the world is flat.

Example 1.22. Investigate the validity of the following argument.

$$\frac{P \wedge \sim P}{\therefore Q}$$

Consider the following truth table:

P	Q	$\sim P$	$P \wedge \sim P$
T	T	F	F
T	F	F	F
F	T	T	F
F	F	T	F

Notice that there are no rows where the hypothesis $P \wedge \sim P$ is true. In order to say that an argument is *invalid*, there must be a row of the truth table in which all the hypotheses are true and the conclusion is false. Since this does not exist, we say that the argument is valid *vacuously*.

Now imagine if we wanted to check the validity of the argument posed at the beginning of this section:

$$B \Rightarrow C$$
$$C \Rightarrow A$$
$$A \Rightarrow S$$
$$\underline{B}$$
$$\therefore S$$

To do so, we would construct a truth table with $2^4 = 16$ rows. That is manageable, but imagine an argument with 10 variables. The truth table we would need to construct would require $2^{10} = 1024$ rows! Certainly there must be another method. When we reasoned through the "logic" of the above argument, we repeatedly used the same argument; if $P \Rightarrow Q$ and P are true, then Q must logically follow. If we can show that this argument is itself valid, for *any* statements P and Q, then we could call upon it to validate "mini-conclusions" from the larger argument above.

The method, called *logical deductions*, appears in the next section, but one of the main tools for them is what are called *rules of inference*.

Definition 1.20. A *rule of inference* is an argument that is valid.

While any argument that is valid could be considered to be a rule of inference, certain ones are natural and occur frequently in our everyday conversations. For example, the assumptions $P \Rightarrow Q$ and P lead to a conclusion understood by our aforementioned ice-creaming loving toddlers. A promise of "if you eat your broccoli, then you get ice cream" is understood to mean "ice cream will be served" once the broccoli is eaten.

There are a handful of other common arguments, each named as follows. P, Q and R are all statements.

Modus Ponens	**Modus Tollens**
$P \Rightarrow Q$	$P \Rightarrow Q$
P	$\sim Q$
$\therefore Q$	$\therefore \sim P$
Transitivity	**Conjunction**
$P \Rightarrow Q$	P
$Q \Rightarrow R$	Q
$\therefore P \Rightarrow R$	$\therefore P \wedge Q$
Generalization	**Specialization**
P (or Q)	$P \wedge Q$
$\therefore P \vee Q$	$\therefore P$ (or Q)
Elimination	**Cases**
$P \vee Q$	$P \vee Q$
$\sim P$ (or $\sim Q$)	$P \Rightarrow R$
$\therefore Q$ (or P)	$Q \Rightarrow R$
	$\therefore R$

We will show that Modus Tollens is valid. Showing that the remainder are valid arguments is left as an exercise.

Example 1.23. The following truth table shows that Modus Tollens is a valid argument.

P	Q	$P \Rightarrow Q$	$\sim Q$	$\sim P$	
T	T	T	F	F	
T	F	F	T	F	
F	T	T	F	T	
F	F	T	T	T	$\leftarrow$

The last row of the truth table is the only one in which both hypotheses, $P \Rightarrow Q$ and $\sim Q$, are true. Note that the conclusion, $\sim P$, is true as well, showing that the argument is indeed valid.

Just as there are common valid arguments arising in everyday language, there are common invalid arguments that show up in our conversations as well, such as the one between Parent A and Parent B prior to Definition 1.19. Sometimes called *logical fallacies* or *logical errors*, we name here two of the most common invalid arguments.

Converse Error	**Inverse Error**
$P \longrightarrow Q$	$P \longrightarrow Q$
Q	$\sim P$
$\therefore P$	$\therefore \sim Q$

With these logical tools in hand, you are not only prepared to address more complicated arguments, forthcoming in the the next section, but you are also better prepared to question and consider arguments in the public discourse. The aforementioned logician Lewis Carroll was quick to point this out, and we conclude our section with a quote from his *Symbolic Logic* [8]:

> *Once master the machinery of Symbolic Logic, and you have a mental occupation always at hand, of absorbing interest, and one that will be of real use to you in any subject you may take up. It will give you clearness of thought, the ability to see your way through a puzzle, the habit of arranging your ideas in an orderly and get-at-able form-and, more valuable than all, the power to detect fallacies, and to tear to pieces the flimsy illogical arguments, which you will so continually encounter in books, in newspapers, in speeches, and even in sermons, and which so easily delude those who have never taken the trouble to master this fascinating Art.*

Exercises

1. Translate each of the following arguments to symbolic arguments and determine, with justification, if they are valid or invalid.

   (a) The moon is made of cheese.
       If the moon is made of cheese, then the world is flat.
       Therefore, the world is flat.

   (b) If France and Portugal face off in the championship match, the television ratings will be at a record level.
       France indeed will play for the championship but Portugal will not.
       Therefore the TV ratings will not be at a record level.

   (c) The symphony would not be cancelled unless it rained.
       If it rains, then we will not have a picnic.
       We will picnic!
       Hence, the symphony will play!

   (d) I drink coffee only if I intend to focus on my research.
       I will do research if and only if I am not in office hours.
       Therefore, I only drink coffee when I am in office hours.

   (e) The governor will not run for re-election if the bill does not pass.
       The public does not want the bill to pass (so it won't) but they do want the governor to run for re-lection.
       The governor will not run for re-election.

   (f) Exams in this class are quite difficult.
       Sawyer performs well on difficult exams, but she does not do well on easier tests.

If Sawyer does well on an exam, Ian does poorly, and vice versa.
Michelle does well on easier exams.

2. Determine if the following arguments are valid or invalid. Justify your conclusions.

   (a) $P \vee Q$
       $P \Rightarrow Q$
       $\therefore Q$

   (b) $P \vee Q \vee R$
       $\sim Q \vee \sim R$
       $\therefore P$

   (c) $R \Rightarrow (P \wedge Q)$
       $\sim Q$
       $\therefore \sim R$

   (d) $P \Leftrightarrow Q$
       $\sim Q$
       $\sim R \Rightarrow P$
       $\therefore R$

   (e) $P \Leftrightarrow Q$
       $P \Rightarrow R$
       $\therefore Q \Rightarrow R$

   (f) $\sim P \vee Q$
       $Q \vee \sim R$
       $\sim (P \vee R)$
       $\therefore Q$

3. Identify which rule of inference each of the following arguments exemplifies.

   (a) This punishment is a consequence of your poor behavior.
       I know I would stay home whenever I got punished.
       Therefore, my poor behavior implies I stay home.

   (b) I decided to buy both a new smartphone and a new tablet.
       Hence, I purchased a new tablet.

   (c) She exited the game at halftime.
       The home team won the game.
       Therefore, she exited the game and the home team won.

   (d) If I hadn't returned home by 6 PM, then I did not have a good run.
       I had a really good run.
       So you know I was home before 6 o'clock.

   (e) Finishing the paper is sufficient for Toby passing the class.
       Toby completed the paper.
       Toby passed the class.

(f) The screen dims if the battery is low.
Bright light also causes the screen to dim.
<u>It's either very bright in here or the phone battery is low.</u>
If I get my phone out, I know the screen will be dim.

4. If the following statements are assumed to be true, determine what conclusion can be drawn. Justify your answer by stating what rule(s) of inference is used to deduce the stated conclusion.

(a) "If everything goes right today, then I will get the job." "I did not get the job."

(b) "Flight 1345, which Sandra is booked on, will arrive in Denver at 10:30 AM if it leaves Seattle on time." "If Sandra is in Denver by 11:00 AM, then she and I will do lunch together."

(c) "You have a choice of a hatchback or a sedan for your rental car." "I do not want a hatchback."

(d) "A temperature over 80 degrees is sufficient for us to go to the water park." "Going to the water park is necessary if it doesn't rain!" "I know for certain that the weather forecast called for temperatures over 80 degrees or no rain."

(e) "If I watched the kids last weekend, then I made $50." "I watched the Dunham kids any time their parents went to a movie." "I did not make any money last weekend."

5. Explain why the following arguments are invalid, stating whether they exhibit the converse or the inverse error.

(a) "If I voted to lower taxes, then you'd all be bringing home more money every month. Your paychecks have increased, so you know I voted to lower your taxes!"

(b) "My doctor says that whenever I begin exercising regularly, I'll lose weight." "You're not exercising at all, so you surely haven't lost any weight."

(c) "The professor told us that we would pass the final only if we studied all weekend." "Did you pass the final?" "No." "You didn't study all weekend, did you?"

(d) "My dad said I could name the kitten Newton if we adopted him last Saturday." "I only met Newton today and you've had him since last Saturday?"

6. Verify that the following rules of inference are indeed valid arguments.

(a) Modus Ponens

(b) Transitivity

(c) Conjunction

(d) Generalization

    (e) Specialization

    (f) Elimination

    (g) Cases

7. Show that the converse and inverse errors are indeed invalid arguments.

8. Give two examples each (different from any presented in this text), in everyday English, of the converse and inverse being used.

9. Explain why the following theorem is true.

Theorem: The argument below is valid if and only $(P_1 \wedge P_2 \wedge \cdots \wedge P_{n-1}) \Rightarrow P_n$ is a tautology.

$$P_1$$
$$P_2$$
$$\vdots$$
$$\frac{P_{n-1}}{\therefore P_n}$$

10. Assuming all variables represent statements, is the following argument valid or invalid? Explain.

$$P \Rightarrow Q$$
$$Q \Leftrightarrow S$$
$$S \veebar T$$
$$T \Rightarrow U$$
$$U \wedge W$$
$$\frac{X \wedge \sim X}{\therefore Y \vee Z}$$

11. Regardless of what statements P, Q, and R are, is the following statement valid or invalid? Explain.

$$P$$
$$Q$$
$$\frac{R}{\therefore P \vee \sim P}$$

1.4 Logical Deductions

Arguments are all around us. We are bombarded with them through advertising ("Use this toothpaste and in just 2 weeks, your teeth will be 50% whiter." "I drink too much coffee and definitely need whiter teeth. If I use this toothpaste, Modus Ponens guarantees me a brighter smile!") and schmoozed into faulty arguments by politicians ("If lions are roaming the streets of our city, then my opponent is our mayor!" "I don't like lions ...I better vote for this guy."). But not all arguments are as "short and sweet" as these simple ones. For example, consider the following symbolic argument.

$$P$$
$$P \Rightarrow Q$$
$$Q \Rightarrow R$$
$$R \Rightarrow S$$
$$\underline{S \Rightarrow T}$$
$$\therefore T$$

A quick examination of the statements makes it fairly obvious that this is a valid argument. The first hypothesis assumes that P is true. Pairing this with the second argument and applying Modus Ponens results in Q being true. Similarly, this with the third hypothesis results in R being true, and consequently S by the same logic, and ultimately T.

Even though this is sound mathematical and logical reasoning, this is not a proper mathematical justification that the argument is valid. To do this, we must use the tools of the previous section and build a truth table. Because the argument contains 5 variables, the truth table will have 32 rows. Regardless of how much fun it is to create truth tables, building one with that many rows is tedious and can eventually lose its appeal. Luckily, there is an easier way, called a *logical deduction*. The importance of this method goes beyond arguments, however. It can be considered as a first glimpse into *proof.*

> **Definition 1.21.** A *logical deduction* is a process for showing an argument is valid. It is a finite sequence of statements, where every term of the sequence is either a hypothesis of the argument or it follows from a single rule of inference or logical equivalence applied to previous terms of the sequence, and the final term of the sequence is the conclusion of the argument. Each statement is accompanied by a justification for why it is allowed to appear in the deduction.

Before proceeding to our first logical deduction, it is critically important to note that logical deductions cannot be used to prove the basic rules of inference are valid ("Modus Ponens follows from ... Modus Ponens"). Truth tables are required to prove their validity, and because each only includes a small number of statement variables, this is not an overwhelming task. Once shown valid, however, they become the workhorses for logical deductions.

> **Example 1.24.** Show that the following argument, where P, Q and R are statements, is valid.

$$P$$
$$P \Rightarrow Q$$
$$\underline{Q \Rightarrow R}$$
$$\therefore R$$

Our logical deduction is as follows.

1.	P	Assumption
2.	$P \Rightarrow Q$	Assumption
3.	$Q \Rightarrow R$	Assumption
4.	Q	Modus Ponens 1, 2
5.	R	Modus Ponens 3, 4

There are certain aspects of Example 1.24 that warrant deeper discussion. First, notice the basic format of a logical deduction. It is a numbered list. The numbering allows for referencing previous lines. That referencing takes place in the right column and that is our second important note. When you present a final draft of a logical deduction, you assume that your work has an audience; you are *communicating* mathematics. At no point do you want your readers to scratch their heads and think, "Where did that line of the deduction come from?" This will be a theme reiterated in the chapters to come.

Mathematical conclusions require mathematical justifications.

Another important note is best highlighted through an example of an incorrect logical deduction. What makes the following an incorrect deduction of the same argument presented in Example 1.24?

1.	P	Assumption
2.	$P \Rightarrow Q$	Assumption
3.	$Q \Rightarrow R$	Assumption
4.	R	Modus Ponens 1, 2, 3

Where is the flaw? Modus Ponens is a rule of inference with only *two* hypotheses. Line 4 of the deduction claims that Modus Ponens was applied with 3 "inputs." This cannot happen. Having three assumptions in that particular rule of inference is not allowed.

Similarly, another mistake would be if the logical deduction had looked as follows.

1.	P	Assumption
2.	$P \Rightarrow Q$	Assumption
3.	$Q \Rightarrow R$	Assumption
4.	R	Modus Ponens 1, 2 and Modus Ponens 3

It appears here that the error of the previous deduction was corrected.

However, in doing so, two different errors were introduced. First, Definition 1.21 allows for only one single rule of inference to be used to obtain any line of the deduction. Line 4 claims to have followed from two applications of Modus Ponens. Secondly, the justification "Modus Ponens 3" is similar to the previous error; Modus Ponens requires two assumptions, not just one.

Example 1.24 was a fairly simple deduction to make. The flow of its arguments seemed somewhat "natural." The next two arguments, and many of those you will encounter in mathematics and in the real world, are not as straightforward.

Example 1.25. Use a logical deduction to show that the following argument, where P, Q and R are statements, is valid.

$$P$$
$$P \Rightarrow Q$$
$$\underline{\sim Q \vee R}$$
$$\therefore R$$

Our deduction is:

1. P	Assumption
2. $P \Rightarrow Q$	Assumption
3. $\sim Q \vee R$	Assumption
4. Q	Modus Ponens 1, 2
5. $\sim(\sim Q)$	Double Negative 4
6. R	Elimination 3, 5

Note the necessity of line 5 of the deduction in Example 1.25. To apply the law of elimination, the negation of one of the terms of the disjunction $\sim Q \vee R$ is required. The statement Q is not itself the negation of $\sim Q$; it is only logically equivalent to the negation by a law of inference. This attention to the details of definitions is crucial to mathematical proofs.

Before proceeding on to more challenging examples, it is worth discussing how logical deductions support a major them of this text: proofs are highly personalizable. What is provided here are just the required pieces, but there are often multiple ways to validate an argument using a logical deduction. Perhaps it is the order of the statements or the way in which you present it. Find and develop *your* logical deduction technique. Once developed, use that same technique consistently. You will be on your way to creating your mathematical voice. To that end, below is an alternative deduction to the argument of Example 1.25.

1.	(a)	P	Assumption
	(b)	$P \Rightarrow Q$	Assumption
		$\therefore Q$	Modus Ponens ((a) and (b))

2.	(a)	$\sim Q \vee R$	Assumption
	(b)	$\sim(\sim Q)$	Double Negative (1)
		$\therefore R$	Elimination ((a) and (b))

Determining how the logical deduction of Example 1.25 "worked" may not have been overly challenging. Perhaps you figured out the sequence of steps in your head: "this implies that, and then that gives me that, and that's the conclusion." As the number of assumptions grow, however, it becomes challenging to work through the thought progression in your head. The next example exhibits this.

Example 1.26. Prove that the following argument, assuming all variables represent statements, is valid.

$$(\sim P \vee Q) \Leftrightarrow R$$
$$S \vee \sim Q$$
$$\sim U$$
$$P \Rightarrow U$$
$$(\sim P \wedge R) \Rightarrow \sim S$$
$$\overline{\therefore \sim Q}$$

As before, we proceed with a logical deduction.

1. $(\sim P \vee Q) \Leftrightarrow R$	Assumption
2. $S \vee \sim Q$	Assumption
3. $\sim U$	Assumption
4. $P \Rightarrow U$	Assumption
5. $(\sim P \wedge R) \Rightarrow \sim S$	Assumption
6. $\sim P$	Modus Tollens 3, 4
7. $((\sim P \vee Q) \Rightarrow R) \wedge (R \Rightarrow (\sim P \vee Q))$	Theorem 1.17 1
8. $(\sim P \vee Q) \Rightarrow R$	Specialization 7
9. $\sim P \vee Q$	Generalization 6
10. R	Modus Ponens 8, 9
11. $\sim P \wedge R$	Conjunction 6, 10
12. $\sim S$	Modus Ponens 5, 11
13. $\sim Q$	Elimination 2, 12

After reading Example 1.26, you may be asking yourself, "How was *that* solution developed?" It should be noted that there is work behind the scenes here that is not obvious in the final product. Constructing a logical deduction is not a one-step problem. It requires thought and trial-and-error and should be approached like a puzzle. The pieces are the assumptions and the tools for solving it are theorems, rules of inference and previously derived steps.

Before proceeding to general tips on how to work through the logical deduction development puzzle, let us think about the scratch work that happened before the writing of the final draft of the deduction in Example 1.26.

Scratch work: Consider what we get to work with and what we are trying to show.

$$(\sim P \vee Q) \Leftrightarrow R$$
$$S \vee \sim Q$$
$$\sim U$$
$$P \Rightarrow U$$
$$\underline{(\sim P \wedge R) \Rightarrow \sim S}$$
$$\therefore \sim Q$$

We have five statements we can assume true, trying to deduce $\sim Q$. If we think about working forwards, straight from the assumptions that we are given, then the third and fourth statements are natural ones to begin with, mostly because they are the simplest and there is a natural conclusion to draw from the two of them together. Assuming $\sim U$ and $P \Rightarrow U$ yields $\sim P$ by Modus Tollens. Does that give us much to work with?

Having $\sim P$ as a true statement is helpful (there are two other assumptions with the variable P in them), but there is not an "obvious" way to "use" $\sim P$ with either the first or last statement. Reaching such a roadblock, we assess how we started the problem: working forwards. Would it have been more helpful to try and work backward?

Our end goal is to deduce $\sim Q$. Looking at the assumptions, knowing $S \vee \sim Q$ means that if we can arrive at $\sim S$, then Elimination will grant us $\sim Q$. Thus, that has become our new goal. Quickly, then, we see that if we can deduce $\sim P \wedge R$, we will have $\sim S$. Hence, we have a *new* new goal: work towards deducing the statement $\sim P \wedge R$. What will this entail? Deducing $\sim P$ (which we know we can do via our working forward approach above) and deducing R.

How can we arrive at the truth of R? The first statement: show $\sim P \vee Q$ and R must follow. But this desired statement, $\sim P \vee Q$, is a conjunction. We already have $\sim P$, thus we have $\sim P \vee Q$. Hence, we know R and consequently $\sim P \wedge R$, and then we simply must put together all of our thoughts.

While there are no definite rules for figuring out "how a logical deduction goes," here are some common tips. These tips are not isolated to logical deductions, however. Keep them in mind when you begin to construct your first mathematical proofs.

Scratch work: as mentioned previously, use scratch work to help derive your solution. Deductions often take effort.

Look for relationships and obvious steps: steps become obvious when you have mastered the tools of the trade. Mastery is more than just memorization, though. It is that deep-down understanding of what the tools are really saying. If you understand why the rules of inference hold, then you are well on your way to mastery. Are there obvious uses of Modus Ponens or Modus Tollens? Are there statements that are closely related somehow?

Work backward: you know where the logical deduction needs to end (the conclusion), so look for statements that may be the penultimate step of the logical deduction. For example, if the conclusion is the statement W and there is an assumption $(S \vee R) \Rightarrow W$, then deriving $S \vee R$ will lead to the conclusion. A next helpful thought would be to look to derive either S or R.

Persist: if you get stuck, take a step backward. If you know an argument is valid, look for other directions. If you are not told that it is valid and you are having trouble validating it, then perhaps it is an invalid argument.

How can a logical deduction be used to show an argument is invalid? Simple: it can't. The only way to show an argument is invalid is to find a particular truth assignment to all variables present in the argument, making all the hypotheses true and the conclusion false. Mathematicians call this *finding a counterexample* to an argument. We will discuss counterexamples more in-depth when we begin developing proof techniques.

In the next example we revisit Lewis Carroll's somewhat confusing argument from Example 1.1 and show that it is indeed valid.

Example 1.27. Show that Lewis Carroll's argument [8] below is valid.

Assumptions

(1) None of the unnoticed things, met with at sea, are mermaids.

(2) Things entered in the log, as met with at sea, are sure to be worth remembering.

(3) I have never met with anything worth remembering, when on a voyage.

(4) Things met with at sea, that are noticed, are sure to be recorded in the log.

Conclusion

I have never met with a mermaid at sea.

To validate this argument we must first translate it to symbolic statements. Define the following variables:

N: it is noticed
M: it is a mermaid
E: it is entered in the log
W: it is worth remembering
S: I have met with it at sea.

With these, our argument becomes

$$\sim N \Rightarrow \sim M$$
$$E \Rightarrow W$$
$$S \Rightarrow \sim W$$
$$N \Rightarrow E$$
$$\therefore S \Rightarrow \sim M.$$

We validate the argument via the following logical deduction.

1. $\sim N \Rightarrow \sim M$	Assumption
2. $E \Rightarrow W$	Assumption
3. $S \Rightarrow \sim W$	Assumption
4. $N \Rightarrow E$	Assumption
5. $\sim W \Rightarrow \sim E$	Theorem 1.14 2
6. $S \Rightarrow \sim E$	Transitivity 3, 5
7. $\sim E \Rightarrow \sim N$	Theorem 1.14 4
8. $S \Rightarrow \sim N$	Transitivity 6, 7
9. $S \Rightarrow \sim M$	Transitivity 1, 8

Logical deductions should be considered your first taste of mathematical proof. While rigid and somewhat cumbersome, they are actual mathematical proofs. They include the most basic concepts that all mathematical proofs must have: assumptions, intermediate steps and conclusions that follow logically from either the assumptions or previously proven results. In the sections and chapters ahead, we will develop proof techniques and spend time refining those that we will have already learned.

Exercises

1. Supply the justification for the steps in each of the following logical deductions.

    (a) Steps 4 and 5:

1. $(P \vee Q) \wedge \sim R$	Assumption
2. $P \Rightarrow S$	Assumption
3. $Q \Rightarrow S$	Assumption
4. $P \vee Q$	
5. S	

(b) Steps 5 through 8:

1. $Q \vee P$	Assumption
2. $R \Rightarrow \sim Q$	Assumption
3. $T \Rightarrow \sim P$	Assumption
4. R	Assumption
5. $\sim Q$	
6. P	
7. $\sim(\sim P)$	
8. $\sim T$	

(c) Steps 5 through 12:

1. $Q \Leftrightarrow R$	Assumption
2. $P \wedge (\sim Q \vee S)$	Assumption
3. $P \Rightarrow \sim S$	Assumption
4. T	Assumption
5. P	
6. $\sim Q \vee S$	
7. $\sim S$	
8. $\sim Q$	
9. $(Q \Rightarrow R) \wedge (R \Rightarrow Q)$	
10. $R \Rightarrow Q$	
11. $\sim R$	
12. $\sim R \wedge T$	

2. Use a logical deduction to show that the following are valid arguments. Assume all variables represent statements.

(a) $P \Rightarrow Q$
$Q \Rightarrow R$
$\underline{\sim R}$
$\therefore \sim P$

(b) $P \wedge (Q \wedge R)$
$\underline{(Q \wedge R) \Rightarrow S}$
$\therefore S$

(c) $Q \wedge P$
$(P \wedge Q) \Rightarrow \sim R$
$\underline{R \vee S}$
$\therefore S$

(d) $(P \vee Q) \vee R$
$\underline{\sim Q \wedge \sim R}$
$\therefore P$

3. Use a logical deduction to show that the following are valid arguments. Assume all variables represent statements.

(a) $P \wedge S$
$\underline{Q \Leftrightarrow \sim R}$
$\therefore P \wedge (Q \Leftrightarrow \sim R)$

(b) $R \wedge \sim Q$
$P \Rightarrow Q$
$\underline{S \Rightarrow \sim(R \wedge \sim P)}$
$\therefore \sim S$

(c) $\sim P \Leftrightarrow (S \vee Q)$
$S \vee (Q \wedge T)$
$\underline{W \Rightarrow P}$
$\therefore \sim W$

(d) $\sim P \vee \sim Q$
$\sim T \Leftrightarrow (P \wedge Q)$
$\sim T \vee S$
$\underline{S \Rightarrow (W \wedge U)}$
$\therefore U$

4. Determine the errors in the following logical deductions. Assume all variables represent statements.

(a)

1. $P \wedge Q$	Assumption
2. $P \Rightarrow R$	Assumption
3. R	Modus Ponens 1, 2

(b)

1. $P \vee Q$	Assumption
2. $P \Rightarrow R$	Assumption
3. $\sim Q$	Assumption
4. R	Modus Ponens 1, 2

(c)

1. $(P \vee Q) \wedge R$	Assumption
2. $P \Rightarrow \sim R$	Assumption
3. R	Specialization 1
4. $P \vee Q$	Generalization 1
5. $\sim P$	Modus Tollens 2, 3
6. Q	Elimination 4, 5

(d)

1. $\sim(P \wedge Q) \Rightarrow T$	Assumption
2. $R \Leftrightarrow (Q \vee \sim S)$	Assumption
3. $R \veebar (W \vee S)$	Assumption
4. S	Assumption
5. $W \vee S$	Generalization 4
6. $\sim R$	Example 1.11
7. $(Q \vee \sim S) \Rightarrow R$	Theorem 1.17
8. $Q \vee \sim S$	Modus Tollens 6, 7
9. $\sim(\sim S)$	Double Negative 4
10. $\sim Q$	Elimination 8, 9
11. $\sim P \vee \sim Q$	Generalization 10
12. $\sim(P \wedge Q)$	De Morgan's Law 11
13. T	Modus Ponens 1, 12

5. Translate each of the following arguments into symbolic form, then, use a logical deduction to validate them.

   (a) If we left at noon, we would not have gotten here before the sun set.
   We arrived while it was still light outside.
   So, we must have left before noon.

   (b) If Olivia or Allen was in charge, then the project was completed.
   The project was not finished.
   Therefore Allen could not have been in charge.

   (c) Andrea playing lead actress is both necessary and sufficient for the show to go on.
   Mrs. McCormack will be angry if and only if Andrea does not play lead actress.
   Thus, if Mrs. McCormack is happy, then the show must have taken place.

   (d) Paul is captain, and Quinn and Rory are alternate captains.
   Trish will be in the starting lineup only if Quinn is not an alternate captain.
   Trish is in the starting lineup if Will is injured.
   We can conclude that Will is not hurt.

6. The following are arguments of Lewis Carroll [8]. Translate them to symbolic statements and use a logical deducation to validate them.

   (a) All babies are illogical.
   Nobody is despised who can manage a crocodile.

Illogical persons are despised.

Therefore, anyone who can manage a crocodile is not a baby.

(b) No interesting poems are unpopular among people of real taste.

No modern poetry is free from affection.

All your poems are on the subject of soap-bubbles.

No affected poetry is popular among people of real taste.

No ancient poem is on the subject of soap-bubbles.

Therefore, your poetry is not interesting.

(c) No shark ever doubts that it is well fitted out.

A fish, that cannot dance a minuet, is contemptible.

No fish is quite certain that it is well fitted out, unless it has three rows of teeth.

All fishes, except sharks, are kind to children

No heavy fish can dance a minuet.

A fish with three rows of teeth is not to be despised.

Thus, no heavy fish is unkind to children.

2

Sets

Much like logic is a fundamental tool to mathematical reasoning, sets are extremely important objects in nearly every area of mathematics. You undoubtedly have encountered them in your previous mathematics courses. In fact, when you first started learning basic mathematics you dealt with them: the collections of numbers (the integers and the real numbers) are sets. Basic arithmetic, even basic counting, exhibits properties of these sets. When you worked with Venn diagrams in your early mathematical years, you were actually exploring set theory.

The history of sets goes back thousands of years. It was the ancient Greeks who investigated properties of various sets of real numbers (such as the positive integers). This was a very specific use of sets; they did not study general *sets* for the sake of studying *sets*. Rather, they, along with those who followed in their footsteps, considered particular sets one at a time.

It was not until the mid-to-late 19th century that focus switched to the theory of sets themselves. It was in his paper titled "On a Property of the Collection of All Real Algebraic Numbers" that Georg Cantor (1845-1918) established set theory as an active area of mathematical research [7], though it was not initially accepted as "proper" mathematics. Why? We shall dive deeper into that question in Chapter 7.

In this chapter, our goal is to develop the notation and theory of sets. This will lead to two particular proof techniques that we will see in Chapter 3. The first, called the *set element method of proof*, is particular to set theory. The second, however, is used throughout mathematics any time we want to show that something is unique.

With this notation and basic theory, we will then be able to address an issue arising in Section 1.1. There, we saw expressions such as

$$(x-1)(x+2) \geq 0$$

are not statements; depending on what value x is assigned, it becomes a statement that may be true or may be false. In Section 2.3, we transform such expressions into statements and spend the remainder of the chapter exploring them. Those lay the groundwork for the claims (and consequently proofs of those claims) to come.

It is worth noting that the set theory defined here is often referred to as *naïve set theory* [23]. Often considered "beginning" set theory, it is a sufficient treatment of the subject for developing most mathematical objects, including those presented in this text. It incorporates the English language (in particular, the words *and, or, all, some,* etc.) into the subject.

The word "naïve" has a somewhat negative connotation in the English language. Various definitions of the word include phrases such as, "unaffected simplicity," "unsophisticated" or "deficient in wisdom." Is this book simply presenting you with a "simple man's set theory," ultimately useless in mathematical practice? No, it is not. What makes the set theory presented here *naïve* is that it leads to paradoxes. The most famous of these, called *Russell's Paradox* (named after Bertrand Russell (1872-1970), one of the great minds to extend Cantor's ideas of logic[1]) simply asks: "is there a set of all sets?" Though we have not defined a *set*, think of it simply as a collection of things. Is there, then, a collection that contains all collections? Likewise, is there a collection of collections that *do not* contain themselves? Both of these questions are considered *paradoxes*.

If there is a *naïve* set theory, there ought to be some other kind of set theory as well. The alternative approach to naïve set theory is called *axiomatic set theory*, and it is a completely formal approach to the subject. Beginning with a slate of axioms (defined in the coming section), the theory is developed through a systematic application of rules and constructions. In this mathematical world, paradoxes do not happen. Axiomatic set theory is what is called a *consistent* theory.

Why then would one even want to develop and consider set theory in the *naïve* sense? It is likely due to the fact that it is the more natural of the two approaches. It is a generalization of the human language, and if the paradoxes are ignored, it creates a system for valid, intuitive and sound mathematics.

2.1   Set Theory Basics

The notion of a *set* is simple yet fundamental to nearly every branch of mathematics. While some texts do not explicitly define one, we do so here.

> **Definition 2.1.** A *set* is a collection of objects called *elements* of the set. If a is an element of the set S, we write "$a \in S$" and sometimes say, "a is in S." If a is not an element of S we write "$a \notin S$."

Why does the term *set* sometimes go undefined? In the above definition, we may ask, "What is a collection?" We then attempt to define it somehow, perhaps as "a compilation of objects..." How then is *compilation* defined?

---
[1]Russell excelled not just in mathematics; he was awarded the Nobel Prize in Literature in 1950.

"An assemblage of objects ... " Soon we find ourselves out of synonyms for *collection*. At some point we must rely on our intuition and understanding of the English language. Thus, we have reached an agreement:

> *You, the educated reader, understand what a collection is and therefore, I, the author, trust that you understand what a set is.*

Leaving certain terms undefined is not new to mathematics, however. Euclid began his *Elements* with undefined terms. He defined a *point* as "that which has no part" and a line or curve as "breadthless length." What, then, is a *part* or *breadth* or *length*? You can see the vicious cycle of undefinability beginning [27, 28, 26]. We accept the definitions for what they are. Definitions are not the only mathematical things sometimes assumed true without validation. Others, such as propositions or results, are assumed true. Regardless of the type of thing assumed true, those that are assumed to be true without validation are known as *axioms*.

Definition 2.2. An *axiom* (or *postulate*) is a mathematical statement assumed to be true without proof.

The appendix titled "Properties of Real Number System" lists those properties of the real number system, along with other mathematical results, taken as axioms for this text.

Before investigating the basics of sets, we remind ourselves of a few sets you are surely familiar with. The real numbers, denoted $\mathbb{R}$, are assumed to be known by the reader. The natural numbers, denoted $\mathbb{N}$, are whole numbers greater than zero. The integers, denoted by $\mathbb{Z}$, are the collection of all natural numbers, their negatives, and 0.[2] Lastly, the collection of all real numbers that can be expressed as a quotient of integers is known as the set of rational numbers, denoted by $\mathbb{Q}$. We may decorate these symbols with $+$, $-$, or $*$ to denote only the elements of the collection that are positive, negative or nonzero, respectively. For example, $\mathbb{Z}^+$ represents the same collection of real numbers as $\mathbb{N}$, whereas $\mathbb{R}^*$ is the set of all nonzero real numbers. The complex numbers, though not appearing in this text, are denoted $\mathbb{C}$.

Example 2.1. List three real numbers that are elements of the set $\mathbb{Q}$ but not elements of $\mathbb{Z}^*$.

The rational numbers $\mathbb{Q}$ are all numbers that can be represented as a quotient of integers, or commonly noted, as fractions of integers. The set $\mathbb{Z}^*$ is the set of all nonzero integers. Thus, any fraction that is not a

---

[2]This symbol comes from the first letter of *zahlen*, the German word for *integer*.

nonzero integer would be such a desired real number. There are infinitely many potential numbers to list, but we choose

$$0, \frac{1}{2} \text{ and } -\frac{9}{4}.$$

There are multiple ways to denote sets. In the real world, sets are often described verbally: "Let S be the set of all female students at University ABC who commute," or, "Consider all students in the classroom." Mathematically, this is not a useful presentation, as we saw in the previous chapter. We move then to a symbolic form of presentation, using the braces { and }. If a set has a finite number of elements, we may simply list all the elements of the set. For example, $E = \{2, 4, 6, 8\}$ is the set named E with four elements: 2, 4, 6, and 8. We can say $4 \in E$ but $5 \notin E$. If the set has an infinite number of elements, such as $\mathbb{Z}$, we can use an ellipsis ($\ldots$) to denote that a certain pattern continues indefinitely.

Example 2.2. Write the following sets using { } notation: (1) the non-negative integers, (2) the set of all integer powers of 2, and (3) the alternating (positive-to-negative) natural numbers.

(1) This set of non-negative integers contains $0, 1, 2$, etc. It can be denoted by

$$\{0, 1, 2, \ldots\}.$$

Note that it would be incorrect to include a "$\ldots$" before 0, as this would indicate the pattern would continue indefinitely "left," including negative integers in the set as well.

(2) The set of all integer powers of 2 contains $\frac{1}{4}$, $\frac{1}{2}$, 1, 2, 4, etc. We can write the desired set as

$$\left\{\ldots, \frac{1}{4}, \frac{1}{2}, 1, 2, 4, \ldots\right\}.$$

You may look at this set and think, "How do I know that the next element after 4 is not 6? The sequence 2 then 4 does not necessarily mean we are taking powers of 2. Perhaps the set is the sequence of even integers." While this questioning is correct, we assume the mathematically-minded reader looks at the *entire* set and observes a pattern.

(3) We must denote the set that includes $1, -2, 3, -4$, etc. There are two ways to do this. The first simply mimics the approach of the previous two sets.

$$\{1, -2, 3, -4, \ldots\}$$

The second option, however, includes an expression "after" the ... to exhibit an arbitrary nth element of the set.

$$\{1, -2, 3, -4, \ldots, (-1)^{n+1}n, \ldots\}$$

Is the set $F = \{8, 4, 2, 6\}$ different from the set $E = \{2, 4, 6, 8\}$? In the definition of a set there is no mention of the order in which elements appear in a set, only that a set is a collection of objects. The order in which those objects appear in the set is not important. However, as we learn to understand what it means to follow mathematical norms, writing a set in the, if possible, obvious way is often best. Thus, you will be hard-pressed to find in any mathematical literature the phrase, "Consider the set $\{8, 4, 2, 6\}$" in place of "Consider the set $\{2, 4, 6, 8\}$."

Knowing the the order in which elements of a set are presented is not important, we turn then to the question, "Must a set have *any* elements?" Nothing in the definition of a set requires this; a set is simply a collection of objects. For example, I very much dislike bugs. Yet, I am proud to say that I *do* have a bug collection. If you were to ask to see my bug collection, I would show you a box with nothing in it. You likely would respond, "That's not a bug collection," yet I would argue that indeed it is, albeit an unimpressive one. It just happens to have *no* bugs in it. In mathematical terminology, I would say that my bug collection is the *empty set*.

Definition 2.3. The *empty set* is the set with no elements and it is denoted $\varnothing$. A set that is not the empty set is called *nonempty*.

Note that we refer to the empty set as *the* empty set. We could have just as easily said, "A set with no elements is called *an* empty set." This would have a different meaning, however. *The* implies that there is only one of these sets, that this set we call the empty set is somehow unique. It may seem intuitively obvious, but mathematicians require more than intuition. In Chapter 3 we will prove the claim that there is one and only one set that has no elements.[3]

Example 2.3. Describe the following four sets: (1) $\varnothing$, (2) $\{\ \}$, (3) $\{\varnothing\}$ and (4) $\{\{\varnothing\}\}$.

---

[3] If you, too, dislike bugs, then your bug collection looks an awful lot like my bug collection. Would we say that they are the *same* collection?

---

(1) This set, $\varnothing$ is the empty set. It is the set with no elements.

(2) This set, { }, is also the empty set. It has no elements.

(3) This set, $\{\varnothing\}$, is *not* the empty set. It is a set with one element: the empty set. To visualize this, imagine a set as a basket and its elements as objects in the basket. The empty set would be an empty basket. The third set listed above, $\{\varnothing\}$, is a basket with an empty basket in it.

(4) Lastly, $\{\{\varnothing\}\}$ is a set with one element, which happens to be the set with the empty set in it. In terms of our basket visualization, it is three nested baskets: a basket that contains a basket that contains an empty basket.

Perhaps a more useful and commonly used notation for sets is what is called *set-builder notation*. For example, if H is the set of all even positive integers less than 45, then we could explicitly list out H. It would be a rather lengthy process and the result is awkward:

$$H = \{2, 4, 6, 8, 10, 12, 14, 16, 18, 20, 22, 24, 26, 28, 30, 32, 34, 36, 38, 40, 42, 44\}.$$

Set-builder notation utilizes variables to expedite the process:

$$H = \{x \in \mathbb{Z}^+ \mid x \text{ is even and } x < 45\}.$$

The vertical bar $\mid$ is read as "such that." What precedes it is a symbolic representation of all values the variable x could *possibly* have. In the set H, the set $\mathbb{Z}^+$ is called the *universal set* for the variable x. We know that regardless of whatever comes after the $\mid$ symbol, H consists only of positive integers. But which ones? That depends on what follows the $\mid$ symbol: the *defining condition*. It is all the properties that any such value x from the universal set must satisfy in order to be in the set.

Example 2.4. List the elements of the set

$$Q = \{x \in \mathbb{R} \mid x(x + 7) = 8\}.$$

First, we know that Q contains only real numbers. To determine which ones, we look at the defining condition. Here, we see that Q consists of all real numbers x that are solutions to

$$x(x + 7) = 8,$$

namely 1 and -8. Thus, the elements of Q are 1 and -8. Using our previous notation, we can write

$$Q = \{1, -8\}.$$

Example 2.5. Show that set-builder notation is not unique by finding two different ways to represent the above set H in set-builder notation.

In trying to rewrite H, we can manipulate either the universal set or the defining condition to find different set-builder representations. If the universal set is changed to $\mathbb{Z}$, then we must change the defining condition; otherwise H would contain every even integer less than 45, including negative numbers and 0. In particular,

$$-4 \in \{x \in \mathbb{Z} \mid x \text{ is even and } x < 45\}$$

but $-4 \notin H$. Thus, we adjust and see we can write H as:

$$H = \{x \in \mathbb{Z} \mid x \text{ is even and } 0 < x < 45\}.$$

We can also rewrite H by only manipulating the defining condition:

$$H = \{x \in \mathbb{Z}^+ \mid x = 2n \text{ for some integer } n \text{ and } x < 45\}.$$

It is important to note that these are just two ways to rewrite H; there are many more.

In the real world, we often consider only some of the elements of a set. Consider this specific situation. For a particular study, a political scientist is researching the most recent U.S. presidential election. She wants data from only the states that were lost by the overall winner of the election. She is looking at only a subcollection of the set of all U.S. states. Rather than consider the set

$$U = \{\text{Alabama, Alaska, } \ldots, \text{ Wisconsin, Wyoming}\},$$

she may only be looking at the set

$$L = \{\text{Colorado, Minnesota, New Mexico, Pennsylvania}\}.$$

These are different sets, yet there is a relationship between the two: every element of L is also an element of U.

Likewise, in mathematical notation, consider the two sets

$$A = \{a, b, c\} \text{ and } B = \{a, b, c, d, e\}.$$

These sets seem to have the same relationship as the sets the political scientist was investigating. Every element of A is also an element of B. In everyday language, we would say that A is a subcollection of the set B.

Definition 2.4. A set S is called a *subset* of a set T if every element of the set S is also an element of the set T. In this case, we write "$S \subseteq T$." If S is not a subset of T, we write "$S \nsubseteq T$." If S is a subset of T and there is an element of T that is not an element of S, we say S is a *proper subset* of T, written "$S \subset T$."

The choice of notation for subset and proper subset is meant to mirror the notation you have come to know about the real number line. We write $3 \leq 4$ and $4 \leq 4$ to mean that both 3 and 4 are less than or equal to 4, respectively. However, we only write $3 < 4$; 3 is strictly smaller than 4. The symbols $\subseteq$ and $\subset$ parallel this idea.

Example 2.6. List all subsets of $T = \{3, 4, 5\}$. Which of those subsets are proper subsets?

In determining subsets of a given set, approach it as building subcollections. We need to construct all subcollections of T, a set with 3 elements. Thus, a subcollection could have up to 3 elements in it.

0 elements: there is only one set with 0 elements:

$$\varnothing.$$

1 element: there are three choices for that one element to be (3, 4 or 5), meaning there are three one-element subsets of T:

$$\{3\}, \{4\}, \{5\}.$$

2 elements: from the 3 elements of T, we must choose 2 to be elements of a constructed subset. There are three ways to do this:

$$\{3, 4\}, \{3, 5\}, \{4, 5\}.$$

3 elements: because T has 3 elements, there is only one way to create a subset with 3 elements, by including all the elements of T in it.

$$\{3, 4, 5\}.$$

Of all these subsets, those with 0, 1 or 2 elements are proper. The set $\{3, 4, 5\}$ is the only non-proper subset of T.

Example 2.7. Which of the following sets are subsets of each other? Which are proper subsets of each other?

1. $\varnothing$

2. $A = \{1, 3, 5\}$

3. $B = \{x \in \mathbb{Z}^+ \mid 1 \le x \le 5\}$

4. $C = \{x \in \mathbb{Z} \mid 0 \le x \le 6 \text{ and } x = 2n + 1 \text{ for some } n \in \mathbb{Z}\}$

5. $D = \{2, -3, 4\}$

By definition, each set is a subset of itself. Since $\varnothing$ has no elements, it is a subset of A, B, C, and D. Moreover, since each of A, B, C, and D are nonempty, $\varnothing$ is a proper subset of each of the other sets (because there is an element in each of them that is not an element of $\varnothing$).

To determine relationships between the other sets, we rewrite B and C by listing their elements: $B = \{1, 2, 3, 4, 5\}$ and $C = \{1, 3, 5\}$. Then it is easy to see that $A \subset B$ (since $2 \in B$ but $2 \notin A$), $A \subseteq C$, and $C \subseteq A$. The set D contains elements not in A, B or C, and vice-versa. Thus, D is not a subset of any of the other sets and none of the other nonempty sets are subsets of D.

Example 2.8. Let A, B, and C be sets. If $A \subseteq B$ and $B \subseteq C$, show that $A \subseteq C$.

Suppose A, B, and C are sets with $A \subseteq B$ and $B \subseteq C$. How can we show that $A \subseteq C$? We need to show that every element of A is also an element of C. We do this by starting with an *arbitrary but specific* element of A: take a in A.

Keep in mind what we are told is true: $A \subseteq B$ and $B \subseteq C$. By using the subset definition, since $a \in A$, we can conclude that $a \in B$. Likewise, by the same definition, we know then that $a \in C$.

If we are completely burrowed down in the problem, we may find ourselves asking, "Where do I go from here? I have no other tools to work with. But always stay keen to what it is you are trying to show. In this case, we want to show that every element of A is also an element of C. That is the conclusion we reached.

If you are familiar with mathematical theory and proofs, you may be thinking, "The argument in Example 2.8 feels like a proof. Why not simply go about *proving* the result, rather than saying we are just *showing* it is true. Our primary goal here is to gain comfort with the language of sets; we refrain from using the language of proofs until Chapter 3 where we will focus on their details.

In Example 2.9, you may be inclined to say that sets A and C are *equal* but you must exercise caution. We have yet to define what actually makes two sets equal, and intuition, again, is not a valid justification. Now that we have defined subsets, however, defining set equality becomes a rather simple task, and it coincides with our intuitive reasoning. We would say that the sets A and C of Example 2.9 are equal because every element in A is also in C, and every element in C is also in A.

> **Definition 2.5.** Sets R and S are said to be *equal*, denoted $R = S$, if $S \subseteq R$ and $R \subseteq S$.

While seemingly obvious, Definition 2.5 is a very important definition. Sets appear throughout mathematics and oftentimes one aims to show that two complicated sets are equal. Definition 2.5 provides the method for doing that: show that the sets are subsets of one another. But how does one go about generally showing a set R is a subset of a set S? It is the same approach taken in Example 2.8.

1. Assume r is an arbitrary but specific element of R.

2. Show that r is an element of S.

Showing that one set is *not* a subset of another is a much simpler task: exhibit it via a counterexample. That is, to show $R \not\subseteq S$, find one particular element of R that is not an element of S. For example, if D and B are the sets from Example 2.9, $D \not\subseteq B$ is justified by simply saying, "$-3 \in D$ but $-3 \notin B$."

Example 2.9. Show that $F \subset E$ if we are given that

$$E = \{x \in \mathbb{Z} \mid x = 2n \text{ for some } n \in \mathbb{Z}\},$$

and

$$F = \{x \in \mathbb{Z} \mid x = 4n \text{ for some } n \in \mathbb{Z}\}.$$

Take note of what we are asked to show: F is a *proper* subset of E. To do this, we show $F \subseteq E$ and that there is an integer $x \in E$ such that $x \notin F$.

To proceed, let $x \in F$. Then, the defining condition for F gives that

$$x = 4n$$

for some $n \in \mathbb{Z}$. In order to conclude that $x \in E$, we must show that x is of the form 2 times *some integer*. Simple algebra does the trick:

$$x = 2(2n).$$

Because $2n \in \mathbb{Z}$ (since $n \in \mathbb{Z}$, applying the closure law axiom of the real numbers), we have shown that $x \in E$ and consequently that that $F \subseteq E$.

Next, we need to show that there is an integer $x \in E$ such that $x \notin F$. Take $x = 6$, which is an element of E since $6 = 2 \cdot 3$. However, 6 is not an integer multiple of 4 since $6 = 4 \cdot (1.5)$ and $1.5 \notin \mathbb{Z}$. Therefore, $6 \notin F$, and we have shown $F \subset E$.

In the latter part of the previous example, observe that the counterexample used to show that F is indeed a proper subset of E consists of one specific integer. There is no need to use general variables when it comes to finding counterexamples to claims. Moreover, the use of variables is often incorrect. Let us see where this would have failed in the previous example.

In finding an element of E that is not an element of F, suppose our argument proceeded as follows:

Let $x \in E$. Then, $x = 2n$ for some $n \in \mathbb{Z}$. Notice that x is not of the form $4m$, where $m \in \mathbb{Z}$. So, $x \notin F$.

Where does this reasoning fail? It fails in that it is proclaiming that no matter the choice of x, we can conclude that x is not an element of F. But what if $x = 8$? This is an integer both of the form $2n$ and of the form $4m$, for some $n, m \in \mathbb{Z}$.

In short, always use specific elements as counterexamples. In disproving a claim that some property is true for *all* things, you need only find one specific thing for which the property does not hold.

The next example highlights Definition 2.5.

Example 2.10. Show the following two sets are equal:

$$A = \{x \in \mathbb{Z} \mid x = 2n \text{ for some } n \in \mathbb{Z}\}$$

and

$$B = \{x \in \mathbb{Z} \mid x = 2n - 2 \text{ for some } n \in \mathbb{Z}\}.$$

To show $A = B$, we must show that A and B are subsets of one another. First, we show $A \subseteq B$. To do so, let $x \in A$. Then,

$$x = 2n$$

for some $n \in \mathbb{Z}$. In order to justify that $x \in B$, keep our eyes on the prize: show

$$x = 2m - 2$$

for some integer m. Then

$$x = 2(n+1) - 2.$$

Since $n + 1 \in \mathbb{Z}$, we have that x is an integer satisfying the defining condition of B, meaning $x \in B$. Thus, $A \subseteq B$.

Similarly, take $x \in B$. Then,

$$x = 2n - 2$$
$$= 2(n - 1),$$

for some $n \in \mathbb{Z}$, and since $n - 1 \in \mathbb{Z}$, we have that $x \in A$. Therefore $B \subseteq A$ and consequently $A = B$.

In this section we have developed the important definitions of sets and subsets and introduced two new proof techniques. In the next section, we investigate how old sets can be combined to form new sets.

Exercises

1. List the elements of the given sets.

   (a) $\{x \in \mathbb{Z} \mid 1 \leq x \leq 2\pi\}$
   (b) $\{m \in \mathbb{Z}^* \mid |m| < 3\}$
   (c) $\{x \in \mathbb{Z}^+ \mid -4 \leq x \leq 4\}$
   (d) $\{4n + 1 \mid n \in \mathbb{Z}, 0 \leq n \leq 3\}$

2. Write each of the following sets using set-builder notation.

   (a) $\{1, 3, 5, 7, 9\}$
   (b) $\{1, 2, 4, 8, 16\}$
   (c) $\{3\pi, 6\pi, 9\pi, 12\pi, 15\pi\}$
   (d) $\{2, -4, 6, -8, 10\}$
   (e) $\{\ldots, -30, -10, 10, 30, \ldots\}$

3. How many elements does each of the following sets have?

   (a) $\{0\}$
   (b) $\{\varnothing, \{\varnothing\}\}$
   (c) $\{\mathbb{Z}, \mathbb{Q}\}$
   (d) $\{\{\{\{1, 2, 3\}\}\}\}$
   (e) $\{1, \{2, \{3, \{\varnothing\}\}\}\}$

4. Determine which of the four sets below are equal and which are subsets of one another.

   (a) $A = \{a, e, i, o, u\}$
   (b) $B = \{a, e, i, o, u, x, y\}$

    (c) $C = \{i, o, u, e, a\}$

    (d) $D = \{x \mid x$ is a vowel in the English alphabet or $x = y\}$

5. Let $A = \{1, 2, 3\}$, $B = \{0, 1, 2, 3\}$ and $C = \{\{1, 2\}, \{3\}\}$. Determine if the following are true or false.

    (a) $1 \in A$

    (b) $\{1\} \in A$

    (c) $\varnothing \subseteq A$

    (d) $\varnothing \subset B$

    (e) $\varnothing \in C$

    (f) $A \subseteq B$

    (g) $A \subset B$

    (h) $A \subseteq C$

    (i) $A = C$

    (j) $\{1, 2\} \in C$

    (k) $\{3\} \subseteq C$

    (l) $\{3\} \in C$

6. List three elements of the following sets or explain why the set does not have three elements.

    (a) {three-letter words in the English language that consist of one vowel and two consonants}

    (b) {symbolic statements with only one variable P}

    (c) {statements in English that are sometimes true and sometimes false}

    (d) $\mathbb{Q}^+$

    (e) $\{x \in \mathbb{Z}^+ \mid x^2 + 5 < 16\}$

    (f) $\{\{1, 2, 3, 4, 5\}\}$

    (g) $\{\varnothing, \{2\}, \pi, \{2\pi, \{3\}\}\}$

    (h) $\{x \in \mathbb{R} \mid x \in \mathbb{Q}, x \notin \mathbb{Z}^+, x > 0\}$

7. Write the empty set three different ways using set-builder notation.

8. Determine which of the following sets are subsets of the others. Of those which are, are they proper subsets?

$$\mathbb{Z}^-, \ \mathbb{Z}^+, \ \mathbb{Z}^*, \ \mathbb{R}, \ \mathbb{R}^+, \ \{x \in \mathbb{R} \mid x^2 < -2\}$$

9. List all proper subsets of

    (a) $\{2, 8, 12\}$

    (b) $\varnothing$

    (c) $\{x \in \mathbb{Z} \mid 0 < x^2 < 4\}$

    (d) $\{\{\{\varnothing\}\}\}$

10. Find a counterexample to show $A \not\subseteq B$:

    (a) $A = \{1, 2, 3\}$
       $B = \{x \in \mathbb{Z} \mid |x| < 3\}$

    (b) $A = \{n \in \mathbb{Z} \mid n = 2m + 1 \text{ for some } m \in \mathbb{Z}\}$
       $B = \{n \in \mathbb{Z} \mid n = 4m + 1 \text{ for some } m \in \mathbb{Z}\}$

    (c) $A = \{n \in \mathbb{Q} \mid n \notin \mathbb{Z}\}$
       $B = \{n \in \mathbb{R} \mid n \text{ cannot be written as a decimal that terminates}\}$

    (d) $A = \{5, 6, 7\}$
       $B = \{\text{all subsets of } A\}$

11. Let $A = \{4n \mid n \in \mathbb{Z}\}$.

    (a) Write A in the form $A = \{x \in \mathbb{Z} \mid \ldots\}$.

    (b) Is A closed under addition? That is, if $x, y \in A$, must $x + y \in A$? Justify.

    (c) Is A closed under multiplication? That is, if $x, y \in A$, must $xy \in A$? Justify.

12. Let $A = \{x \in \mathbb{Z}^+ \mid \sqrt{x} \in \mathbb{Z}\}$ and $B = \{x^2 \mid x \in \mathbb{Z}\}$. Are A and B equal? If not, is either a subset of the other? Justify.

13. Explain why the following statement is true or find a counterexample to show it is false:

    Every nonempty set A is equal to the set of all subsets of A.

14. If a set S has n elements, make a conjecture about how many subsets S has.

15. For every integer n, let $A_n = \{n, n^2, n^3\}$.

    (a) Find the sets A_1, A_2 and A_{-3}.

    (b) How many elements (which may depend on n) are in A_n?

    (c) Let m and n be different integers. Explain why $A_m \neq A_n$.

16. For any $i \in \mathbb{Z}^+$, let D_i be the set

$$D_i = \{in \mid n \in \mathbb{Z}\}$$

    (a) Find 4 elements in each of D_2, D_3 and D_4.

    (b) Find 3 elements that D_2 and D_5 have in common.

    (c) In plain English, describe the set D_i.

    (d) What relationship between the positive integers i and j will guarantee that $D_i \subseteq D_j$?

17. If D is the set of all differentiable real-valued functions and C is the set of all continuous real-valued functions, is $C \subseteq D$? Is $D \subseteq C$? Does $C = D$? Justify.

18. Let P be the set defined as

$$P = \{\{L_1, L_2\} \mid L_1 \text{ and } L_2 \text{ are distinct parallel lines in } \mathbb{R}^2\}.$$

   (a) List 3 elements of P.
   (b) Is $\varnothing \subseteq P$? Is $\{\varnothing, \varnothing\} \in P$?
   (c) For any line L_1 in $\mathbb{R}^2$, must there exist a line L_2 in $\mathbb{R}^2$ such that $\{L_1, L_2\} \in P$? If so, how can such a line be found?
   (d) If $\{L_1, L_2\} \in P$ and $\{L_2, L_3\} \in P$ (with $L_1 \neq L_2$), must $\{L_1, L_3\}$ be an element of P?

19. For every $i \in \mathbb{R}$, let

$$S_i = \{S \mid S \subset \{\varnothing, \{-i, i, \{-i, i\}\}\}\}.$$

   (a) Find S_1, S_2, and S_π.
   (b) Do S_i and S_j have the same number of elements, for any real numbers i and j?
   (c) List all the subsets of S_i.

2.2 Properties of Sets

When mathematicians define new *things*, they first ask what makes two of those new *things* equal. Then they ask how those *things* can be combined to create more *things*. Set theory is no different. Definition 2.5 accomplished the goal of determining when two sets are equal. Now we aim to combine sets in various ways to create new sets. As we define these set operations, you should begin to notice parallels between set theory and symbolic logic. If you find yourself asking, "Why," consider this:

For any set A, it is either *true* or *false* to say that that $x \in A$.

Also, note that a universal set is often referenced. This set represents every possible element that exists (hence the name *universal*). For example, if A is the set of all solutions to $x^2 + x = 0$, we need to know what universal set A belongs to. If the universal set is $\mathbb{R}$, then $A = \{-1, 0\}$. If the universal set is $\mathbb{Z}^+$, then $A = \varnothing$. However, if the universal set is the set of all non-negative integers, then $A = \{0\}$. Always be cognizant of the universe a particular set is assumed to be in. This will usually be explicitly told or context will make it clear.[4]

---

[4]For example, an algebra student asked to solve $x^2 + x = 0$ will likely receive a low score if he says no solutions exist. Technically, the student is correct (but would be better served by saying "no solutions exist *in the positive integers*"). Context of the problem is that the universal set is considered to be $\mathbb{R}$.

We begin by combining sets using terms that should sound familiar from your previous investigations of Venn diagrams.

Definition 2.6. Let A and B be sets, both belonging to some universal set U.

(1) The *union* of A and B, denoted $A \cup B$, is the set of all elements in at least one of A or B. Symbolically,

$$A \cup B = \{x \in U \mid x \in A \text{ or } x \in B\}.$$

(2) The *intersection* of A and B, denoted $A \cap B$, is the set of all elements in both A and B. Symbolically,

$$A \cap B = \{x \in U \mid x \in A \text{ and } x \in B\}.$$

(3) The *difference* A minus B, denoted $A - B$, is the set of all elements of A not in B. Symbolically,

$$A - B = \{x \in U \mid x \in A \text{ and } x \notin B\}.$$

(4) The *complement* of A, denoted A^C, is the set of all elements in the universe U not in A. Symbolically,

$$A^C = \{x \in U \mid x \notin A\}.$$

These set operations can be visualized using Venn diagrams, as in Figure 2.1. While Venn diagrams are wonderful tools for determining if set theoretic results hold, they do not serve as rigid justification (or disproof) of results. Example 2.11 highlights this.

Example 2.11. Show that for sets A and B, $A \cap B \subseteq A$.

Figure 2.1(b) exhibits this containment. The shaded region, depicting $A \cap B$, is clearly contained in A. But this is not a proof. Perhaps the picture is misleading and is drawn in such a way that it supports the result. A precise mathematical approach is to use the method for showing one set is a subset of another discussed in Section 2.1.

Consider a general $x \in A \cap B$. By Definition 2.6, $x \in A$, giving us that

$$A \cap B \subseteq A.$$

However, by taking a general $x \in A \cap B$, we are assuming that such an x exists. If A and B have no elements in common, then vacuously, every element of $A \cap B$ is also an element of A, yielding the same result.

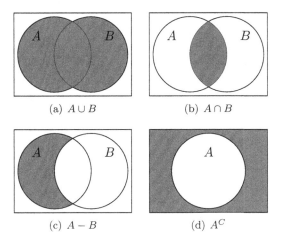

(a) $A \cup B$ (b) $A \cap B$

(c) $A - B$ (d) A^C

FIGURE 2.1
Set operations, represented as shaded areas

Often, when mathematicians are asked to show a set A is a subset of another, they begin by saying, "Let $x \in A$," much like we did in the previous example. Yet, the special case where A actually has no elements is not usually pointed out (as it was in Example 2.11). As you will come to learn, understanding your audience is key to writing a good mathematical solution or proof. More often than not, it is assumed that your reader understands that if A were empty, the result holds vacuously. Additionally, you will come to learn that actually *including* this trivial situation can serve to clutter up your solution. Finding the balance between too little and too much is critical, and we will work diligently to understand where that fine line lies.

The special case in Example 2.11 of two sets having no elements in common is an important term in set theory. Two such sets are called *disjoint*.

Definition 2.7. Sets A and B are called *disjoint* if $A \cap B = \varnothing$.

The result in Example 2.11 is just one property that holds for *any* sets A and B. Results such as these are called *set identities* or *laws*. Theorem 2.8 provides many more.

Theorem 2.8. *Let A, B, and D be sets in a universal set U. The following hold.*

1. *Inclusion of Intersection*	$A \cap B \subseteq A$
	$A \cap B \subseteq B$
2. *Inclusion of Union*	$A \subseteq A \cup B$
	$B \subseteq A \cup B$
3. *Transitivity of Subsets*	If $A \subseteq B$ and $B \subseteq D$, then $A \subseteq D$.
4. *Commutativity*	$A \cup B = B \cup A$
	$A \cap B = B \cap A$
5. *Associativity*	$(A \cup B) \cup D = A \cup (B \cup D)$
	$(A \cap B) \cap D = A \cap (B \cap D)$
6. *Distribution*	$A \cup (B \cap D) = (A \cup B) \cap (A \cup D)$
	$A \cap (B \cup D) = (A \cap B) \cup (A \cap D)$
7. *Identity*	$A \cup \varnothing = A$
	$A \cap U = A$
8. *Complement*	$A \cup A^C = U$
	$A \cap A^C = \varnothing$
9. *Double Complement*	$(A^C)^C = A$
10. *Idempotent*	$A \cup A = A$
	$A \cap A = A$
11. *Universal Bound*	$A \cup U = U$
	$A \cap \varnothing = \varnothing$
12. *Universal Complements*	$U^C = \varnothing$
	$\varnothing^C = U$
13. *Set Difference*	$A - B = A \cap B^C$

We verify here one of the Distribution properties, making a key observation, and leave most of the remainder as exercises in Section 3.3 (where we learn a proof technique called the *set element method*).

Example 2.12. For sets A, B, and D in a universal set U, show that

$$A \cup (B \cap D) = (A \cup B) \cap (A \cup D).$$

What is it we ultimately must show? We have to show two *sets* are equal. Definition 2.5 provides an outline for doing this: show that

$$A \cup (B \cap D) \subseteq (A \cup B) \cap (A \cup D)$$

and

$$(A \cup B) \cap (A \cup D) \subseteq A \cup (B \cap D).$$

Using our arbitrary-but-specific element approach, we proceed by showing the first of these.

Let $x \in A \cup (B \cap D)$. Definition 2.6 tells us $x \in A$ or $x \in B \cap D$. If $x \in A$, then $x \in A \cup B$ by the Inclusion of Union law of Theorem 2.8 (we are assuming any results prior to the one we are showing have been shown true). Likewise, we have that $x \in A \cup D$.

Now suppose $x \in B \cap D$. By Definition 2.6, then, $x \in B$ and $x \in D$. Knowing that x is an element of B, we can say x is an element of A or B, and likewise, x is an element of A or D. The definition of set union yields $x \in A \cup B$ and $x \in A \cup D$, and consequently, $x \in (A \cup B) \cap (A \cup D)$. This shows

$$A \cup (B \cap D) \subseteq (A \cup B) \cap (A \cup D).$$

Now we go about showing

$$(A \cup B) \cap (A \cup D) \subseteq A \cup (B \cap D).$$

Let $x \in (A \cup B) \cap (A \cup D)$. Then, again by Definition 2.6, both $x \in A \cup B$ and $x \in A \cup D$. Therefore, $x \in A$ or $x \in B$. Consider both possibilities.

If $x \in A$, then $x \in A \cup B$ and $x \in A \cup D$ (Theorem 2.8, Inclusion of Union). If $x \notin A$, then we know $x \in B$ (since $x \in A$ or $x \in B$). Also, since $x \in A \cup D$ means $x \in A$ or $x \in D$, it must be that $x \in D$. Thus, $x \in B \cap D$, and consequently, $x \in A$ or $x \in B \cap D$, or equivalently, $x \in A \cup (B \cap D)$. Thus, we have shown that

$$(A \cup B) \cap (A \cup D) \subseteq A \cup (B \cap D).$$

Having shown the sets are subsets of each other, we have

$$A \cup (B \cap D) = (A \cup B) \cap (A \cup D).$$

De Morgan's Law of symbolic logic has a parallel in set theory, known by the same name. This should come as no surprise, as unions and intersections of sets are defined using the words "and" and "or," respectively. Their proofs are left as exercises in Section 3.3. You are encouraged to intuitively justify these results with Venn diagrams, however.

Theorem 2.9. (De Morgan's Laws): *If A and B are sets, then*

    1. $(A \cap B)^C = A^C \cup B^C$, *and*

    2. $(A \cup B)^C = A^C \cap B^C$.

Associativity of unions and intersections allow for writing arbitrary unions and intersections without the need for grouping symbols: $A \cap (B \cap C)$ can be written simply as $A \cap B \cap C$. Such notation could get overwhelming if we were to consider the intersection of 10 sets, for example. To address this problem, we introduce a shorthand notation for an arbitrary finite union or intersection. If A_i is a set for $i = 1, 2, \ldots, n$ for some $n \in \mathbb{Z}^+$, then

$$A_1 \cup A_2 \cup \cdots \cup A_n = \bigcup_{i=1}^{n} A_i,$$

and

$$A_1 \cap A_2 \cap \cdots \cap A_n = \bigcap_{i=1}^{n} A_i.$$

Example 2.13. For $i \in \mathbb{Z}^+$, let

$$A_i = \{x \in \mathbb{R} \mid -\frac{1}{i} \le x \le \frac{1}{i}\}.$$

That is, $A_i = \left[-\frac{1}{i}, \frac{1}{i}\right]$, in standard interval notation. Find

    1. $\displaystyle\bigcup_{i=1}^{6} A_i$

    2. $\displaystyle\bigcap_{i=1}^{6} A_i$.

(1) Observe that $A_1 \cup A_2 = [-1, 1]$. Generalizing,

$$\bigcup_{i=1}^{6} A_i = [-1, 1] = A_1.$$

(2) Likewise, $A_1 \cap A_2 = \left[-\frac{1}{2}, \frac{1}{2}\right]$. Thus,

$$\bigcap_{i=1}^{6} A_i = \left[-\frac{1}{6}, \frac{1}{6}\right] = A_6.$$

Suppose in the previous example that we were asked not for the union or intersection of a finite number of the sets A_i but for the union or intersection of *all* sets A_i. The notation for an infinite union or intersection is a natural generalization of arbitrary finite unions or intersections. We state the definition in terms of index sets.[5] While an index set can be defined more precisely in terms of functions, we provide the following definition.

Definition 2.10. An *index set* is a set whose elements label the elements of another set.

Example 2.14. Consider the set

$$\{A_1,\, A_2,\, A_3,\, A_4\},$$

with A_i coming from Example 2.13. This collection of sets is indexed by the set $\{1,2,3,4\}$. The set

$$\{A_1,\, A_2,\, A_3,\, \ldots\}$$

is indexed by $\mathbb{Z}^+$. If the A_i are generalized to any positive real number i, then $\mathbb{R}^+$ is the index set for

$$\{A_i \mid i \in \mathbb{R}^+\}.$$

It is important to note some of the differences in various index sets. Consider the three in Example 2.14. The first of these, $\{1,2,3,4\}$, is a finite set, while the second, $\mathbb{Z}^+$, is not[6]. The third index set, $\mathbb{R}^+$, is also infinite but it has a very different sort of "infinite-ness." There is a natural way to list the elements of the set indexed by $\mathbb{Z}^+$: A_1, A_2, A_3, ... If we try to do the same thing for the set

$$\{A_i \mid i \in \mathbb{R}^+\},$$

we find ourselves a bit perplexed. Where do we begin? There is no "first" positive real number. In fact, if we list two of these sets successively, $A_{.5}$ and $A_{.6}$, there are still infinitely many of the indexed sets "between" these. We will address this notion of infinite sets in Chapter 7.

The next example highlights the notation for infinite unions and intersections.

---

[5]It is common for mathematicians to talk about objects being defined "naturally." This is simply a way to say, "This thing is being defined intuitively. It works like you think it ought to work."

[6]We have not yet defined what it means for a set to be finite or infinite, but it is assumed the reader has an intuitive feeling for the terms.

Example 2.15. Let A_i be the sets of Example 2.13, where $i \in \mathbb{R}^+$. Then,

$$A_1 \cap A_2 \cap A_3 \cap \cdots = \bigcap_{i=1}^{\infty} A_i = \{0\},$$

while

$$\bigcup_{i \in \mathbb{R}^+} A_i = \mathbb{R}.$$

Many of the properties that hold for the union or intersection of two sets carry over naturally to results about unions or intersections of any finite number of sets. Proving these results, however, requires a proof technique known as *mathematical induction*. In Section 3.3, we use this technique to verify such generalizations (such as De Morgan's Law on a general finite number of sets, rather than simply two, as it was stated in Theorem 2.9).

We continue now in creating new sets from old. Subsets of a set give rise to numerous important definitions (including set equality, as previously defined). Here, we use the notion of a subset to create a new set from one that is given: a set whose elements are themselves sets.

Definition 2.11. Let A be a set. The *power set of A*, denoted $\mathcal{P}(A)$, is the set of all subsets of A.

Example 2.16. Find $\mathcal{P}(A)$, if A is the given set.

1. $A = \{1\}$
2. $A = \{1,3\}$
3. $A = \{1,3,5\}$

Finding all the elements of $\mathcal{P}(A)$ equates to finding all subsets of A. We developed an approach this problem previously: determine all subsets with 0 elements, all subsets with 1 element, all subsets with 2 elements, etc. Thus,

1. $\mathcal{P}(\{1\}) = \{\varnothing, \{1\}\}$,
2. $\mathcal{P}(\{1,3\}) = \{\varnothing, \{1\}, \{3\}, \{1,3\}\}$, and
3. $\mathcal{P}(\{1,3,5\}) = \{\varnothing, \{1\}, \{3\}, \{5\}, \{1,3\}, \{1,5\}, \{3,5\}, \{1,3,5\}\}$.

Example 2.17. Is the power set of the empty set itself empty?

The power set of the empty set is the set of all subsets of the empty set. Thus, if we were to answer the initial question affirmatively, that *yes*, the power set of the empty set *is* empty, we would be saying that the empty set has no subsets. Is that true? If not, what are the subsets of the empty set? If $A \subseteq \varnothing$, can A have any elements? If $x \in A$, then we would necessarily have $x \in \varnothing$, which is not possible. But this does not mean that A cannot exist; it only means that the set A can have no elements. There is such a set: the empty set.

$$\mathcal{P}(\varnothing) = \{\varnothing\}$$

Two particular subsets of $\mathcal{P}(\{1,3,5\})$ from Example 2.16 have very different properties and inspire the next two definitions. The union of the elements of $\{\{1\}, \{3,5\}\}$ and the union of the elements of $\{\{1\}, \{3\}, \{1,5\}\}$ both equal A. The sets in the first collection have no overlap, however; 1 is an element of two of the elements of the latter set. This important idea is summed up in the following two definitions.

Definition 2.12. A collection of sets A_i, where $i \in I$ for some index set I, is said to be *pairwise* (or *mutually*) *disjoint* if $A_i \cap A_j = \varnothing$ for $i, j \in I$ and $i \neq j$.

The choice of words in the previous definition highlights its intuitive meaning. The word *pairwise* means to consider elements of the larger set two at a time. If any pair of sets are not disjoint, then the collection is not pairwise disjoint. This is *very* different from saying the collection of sets is simply disjoint. Such a claim means that there is no *single element* common to every set in the collection. The three sets $\{1,2\}$, $\{2,3\}$, and $\{3,4\}$ are disjoint but they are not pairwise disjoint.

Example 2.18. The first collection A of sets is pairwise disjoint, while the second collection B is not.

$$A = \{\mathbb{Z}^+, \{n + .25 \mid n \in \mathbb{Z}^+\}, \{n + .5 \mid n \in \mathbb{Z}^+\}, \{n + .75 \mid n \in \mathbb{Z}^+\}\}$$

$$B = \{\mathbb{Q}^+, \{0, 2, 4, \dots\}, \{1, 3, 5, \dots\}, \{\pi, 2\pi, 3\pi, \dots\}\}$$

In addition to being pairwise disjoint, a second definition is exhibited by

$\{\{1\}, \{3,5\}\}$ relative to the set $A = \{1,3,5\}$. It is a property that the pairwise disjoint set $\{\{1\}, \{3\}\}$ does *not* have. If we take the union of all the elements of this first set, the result is the *entire* set A; this does not happen in the latter case. Such a collection of pairwise disjoint subsets of A is called a *partition* of A.

Definition 2.13. A collection $\{A_i \mid i \in I\}$ (for some index set I) of nonempty subsets of a set A is said to *partition* A if

$$\bigcup_{i \in I} A_i = A,$$

and if the collection is pairwise disjoint. The sets A_i are called the *partitioning sets* of A.

The concept of a partition is one we encounter throughout our everyday lives. In a class, a teacher may say to her students, "Break into groups." The groups partition the class: every group has at least one student in it, every student is in a group and no student is in more than one group. The groups are the partitioning sets. The way a set can be partitioned is not unique, however. If the teacher has the students break into groups another day, they need not form the same groups. In doing so, they have formed a different partition.

Can every set be partitioned? Almost. There is only one set that cannot be partitioned: the empty set. The partitioning sets that partition a set S must be nonempty, yet the only subset of the empty set is the empty set (see Example 2.17). Hence, no partition of it exists. If $S \neq \varnothing$, we can partition S into single element subsets $\{s_i\}$, $i \in I$ for some index set I, where $S = \{s_i \mid i \in I\}$. But sets with more than one element have more than one partition, as seen in the Exercises and the following example.

Example 2.19. Find three partitions of $\mathbb{R}$, one of which consists of two sets, one consisting of three sets and another consisting of infinitely many sets.

Note that while there are infinitely many correct solutions, each of the following satisfies the definition being a partition.

(1) Two sets: $\mathbb{R}$ is partitioned by $\mathbb{Q}$ and $\mathbb{R} - \mathbb{Q}$ (the rational numbers and the irrational numbers).

(2) Three sets: $\mathbb{R}$ is partitioned by $\mathbb{R}^-$, $\{0\}$, and $\mathbb{R}^+$.

> (3) Infinitely many sets: the collection of all intervals of the form $[n, n+1)$ (where $n \in \mathbb{Z}$) partitions $\mathbb{R}$.

We shift back to constructing new sets from existing ones. Our last definition allows us to construct sets whose elements are of a completely different form than those from which they are formed. It is a concept you have encountered throughout your algebraic upbringings. Properties of the real number line, $\mathbb{R}$, lead to properties of the Cartesian plane, $\mathbb{R} \times \mathbb{R}$.[7] The latter consists of ordered pairs of real numbers, a completely different type of mathematical object than a single real number; consider 1 versus $(1, 2)$.

Definition 2.14. Let A_1, A_2, ..., A_n be sets (which need not belong to the same universal set). Then,

$$A_1 \times A_2 \times \cdots \times A_n = \{(a_1, a_2, \ldots, a_n) \mid a_i \in A_i, 1 \leq i \leq n\}$$

is called the *Cartesian product* of the sets A_1 through A_n. We say that the Cartesian product $A_1 \times A_2 \times \cdots \times A_n$ is the set of *ordered n-tuples*. The *ith coordinate* of the n-tuple $(a_1, a_2, \ldots, a_n)$ is a_i. Two n-tuples $(a_1, a_2, \ldots, a_n)$ and $(b_1, b_2, \ldots, b_n)$ are equal if and only if $a_i = b_i$ for $i = 1, 2, \ldots, n$.

Example 2.20. If $A = \{1, 4\}$ and $B = \{x \in \mathbb{Z} \mid x^2 = 4\}$, list the elements of

1. $A \times B$,
2. $B \times A$, and
3. $(A \times B) \times A$.

Note that $B = \{-2, 2\}$. Then,

$$A \times B = \{(1, -2), (1, 2), (4, -2), (4, 2)\}.$$

The Cartesian product $B \times A$ is a different set:

$$B \times A = \{(-2, 1), (-2, 4), (2, 1), (2, 4)\}.$$

---

[7]The Cartesian product is named after René DesCartes, French philosopher and mathematician (1596 - 1650). Perhaps best known for his philosophical quote, "I think, therefore I am," DesCartes made significant advances in mathematics. He is credited with developing the rectangular coordinate system (the xy-plane), also known as the Cartesian plane [20]. Formally, it is the Cartesian product $\mathbb{R} \times \mathbb{R}$.

Since we have already found $A \times B$, the set $(A \times B) \times A$ is the set of all ordered pairs whose first coordinate is an element of $A \times B$ and whose second coordinate is an element of A. In particular,

$$(A \times B) \times A = \{((1, -2), 1), ((1, -2), 4), ((1, 2), 1), ((1, 2), 4),$$
$$((4, -2), 1), ((4, -2), 4), ((4, 2), 1), ((4, 2), 4)\}.$$

Note the necessity of the parentheses in the elements of $(A \times B) \times A$. It is true that

$$((1, -2), 1) \in (A \times B) \times A,$$

but

$$(1, -2, 1) \notin (A \times B) \times A.$$

Hidden in Example 3.16 is a counterexample to the claim that for two general sets A and B, the sets $A \times B$ and $B \times A$ are equal. When would $A \times B = B \times A$? If $A = B$, then clearly the result holds. It turns out that this is not only sufficient for the equality to hold but it is also necessary. The proof of this result appears in Section 3.2.

Exercises

1. If A, B, and D are sets in a universe U, sketch a Venn diagram with a shaded region representing the following sets.

    (a) $A - B^C$
    (b) $A \cup (B \cap D)$
    (c) $(A^C \cap B^C)^C$
    (d) $A^C - (B^C \cap D)$
    (e) $(A - B)^C - (B^C \cup D)$

2. Sketch Venn diagrams justifying De Morgan's Laws for sets.

3. Let $A = \{a, b, c, d, e, f, g\}$ and $B = \{a, e, i, o, u\}$ in the universe $U = \{$letters of the English alphabet$\}$. Find the following sets.

    (a) $A \cup B$
    (b) $A \cap B$
    (c) A^C
    (d) $A - B$
    (e) $B - A$

4. In the universe $U = \{x \in \mathbb{Z} \mid x^2 < 9\}$, let $A = \{x \in \mathbb{Z}^+ \mid |x| \leq \pi\}$ and $B = \{x \in \mathbb{Z} \mid -|x| \leq \pi\}$. Find:

    (a) all elements of U

    (b) all elements of A

    (c) A^C

    (d) all elements of B

    (e) B^C

5. Let $A = \{4, 5\}$ and $B = \{\varnothing, 1, 4\}$. Determine the following sets.

    (a) $A \times B$

    (b) $B \times A$

    (c) $A \times A$

    (d) $B \times B$

    (e) $\mathcal{P}(A)$

    (f) $\mathcal{P}(B)$

6. Let $A = \{a\}$, $B = \{b, c\}$ and $D = \{d\}$. Find each of the following.

    (a) $A \times (B \times D)$

    (b) $(A \times B) \times D$

    (c) $\mathcal{P}(A \times B)$

    (d) $\mathcal{P}(A \times B \times D)$

7. Let $A = \{1\}$, $B = \{2, 3\}$, and $C = \{i\}$. Find the given sets.

    (a) $A \times B \times C$

    (b) $A \times (A \times B)$

    (c) $(A \times A) \times B$

    (d) $\mathcal{P}(A) \times \mathcal{P}(B)$

    (e) $\mathcal{P}(\varnothing) \times \mathcal{P}(\mathcal{P}(A))$

8. Let $A = \{2, 3, 5, 7\}$ and $B = \{2, 3, 6, 8\}$ be sets in the universal set $U = \{x \in \mathbb{Z} \mid 1 \leq x \leq 10\}$. Show that the following cases of Theorem 2.8 are true for these sets.

    (a) Inclusion of Intersection

    (b) Inclusion of Union

    (c) Complement

9. Show each of the following is false. In each, A, B and D are sets in a universe U.

    (a) If $A \subseteq B \cup D$, then $A \subseteq B$ or $A \subseteq D$.

    (b) For any set A, $A \subseteq A \times A$.

    (c) If $A^C \subseteq B$, then $A \subseteq B^C$.

    (d) $(A \cap B)^C = A^C \cap B^C$.

    (e) $(A \cup B)^C = A^C \cup B^C$.

10. Show each of the following is true. In each, A, B and D are sets in a universe U.

(a) If $A \subseteq B \cap D$, then $A \subseteq B$ and $A \subseteq D$.

(b) If $A \subseteq D$ and $B \subseteq D$, then $A \cup B \subseteq D$.

(c) If $A \subseteq B$ and $A \subseteq D$, then $A \subseteq B \cap D$.

(d) If $A \subseteq B$, then $A \times A \subseteq B \times B$.

11. Determine, with justification, if each of the following is true or false. In each, A, B and D are sets in a universe U.

(a) $(A \cap B) \cup (A \cap B^C) \subseteq A$.

(b) $(A \cup B) \cap D = A \cup (B \cap D)$.

(c) $(B - A) \cup (D - A) \subseteq (B \cup D) - A$.

(d) If $A - B = \varnothing$, then $A = B$.

12. Let $B_i = (-i, i)$ for every $i \in \mathbb{Z}^+$ (that is, B_i is standard interval notation for the set $\{x \in \mathbb{R} \mid -i < x < i\}$). Find:

(a) $\displaystyle\bigcap_{i=1}^{4} B_i$

(b) $\displaystyle\bigcup_{i=1}^{4} B_i$

(c) $\displaystyle\bigcap_{i=1}^{\infty} B_i$

(d) $\displaystyle\bigcup_{i=1}^{\infty} B_i$

13. Repeat the previous problem if B_i is defined, for every $i \in \mathbb{Z}^+$, as

(a) $B_i = \left[1 - \dfrac{2}{i}, 1 + \dfrac{2}{i}\right]$.

(b) $B_i = (0, 2i]$.

(c) $B_i = \left[-i, \left(\dfrac{1}{2}\right)^i\right)$.

14. Suppose $D_i = \{in \mid n \in \mathbb{Z}\}$, where $i \in \mathbb{Z}^+$.[8] Find:

(a) $\displaystyle\bigcap_{i=1}^{4} D_i$

(b) $\displaystyle\bigcup_{i=1}^{4} D_i$

(c) $\displaystyle\bigcup_{i=1}^{\infty} D_i$

---

[8] The set D_i is sometimes referred to as $i\mathbb{Z}$.

15. Determine which of the following collections of sets are a partition of $\mathbb{Z}^+$.

    (a) The set of odd positive integers, the set of even positive integers
    (b) $\{0, 1, 2, 3\}$, $\{3, 4, 5, \ldots\}$
    (c) $\{2^n \mid n \in \mathbb{Z}^+\}$, $\{3^n \mid n \in \mathbb{Z}^+\}$, $\{5^n \mid n \in \mathbb{Z}^+\}$
    (d) $\{1, 2, \ldots, 9\}$, $\{10, 11, \ldots, 98, 99\}$, $\{100, 101, \ldots, 998, 999\}$, $\ldots$
    (e) The sets A_i, $i \in \{1, 2, \ldots, 9\}$ where
        $A_i = \{x \in \mathbb{Z}^+ \mid$ the units digit of x is $i\}$.

16. Provide three examples of how the population of the living people in the world is naturally partitioned.

17. Provide an example of how the population P of living people in the world is broken into three nonempty subsets A_1, A_2, A_3 such that $A_1 \cup A_2 \cup A_3 = P$ but three sets do not form a partition of P.

18. Suppose a set S has at least two elements. Partition S two different ways.

19. Explain why $\mathcal{P}(A \times A) \neq \mathcal{P}(A) \times \mathcal{P}(A)$, in both the case when A is nonempty and when $A = \varnothing$.

20. Let $S = \{x \in \mathbb{Z} \mid |x| \leq 2\}$.

    (a) Find 3 nonempty subsets of S whose union equals S yet they do not form a partition of S.
    (b) Find 3 disjoint, nonempty subsets of S that do not partition S.
    (c) Find a partition of S consisting of 3 partitioning sets.

21. Repeat the previous problem but with $S = \{a, b\} \times \{a, b\}$.

22. Find a partition consisting of at least two partitioning sets for each of the following sets.

    (a) $\mathbb{R}^*$
    (b) $\mathbb{Z} \times \mathbb{Z}$
    (c) $\mathbb{R} \times \mathbb{R} \times \mathbb{R}$
    (d) $\mathcal{P}(\mathbb{Z}^+)$

23. The *disjunctive union* of two sets A and B in a universe U is defined as

$$A \oplus B = \{x \in U \mid x \in A \cup B \text{ and } x \notin A \cap B\}.$$

    (a) Draw a Venn diagram illustrating $A \oplus B$ for general sets A and B.
    (b) Find $A \oplus B$ for the sets A and B of Exercise 3.
    (c) Find $A \oplus B$ for the sets A and B of Exercise 8.
    (d) For any set A in a universe U, find

    i. $A \oplus A$.

    ii. $A \oplus \varnothing$.

    iii. $A \oplus U$.

24. Suppose a set A is partitioned by A_1, A_2 and A_3. If A_1 is partitioned by B_1, B_2 and B_3, is A partitioned by B_1, B_2, B_3, A_1, A_2 and A_3? If not, is A partitioned by some subcollection of these 6 sets?

2.3   Quantified Statements

In Chapter 1 we saw that sentences such as, "$x^2 = 9/4$," and, "Today is Monday" are not statements because they are sometimes true (for example, when $x = -3/2$ and when today is January 6, 2014), and sometimes false (when $x = 1$ and when today is January 10, 2014). The culprit in both cases is the presence of a variable (x, "Today"). In this section, we will explore ways to convert this type of sentence into a statement. One of the main tools for doing so will be sets.

> **Definition 2.15.** A sentence that has at least one variable in it is called a *predicate*. When specific values are substituted for every variable, it becomes a statement. Those values that may be substituted for a particular variable are the elements of the *domain set* of that variable. The subset of D consisting of all such values that make the predicate true is called the *truth set* for the predicate.

It is important to note that predicates are *not* statements. The two sentences given in the open lines of this section are predicates but are not statements. When the variable(s) in a predicate are replaced by specific values, then the predicates become statements.

> **Example 2.21.** Let $P(x)$ be the predicate, "$|x| = x$." The truth set of $P(x)$ depends on how the domain D of x is defined.
>
> (1) If $D = \{-2, -1, 1, 2\}$, then the truth set of $P(x)$ is $\{1, 2\}$, because $P(1)$ and $P(2)$ are true but $P(-2)$ and $P(-1)$ are false.
>
> (2) If $D = \mathbb{R}$, then the truth set of $P(x)$ is the set of all non-negative real numbers; for any $x \in \mathbb{R}$, $x \geq 0$, $P(x)$ is true since $|x| = x$. If $x < 0$, then $P(x)$ is false, since $|-x| = -x$.

(3) If $D = \mathbb{Z}^-$, then the truth set of $P(x)$ is empty, since $P(x)$ is false for every negative integer x.

Now, sentences like "$x^2 = 9$" can be amended, using truth sets to turn them into statements: "For every $x \in \{-3, 3\}$, $x^2 = 9$," or, "For every $x \in \mathbb{R}^+$, $x^2 = 9$." The first sentence is *always* true. The values of x are restricted to be -3 or 3, both of which are in the truth set of the predicate $x^2 = 9$. The second sentence, however, is not so clearly a statement. You may be thinking, "If $x = 1$, then $x^2 = 9$ is false, but if $x = 3$, then $x^2 = 9$ is true. This sentence is sometimes true and sometimes false, making it not a statement." It is the phrase "for every" that makes this a statement though. For example, the sentence "Every member of the basketball team is over seven feet tall" is a statement, even if the team has a short player on it. In this case it is just a false statement.

Note that the statement, "For every $x \in \{-3, 3\}$, $x^2 = 9$" is also a true statement. Assign x any value in the set $\{-3, 3\}$, and it makes the predicate "$x^2 = 9$" true. It does not matter that there are other values that x could be assigned to make the predicate true (namely $x = 3$). It only depends on how the domain set is defined.

Similarly, rather than using the word "every," which is making a conclusion about all elements of a particular set, we just as easily could have made a claim that there is *some* element of the set making the predicate true. "There is a member of the basketball team that is over seven feet tall," and, "There is $x \in \mathbb{R}^+$ making $x^2 = 9$."

These are examples of *quantified statements*, defined below.

Definition 2.16. A *universal statement* is a statement of the form

$$\forall x \in D, P(x),$$

where $\forall$ stands for "for all," $\in$ is the usual set theoretic "is an element of" symbol and $P(x)$ is a predicate with domain set D. The universal statement is true if and only if $P(x)$ is true for every x in the set D.

Definition 2.17. An *existential statement* is a statement of the form

$$\exists x \in D \mid P(x),$$

where $\exists$ stands for "there exists", $\mid$ is read as "such that" and $P(x)$ is a predicate with domain D. It is true if and only if $P(x)$ is true for at least one x in the set D.

Definition 2.18. The symbols ∀ and ∃ are called *quantifiers*. Both universal and existential statements are statements; we call them *quantified statements*.

There are many ways quantified statements appear in our everyday language. Key words to look for include *any*, *all*, and *some*.

Example 2.22. Verbalize each of the following quantified statements three different ways.

(1) $\forall x \in \mathbb{Z}$, $x = 1 \cdot x$.
(2) $\exists\, x \in \mathbb{R}$ such that $x = x^2$.

Verbalizations of the universal statement include the following. Note that each includes some variant of the phrase "for all."

(1) One times any integer equals that integer.
(2) The product of an integer and one is the integer.
(3) If x is an integer, then $x = x \cdot 1$.

The existential statement necessitates different ways to say "for some."

(1) Some integer is the square of itself.
(2) One can find an integer that when squared is itself.
(3) There are integers that equal themselves when squared.

Example 2.23. Write the following mathematically as quantified statements, defining any sets or predicates used and using the appropriate symbols.

(1) All the National Parks are beautiful.

Define N as the set of all National Parks. Let $B(x)$ be the predicate, "x is beautiful." Then, the statement can be written as,

$$\forall x \in N,\ B(x).$$

We also could have let T be the set of all things and B the set of all beautiful things. Then, the statement can be represented as,

$$\forall x \in T,\ \text{if } x \in N,\ \text{then } x \in B.$$

(2) There are people who refuse to speak when spoken to.

This statement is existential. It is a claim about some, but not all, people. Let P be the set of all people, and let $S(x)$ be the predicate, "x is spoken to," and $T(x)$ the predicate, "x speaks." Then, the statement can be expressed as,

$$\exists\, x \in P \mid S(x) \Rightarrow \sim T(x).$$

Example 2.24. Determine which of the following universal statements are true.

(1) $\forall a \in \{-3, -1, 0, 1, 3\}$, $|a + 5| < 10$.

To check the truth of this statement, we must determine if the predicate is true when every element of the domain set is substituted for the variable in it: $|-3 + 5| = 2$, $|-1 + 5| = 4$, $|0 + 5| = 5$, $|1 + 5| = 6$ and $|3 + 5| = 8$. Since every one of these resulting values is less than 10, the universal statement is true.

(2) $\forall x \in \mathbb{Z}$, $x^2 > 0$.

To be true, the square of *every* integer must be strictly greater than zero. At first glance, you may be thinking, "Of course the square of any integer is positive: $1^2 = 1$, $2^2 = 2$, $(-3)^3 = 9$, etc. Use caution when determining the truth of universal statements. Consider all possible elements of the domain set. Perhaps partitioning them by some characteristic property will help you determine the truth of the statement. Here, $\mathbb{Z}$ is not partitioned by the sets of positive and negative integers; 0 is neither positive nor negative. It is, however, the domain element showing that the universal statement is false: $0^2 = 0$.

(3) $\forall x \in \mathbb{R}$, if $x^2 < 0$, then $x = 10$.

We may be tempted to say this statement is false; 10^2 is clearly not negative! But consider the form of the predicate here. It is a conditional statement and a conditional statement is false only when the hypothesis is true and the conclusion is false. The hypothesis of this conditional statement, $x^2 < 0$, is false for every element in the domain set. Thus, for any $x \in \mathbb{R}$, the predicate if $x^2 < 0$, then $x = 10$ is true, making the universal statement true.

In Example 2.24(2), $x = 0$ is an element of the domain that makes the predicate false, making the universal statement false. We call such values *counterexamples*.

> **Definition 2.19.** If the universal statement $\forall x \in D$, $P(x)$ is false, then any value of $x \in D$ that is not in the truth set of $P(x)$ is called a *counterexample* to the universal statement.

Single specific elements of the domain set can be used to show a universal statement is false. The opposite is true for an existential statement. Finding one element of the domain set making the predicate of an existential statement true proves that the entire existential statement itself is true.

Example 2.25. Which of the following existential statements are true?

(1) $\exists\, x \in \mathbb{R} \mid x^2 \leq 0$.

This statement is true because for the same reason that (2) in Example 2.24 is false: $0^2 \leq 0$.

(2) $\exists\, x \in \{2, 3, 4\} \mid x = 2n + 1$ for some $n \in \mathbb{Z}$.

Note that $3 = 2(1) + 1$. Because of this, 3 is the domain value making the predicate of the existential statement true, meaning the statement itself is true.

(3) $\exists\, x \in \mathbb{Q} \mid x$ cannot be written as a quotient of two integers.

By definition, if $x \in \mathbb{Q}$, then x can be written as a quotient of two integers. Thus, this statement is false.

Just like there are multiple ways to informally verbalize quantified statements, there are often numerous ways to mathematically express a quantified statement. The statements $\forall x \in D$, $P(x)$ and $\exists\, x \in D \mid P(x)$ can be manipulated without changing the meaning of the statement. If $Q(x)$ is a predicate with truth set D, then $\forall x \in D$, $P(x)$ can be rewritten as $\forall x$, $Q(x) \Rightarrow P(x)$. Likewise, $\exists\, x \in D \mid P(x)$ has the same meaning as $\exists\, x \mid Q(x)$ and $P(x)$, where $Q(x)$ is a predicate with truth set D.

Example 2.26. The statement $\forall x \in \mathbb{Z}$, $x = 1 \cdot x$ can be rewritten mathematically:

(1) If $x \in \mathbb{Z}$, then $x = 1 \cdot x$.
(2) $\forall x$, if $x \in \mathbb{Z}$, then $x = 1 \cdot x$.

The statement

$$\exists\, x \in \mathbb{R} \mid x \notin \mathbb{Q}$$

has the same meaning as the statement

$$\exists\, x \mid x \in \mathbb{R} \text{ and } x \notin \mathbb{Q}.$$

Note that it does not make sense to talk about a single counterexample to an existential statement. Rather, to show an existential statement of the form $\exists\, x \in D \mid P(x)$ is false, we must show that there is no single element of D that is in the truth set for $P(x)$. For example, if a friend says to you, "There is a student in our math class who was born on Halloween," how would your friend that he is wrong? It would not suffice to show just one or two of your classmates have birthdays not on October 31. You would show that *every* classmate has a birthday on some day other than October 31.

Symbolically, the idea of negating quantified statements is summarized as follows.

1. $\sim(\forall x \in D,\, P(x)) \equiv \exists\, x \in D \mid \sim P(x)$.

2. $\sim(\exists\, x \in D \mid P(x)) \equiv \forall x \in D,\, \sim P(x)$.

These equivalences say that the negation of an existential statement is a universal statement and vice versa. Informally, we can think about negating a universal or existential statement by switching its quantifier and negating its predicate.

Example 2.27. Negate each of the following quantified statements.

(1) Some member of our class earned a perfect score on the exam.

This is an existential statement. Its negation is a universal statement: "Every person in our class earned less than a perfect score on the exam."

(2) $\forall x \in \{1, 2, 3, 5\}$, x is prime.

The negation of this universal statement is the existential statement

$$\exists\, x \in \{1, 2, 3, 5\} \mid x \text{ is not prime.}$$

(3) $\exists\, x \in \mathbb{Z} \mid$ if $x > 1$, then x does not have a prime divisor.

Negating this statement is somewhat more complicated. Note that the predicate of the quantified statement is an implication, and recall that the negation of an implication is logically equivalent to a particular conjunction. Thus, we have the negation of this statement as

$$\forall x \in \mathbb{Z},\ x > 1 \text{ and } x \text{ has a prime divisor.}$$

We conclude this section by introducing a third quantifier related to the existential quantifier. The existential statement, "$\exists\, x \in D \mid P(x)$" means that there is *at least* one domain element that is in the truth set of $P(x)$. Often, though, we want to say that there is one *and only one* such element, that that element is unique. The quantifier defined below accomplishes this for us.

Definition 2.20. A *unique existential statement* is a statement of the form

$$\exists!\, x \in D \mid P(x),$$

where $\exists!$ stands for "there exists a unique" and $P(x)$ is a predicate with domain D. It is a statement that is true if and only if $P(x)$ is true for exactly one x in the set D.

Example 2.28. The statement, "There is only one even prime integer," can be formally written as

$$\exists!\, x \in \mathbb{Z} \mid x \text{ is even and } x \text{ is prime.}$$

Though we have yet to explicitly define what it means for an integer to be prime, it is a familiar concept. We know that 2 is prime, and we know that every even number greater than 2 has 2 as a divisor, and consequently, is not prime. Hence, we have that the above statement is true.

To determine what the negation of a general unique existential statement is, let us consider a real-world example. After taking a 10-question true/false quiz, your friend turns to you and says, "There was only one question whose answer was true." Your friend just presented you with a unique existential statement. You, being the expert in this class, must inform your friend he was wrong. In what situations would you do this?

If there were two or more questions whose answer was true, then you would tell your friend he was wrong. But this is not the only situation in which your friend would be wrong. If *no* questions on the quiz had true as a correct answer, then you would be sharing bad news with your friend. Symbolically, we have:

$$\sim (\exists! \, x \in D \mid P(x))$$

is logically equivalent to

$$[\, (\exists \, x, y \in D \mid x \neq y \text{ and } P(x) \wedge P(y)), \text{ or, } \forall x \in D, \, \sim P(x)].^{9}$$

In summary, quantified statements are indeed statements. Their truth value depends upon the type of quantifier being used in the statement and the existence (or non-existence) of elements of the domain that make the predicate of the quantified statement true. The reasoning behind this is exactly how we have come to use quantified statements in our everyday language. Remember, the goal of the fathers of symbolic logic was not to change the logic we use on a daily basis but rather to represent it mathematically. In the next section, we construct statements with multiple quantifiers, because such statements are regular parts of our everyday speech. Just like here, their symbolism will mirror the way they show up in our daily language.

Exercises

1. Let $P(x)$ be the predicate, "$\sqrt{x} \in \mathbb{Z}^+$." Determine the truth value of the following.

   (a) $P(1)$
   (b) $P(3)$
   (c) $P(\pi^2)$
   (d) $P((-2)^2)$

2. Let $P(x, y)$ be the predicate, "x is a prime factor of y." Determine the truth value of the following.

   (a) $P(2, 10)$
   (b) $P(12, 3)$
   (c) $P(5, 5)$
   (d) $P(10, 20)$
   (e) $P(1, 3)$

3. Let $Q(x, y, z)$ be the predicate, "The product xy is in the domain of the function z." Determine the truth value of the following.

---

[9] This may appear to be a statement with multiple quantifiers, and indeed it is. However, its truth is understood in this context. It is the conjunction of two quantified statements. In the coming section, we will discuss how this differs from a general *statement with multiple quantifiers*.

(a) $Q(2, \pi, \cos(x))$

(b) $Q(\frac{1}{2}, \pi, \tan(x))$

(c) $Q(-2, -3, \sqrt{x})$

(d) $Q(5, -e, \ln(x))$

4. Determine the truth set of each of the following predicates (a) if the domain of x is $\mathbb{Z}$, and (b) if the domain of x is $\mathbb{R}$.

 (a) $x^2 = x$

 (b) $x^2 + x - 3 = 0$

 (c) $x^2 + x + 3 = 0$

5. Write each of the following as quantified statements, defining any sets that you use.

 (a) All trees are beautiful.

 (b) Everyone is treated equally.

 (c) I need someone to help me with these problems.

 (d) There's a cool cat in the alley.

6. Determine which of the following universal statements are true. For those that are true, explain why, and for those that are false, provide a counterexample.

 (a) $\forall x \in \{1, 2, 3, 4, 5\}$, $x^2 \le 25$.

 (b) $\forall x \in \{x \in \mathbb{Z} \mid |x| < 4\}$, $(x + 1)(x - 1) \ge 0$.

 (c) $\forall x \in \{\text{polygons that are squares}\}$, x is a rectangle.

 (d) $\forall x \in \mathbb{Z}$, $x \cdot \dfrac{1}{x} = 1$.

 (e) $\forall x \in \{x \in \mathbb{R} \mid |x| < 0\}$, $\sin(x) > 1$.

7. Determine which of the following existential statements are true. For those that are true, find an element of the domain making the predicate true. For those that are false, explain why no such element exists.

 (a) $\exists x \in \mathbb{Q}$ such that $\dfrac{2}{x} \notin \mathbb{Q}$.

 (b) $\exists x \in \mathbb{Z}^+$ such that $x^2 \le x$.

 (c) $\exists x \in \mathbb{Z}^-$ such that $x^2 \le x$.

 (d) If C is the set of continuous functions, then $\exists\, f(x) \in C$ such that $f(x)$ is not differentiable.

 (e) $\exists x \in \mathbb{R}$ such that $x^5 + 2x + 1 = 0$.

8. Determine which of the following quantified statements are true.

 (a) $\forall x \in \mathbb{Z}$, $x = 3 + y$ for some $y \in \mathbb{Z}$.

 (b) $\exists x \in \mathbb{Z} \mid x = a^2 + 2$ for some $a < 0$.

 (c) $\forall x \in \mathbb{Z}$, $x = a^2 + 2$ for some $a < 0$.

(d) $\exists\, x \in \mathbb{R}$ such that if $\sqrt{x} < 0$, then $x = 11^{210}$.

(e) $\forall x \in \mathbb{Q}$, cats are reptiles if and only if $x^{20} < -\pi$.

(f) $\forall x \in \{x \in \mathbb{Z} \mid x^2 + x + 1 = -2\}$, $x = 10$ or $x > 100$.

9. For each of the following, find a domain D for which the quantified statement is true and a domain D for which the quantified statement is false.

(a) $\forall x \in D$, $x + 1 \geq 2x$.

(b) $\exists\, x \in D$ such that $x^2 - 1 = 0$ has a unique solution.

(c) $\forall x \in D$, $x^2 - 1 - 0$.

10. Rewrite each of the following without using quantifiers.

(a) $\forall x \in \mathbb{R}$, $\sqrt{x^2} = |x|$.

(b) $\forall x \in \mathbb{R}$, $x = \dfrac{x}{1}$.

(c) $\forall x \in \mathbb{R}$, if $x^2 > 4$, then $|x| > 2$.

11. Negate the following universal statements.

(a) $\forall x \in S$, $|x| \geq 10$.

(b) $\forall x \in D$, $x \neq a$.

(c) $\forall x \in A$, $x = x_1$ or $x = x_2$.

(d) $\forall x \in \{a, b, c\}$, if x has Property A, then x has Property B.

(e) $\forall x \in N$, $\sim P(x) \Leftrightarrow Q(x)$.

12. Negate the following existential statements.

(a) $\exists\, x \in S$ such that $x > 0$.

(b) $\exists\, x \in \mathbb{R} \mid x(x + 1) = -10$.

(c) $\exists\, x \in \mathbb{Z}^+$ such that $0 \leq x^2 \leq 1$.

(d) $\exists\, x \in D \mid P(x) \wedge \sim Q(x)$.

(e) $\exists\, x \in \mathbb{Q}$ such that if $P(x)$, then $x < 3$.

13. Negate the following unique existential statements.

(a) $\exists!\, x \in D$ such that $x < 2$.

(b) $\exists!\, x \in S$ such that $P(x) \wedge Q(x)$.

(c) $\exists!\, x \in P$ such that $(P(x) \wedge Q(x)) \Rightarrow (R(x) \vee T(x))$.

14. Find a counterexample showing that each of the following universal statements is false.

(a) $\forall x \in \mathbb{R}$, $x < x^2$.

(b) $\forall n \in \mathbb{Z}^+$, at least one of n, $n + 1$ or $n + 2$ is prime.

(c) $\forall x \in \mathbb{Q}$, $-x \leq x$.

(d) $\forall m \in \mathbb{R}$, if $2m^2 + 5m + 3 = 0$, then $m \in \mathbb{Z}$.

15. Explain why each of the following universal statements is true.

(a) $\forall x \in \{x \in \mathbb{R} \text{ such that } x^4 + x^2 < 0\}$, $x - 9 = 2x^2$.

(b) $\forall x \in \mathbb{Z} - \mathbb{R}$, $100 \leq \sqrt{x}$.

(c) $\forall x \in \varnothing$, x is a purple cow whose square root is negative.

(d) $\forall x \in \mathbb{R}$, if $x \in \mathbb{Z}$ and $x \notin \mathbb{Q}$, then $\tan(x) = 0$.

(e) Every triangle that includes an angle greater than 180 degrees is a right triangle.

16. Explain why the following statement is true:

   Every cat that lives underwater is a tuxedo cat named Gordon.

17. Suppose you know that the following universal statement is true:

$$\forall x \in D, \ P(x) \Rightarrow Q(x),$$

where D is some set and $P(x)$ and $Q(x)$ are predicates. Which of the following must be true?

(a) $\forall x \in D$, $Q(x) \Rightarrow P(x)$.

(b) $\forall x \in D$, $\sim P(x) \Rightarrow \sim Q(x)$.

(c) $\forall x \in D$, $\sim Q(x) \Rightarrow \sim P(x)$.

(d) $\forall x \in D$, $P(x)$ is a necessary condition for $Q(x)$.

(e) $\forall x \in D$, $P(x)$ is a sufficient condition for $Q(x)$.

(f) $\forall x \in D$, $P(x)$ only if $Q(x)$.

(g) $\forall x \in D$, $Q(x)$ only if $P(x)$.

2.4 Multiple Quantifiers and Arguments with Quantifiers

Suppose that you and a friend are in two different sections of a college history course. Each of your classes requires a semester-long research project on the War of 1812. In your section, the professor says to the entire class, "Every one of you has to research some topic on the War of 1812." In your friend's section of the course, taught by a different professor, everyone is told, "There is some topic on the War of 1812 that every one of you must research."

Think about how you and your friend will approach obtaining a topic for your papers. The assignment is completely different for each of you. In your section, each student has a choice of topic to research. Everyone may decide on completely different topics. However, in your friend's class, there is no option to choose a topic. "There is some topic" means that each student in the course will research the same exact topic.

How can this be? The instructions given by the professors were very similar but their meanings are completely different. Even though both of these statements involve two quantifiers, ∀ and ∃, it is the order in which they appear in the statements that affects how they are interpreted.

In Section 2.3 we investigated statements with a single quantifier and a predicate with only one variable present. But such statements are not the only type of quantified statements appearing in our everyday conversations, as illustrated in the preceding paragraphs. Often in our everyday speech we verbalize sentences with more than one variable, and consequently, more than one quantifier. These are indeed statements, called *statements with multiple quantifiers*.

Definition 2.21. Let $P(x,y)$ be a predicate with domain $x \in D$, $y \in E$. The statement

$$\forall x \in D, \exists y \in E \mid P(x,y)$$

is true if for any given value of $x \in D$, there is some particular $y \in E$ (dependent upon x) making $P(x,y)$ true.

In general, to show the statement

$$\forall x \in D, \exists y \in E \mid P(x,y)$$

is true, we must show that for an *arbitrary but particular* $x \in D$ that there is some $y \in E$ so that $P(x,y)$ holds. That is, we proceed as follows.

1. Let $x \in D$.
2. Determine what choice of $y \in E$, dependent upon the arbitrary value of x, makes $P(x,y)$ true.

Let us demonstrate this process with a straightforward example.

Example 2.29. Show that the following statement is true:

$$\forall x \in \mathbb{Z}, \exists y \in \mathbb{Z} \mid x - y = 10.$$

The result is clearly true; given any integer, there is another integer that differs from it by 10. That is, if I give you an integer, you would subtract 10 from it to determine the other integer. And this is *precisely* how you show statements of this form are true! To make the process more precise:

Let $x \in \mathbb{Z}$. By closure properties of the integers, $x - 10 \in \mathbb{Z}$. Then,

$$x - (x - 10) = 10,$$

showing that the given statement is true.

Compare this process to the example at the beginning of this section. Your professor said, *"Every* student must do *some* research project..." The interpretation that each student has flexibility in choosing the topic for the project agrees with Definition 2.21. Select an arbitrary student in the class and there will be some research topic, dependent upon the choice of the student. But this is very different from your friend's experience.

Definition 2.22. Let $P(x, y)$ be a predicate with domain $x \in D$, $y \in E$. The statement,

$$\exists\, x \in D,\ \forall y \in E,\ P(x, y)$$

is true if there is a particular $x \in D$ that makes $P(x, y)$ true for every $y \in E$.

To show a statement of the form

$$\exists\, x \in D,\ \forall y \in E,\ P(x, y)$$

is true we must show that there is one single $x \in D$ so that no matter what choice $y \in E$ is made, $P(x, y)$ is true. The arbitrary selection of an element plays a role in verifying the truth of the statement; there is no dependence of one variable choice upon another. More specifically, we show such a statement is true as follows.

1. Determine a specific choice of $x \in D$ so that ...

2. no matter the choice of $y \in E$, $P(x, y)$ is true.

The next example demonstrates this process.

Example 2.30. Show that the following statement is true:

$$\exists\, x \in \mathbb{Z} \mid \forall y \in \mathbb{Z},\ x + y = y.$$

This statement is saying that there is some integer so that when it is added to any other integer, the result is the other integer. Clearly this is true: 0 satisfies this property (and is called the *additive identity* because it does). As before, this thought process demonstrates the precise method for showing the statement is true:

Let $y \in \mathbb{Z}$. Because $0 + y = y$, the result holds.

In the previous example, our precise justification of the result begins by arbitrarily selecting an integer y. However, we mentioned prior to the example that the process does not begin by arbitrarily choosing an element but rather begins by finding a specific element. Observe that this is actually what we did. We *found* 0. The phrasing, "Let $y \in \mathbb{Z}$," is simply a way to present the result. There is no need to say, "$0 \in \mathbb{Z}$."

Similarly, if we again consider the War of 1812 research example preceding these definitions, we see that the mathematical definition (Definition 2.22) agrees with our intuitive interpretation of what your friend's professor said: "There is *some* topic that *all* students will research . . . ." The professor chooses a topic, then, regardless of the student in that course, he or she will write a paper on that topic.

> **Definition 2.23.** Statements like those in the preceding two definitions are called *statements with multiple quantifiers.*

It should be noted that a statement with multiple quantifiers is not necessarily just a statement that, when looking at it, has multiple quantifiers present in it.[10] As defined, a statement with multiple quantifiers must have a quantifier falling *within the scope* (that is, within the extent of the effect of a quantifier) of another quantifier. For example,

$$\forall x \in D, \exists\, y \in E \mid P(x,y)$$

is interpreted as

$$\forall x \in D \; (\exists\, y \in E \mid P(x,y)).$$

The existential quantifier is "inside" the universal quantifier; for every $x \in D$, *something* happens. It just so happens that the *something* involves an existential quantifier. This is very different from saying,

$$\forall x \in D, \; P(x) \land \exists y \in E \mid Q(y).$$

This is simply the conjunction of two quantified statements, the first of which is a universal statement (every $x \in D$ satisfies $P(x)$) and the seconds an existential statement (some $y \in E$ satisfies $Q(y)$). This statement is interpreted as

$$(\forall x \in D, \; P(x)) \land (\exists y \in E \mid Q(y)).$$

In everyday language, such a statement would be something along the lines of, "Every student must select a book, and some teacher will win a prize." The students selecting books have no bearing on which teacher wins a prize.

Let us solidify these definitions with a pair of examples.

---

[10]Some texts define statements with multiple quantifiers in the latter term and use the term "statements with nested quantifiers" as in Definition 2.23.

Example 2.31. Let $A = \{1, 2, 3\}$ and $B = \{2, 4, 6, 8, 9\}$. The following statements are true.

(1) $\forall x \in A,\ \exists\, y \in B \mid 2x = y$.
(2) $\exists\, x \in A \mid \forall y \in B$, x is a divisor of y.

The first statement is true because we can consider each element x of A and determine a corresponding element y of B so that $2x = y$. Namely, $2(1) = 2$, $2(2) = 4$ and $2(3) = 6$.

For the second statement to be true, there must be a *single* element x of A that is a divisor of *every* element y of B. In this case there is only a single element of A that divides each element of B (and that element is 1). But, in order for this statement to be true, there need only be one such element.

Example 2.32. Determine if the following statements are true or false.

(1) $\forall$ nonzero real numbers u, $\exists$ a real number $v \mid uv = 1$.
(2) $\exists\, u \in \mathbb{R}^* \mid \forall v \in \mathbb{R}$, $uv = 1$.
(3) $\forall x \in \mathbb{R},\ \exists\, n \in \mathbb{Z} \mid x < n$.
(4) $\exists\, x \in \mathbb{R} \mid \forall n \in \mathbb{Z}$, $x < n$.

For each, refer to Definitions 2.21 and 2.22.

(1): Take $u \in \mathbb{R}^*$. Can we find $v \in \mathbb{R}$ so that $uv = 1$? Of course; every nonzero real number has a multiplicative inverse. If we define $v = \frac{1}{u}$, then $uv = u(\frac{1}{u}) = 1$, as desired. Thus, the statement is true.

(2): Notice the different interpretation of this statement versus the first one. For this statement to be true, there must be a single real number u whose product with *any* real number v is 1. Clearly this is false. If $u \cdot 2 = 1$, we know $u = \frac{1}{2}$. Yet $\frac{1}{2} \cdot 3 \neq 1$.

(3): In checking the truth of this statement, we are asking if, given any real number x, is there an integer n greater than x? This is true via the Archimedean law of the real numbers. We can explicitly define n in terms of the greatest integer function: $n = \lceil x \rceil$.

(4): For this to be true there must be a single real number smaller than every integer. Since the integers are not bounded below, this statement is false.

Statements with multiple quantifiers appear early on in calculus. The definition of a limit is one such statement. We say,

$$\lim_{x \to a} f(x) = L$$

if

$\forall \epsilon > 0$, $\exists \delta > 0$ such that $|f(x) - L| < \epsilon$ whenever $0 < |x - a| < \delta$.

To show, using the definition, that $\lim_{x \to a} f(x) = L$, you first assume $\epsilon > 0$ (an arbitrary but fixed value for ϵ). Then, you determine how to define δ, often in terms of ϵ, so that $0 < |x - a| < \delta$ implies $|f(x) - L| < \epsilon$.

Statements with multiple quantifiers need not be just of the form $\forall x, \exists y$ or $\exists x \mid \forall y$. They may include multiple $\forall$ symbols, just $\exists$ symbols or some other combination of the two.

Example 2.33. Interpret the following statements with multiple quantifiers.

(1) $\forall x \in \mathbb{R}, \forall y \in \mathbb{R}, x + y = y + x$.
(2) $\exists x \in \mathbb{Z}, \exists y \in \mathbb{Z} \mid x^2 + y = 5$.
(3) $\forall x \in \mathbb{R}, \forall y \in \mathbb{R}$ with $x < y$, $\exists z \in \mathbb{Q} \mid x < z < y$.

(1): This statement claims that for every real number x, no matter the choice of $y \in \mathbb{R}$, $x + y = y + x$. In other words, that addition of real numbers is commutative. This is a true statement.

(2): If this statement is true, then there is some integer x and some integer y so that $x^2 + y = 5$. This is true: $x = 2$, $y = 1$.

(3): What is this statement saying? The first quantifier says let x be any arbitrary but fixed real number. Then, let y be an arbitrary real number greater than x. With these two fixed values, the last quantifier states that there exists a rational number z strictly between x and y. In short, is there a rational number between any two real numbers? Yes; the statement is true.

The unique existential quantified statement

$$\exists! \, x \in D \mid P(x)$$

can be written as a statement with multiple quantifiers. The following example solidifies this fact using our new tools.

Example 2.34. Rewrite $\exists! \, x \in D \mid P(x)$ using only $\exists$ and $\forall$.

Recall that the statement

$$\exists! \, x \in D \mid P(x)$$

means that the truth set of $P(x)$ contains exactly one element of D. In other words, *there exists* an $x \in D$ making $P(x)$ true, *and*, if any element of D does indeed make $P(x)$ true, then that element of D must actually be x. The equivalent statement is

$$\exists x \in D \mid (\, P(x) \wedge \forall y \in D, P(y) \Rightarrow (y = x)).$$

Note the necessity of the parentheses around the statement following the $\mid$ symbol. This guarantees that the universal quantifier falls within the scope of the existential quantifier.

Note that the quantifier $\forall$ must be included in the predicate of the initial $\exists$ quantifier of the last statement of Example 2.34. Otherwise, the following is not actually a statement:

$$\exists x \in D \mid P(x) \text{ and } \forall y \in D, P(y) \Rightarrow (y = x).$$

What makes this not a statement? In the universal statement

$$\forall y \in D, P(y) \Rightarrow (y = x)$$

the variable x is undefined. By stating

$$\exists x \in D \mid P(x)$$

we are saying that there is some element in D, temporarily called x, that is in the truth set of $P(x)$. Beyond this existential statement, the notion of x carries no meaning. It is outside the scope of the quantifier.

In calculus, how do we go about, using the definition, showing that a particular limit does *not* equal a particular value? For example, how can the definition of a limit be used to show

$$\lim_{x \to 1} 3x + 2 \neq 4?$$

We look for a particular real number ϵ (in this case $\epsilon = .5$ would suffice) so that no matter what positive real value δ takes,

$$0 < |x - 1| < \delta \text{ but } |(3x + 2) - 4| \geq \epsilon.$$

Logically, what are we doing? We are negating a statement with multiple quantifiers. Negating a statement with multiple quantifiers involves nothing more than repeatedly negating quantified statements. Treat the quantifiers one at a time:

$$\sim(\forall x \in D, \exists\, y \in E \mid P(x,y)) \equiv \exists\, x \in D \mid \sim(\exists\, y \in E \mid P(x,y))$$
$$\equiv \exists\, x \in D \mid (\forall y \in E, \sim P(x,y))$$

$$\sim(\exists\, x \in D \mid \forall y \in E, Q(x,y)) \equiv \forall x \in D \sim(\forall y \in E, Q(x,y))$$
$$\equiv \forall x \in D, (\exists\, y \in E \mid \sim Q(x,y))$$

Negating statements with more than two quantifiers follows this same process, treating them one at a time.

Example 2.35. Negate the following statement.

$$\forall x \in \mathbb{R}, \forall y \in \mathbb{R} \text{ with } x < y, \exists\, z \in \mathbb{Q} \mid x < z < y.$$

To negate the statement, we proceed as above, treating each quantifier individually. The resulting negation is:

$$\exists\, x \in \mathbb{R}, \exists\, y \in \mathbb{R} \text{ with } x < y \mid \forall z \in \mathbb{Q}, z < x \text{ or } y < z.$$

Note that the resulting negation has "or" in it. This is because the expression

$$x < z < y$$

is shorthand for a conjunction:

$$x < z \text{ and } z < y.$$

As such, De Morgan's Law yields the result.

We now turn our attention to arguments with quantified statements. Consider the following.

> All cats make me smile.
> Petey is a cat.
> ∴ Petey makes me smile.

Does this argument intuitively seem valid? Of course. It is common to our everyday conversations, yet, it is structurally very different from arguments we previously investigated. The first hypothesis is a quantified statement. None of the rules of inference or previous theorems apply to such arguments. The key tool to deal with these types of arguments is the rule defined below.

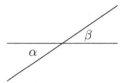

FIGURE 2.2
Vertical angles are equal

Definition 2.24. The rule of *Universal Specification* states that if something is true about every element of a set S, then that thing is true for any specific element of S. Symbolically, for a set S and predicate $P(x)$ defined on S, the rule is

$$\forall x \in S,\, P(x)$$
$$\underline{s \in S}$$
$$\therefore P(s).$$

Universal Specification gives rise to the following three universal rules of inference, each highlighted with an example.

Definition 2.25. If D is a set and both $P(x)$ and $Q(x)$ are predicates whose domains contain D, then the following argument, called *Universal Modus Ponens*, is valid.

$$\forall x \in D,\, P(x) \Rightarrow Q(x)$$
$$\underline{P(a) \text{ for some } a \in D}$$
$$\therefore Q(a).$$

Example 2.36. Book I of Euclid's *Elements* contains numerous geometric results [27]. One such result, known as Proposition I.15, deals with vertical angles and is the first hypothesis of the argument below.

> If straight lines cut one another, they make the vertical angles equal to one another.
> Lines L_1 and L_2 (as in Figure 2.4) intersect.
> $\therefore$ Angles α and β are equal.

Definition 2.26. If D is a set and both $P(x)$ and $Q(x)$ are predicates whose domains contain D, then the following argument, called *Universal Modus Tollens*, is valid.

$$\forall x \in D,\ P(x) \Rightarrow Q(x)$$
$$\underline{\sim Q(a) \text{ for some } a \in D}$$
$$\therefore \sim P(a)$$

Example 2.37. Consider this argument.

> If a student graduates as a chemistry major, then he or she must have passed Calculus II.
> Brayden never passed Calculus II.
> ∴ Brayden did not major in chemistry.

This is an example of Universal Modus Tollens being used in our everyday language. The first hypothesis is a universal statement:

$$\forall x \in S,\ \text{if } x \in C, \text{ then } x \in P,$$

where S represents the set of all students, C represents the set of all chemistry majors, and P represents the set of all persons passing Calculus II.

Definition 2.27. If D is a set and $P(x)$, $Q(x)$, and $R(x)$ are predicates whose domains contain D, then the following argument, called *Universal Transitivity*, is valid.

$$\forall x \in D,\ P(x) \Rightarrow Q(x)$$
$$\underline{\forall x \in D,\ Q(x) \Rightarrow R(x)}$$
$$\therefore \forall x \in D,\ P(x) \Rightarrow R(x)$$

Example 2.38. The following hypothetical discussion exhibits universal transitivity.

Ben: The handbook says if a student's GPA is below 1.0, that student can no longer be enrolled at the school.

Sam: I read here that if a student is not enrolled at the school, then that student cannot live in college housing.

Ben: So, if a student has a GPA below 1.0, then that student cannot live in the residence halls. Am I right?

> *Sam*: Ben, you sure understand Universal Transitivity well!

Rules of inference for quantified statements validate very specific types of arguments (namely, those precisely fitting the form of Definitions 2.25, 2.26, and 2.27). What, then, makes a general argument with quantifiers valid? We say an argument containing quantified statements is *valid* if the argument is valid in the sense of Definition 1.19 no matter what domain value is substituted for each variable present in the argument.

As with arguments in Section 1.3, there are two common invalid arguments that oftentimes appear in our everyday conversations. These are natural extensions of the previously defined Converse and Inverse Errors.

Definition 2.28. Let $P(x)$ and $Q(x)$ be predicates whose domains contain set D. Then the following two arguments, known respectively as the *Universal Converse Error* and the *Universal Inverse Error*, are invalid.

$$\forall x \in D,\ P(x) \Rightarrow Q(x)$$
$$\underline{Q(a),\text{ for some } a \in D}$$
$$\therefore P(a)$$

$$\forall x \in D,\ P(x) \Rightarrow Q(x)$$
$$\underline{\sim P(a),\text{ for some } a \in D}$$
$$\therefore \sim Q(a)$$

How can we go about arguing that these arguments are indeed invalid? We do so by assuming they are valid. Take the first argument. For *any* arbitrary but fixed $a \in D$, the following argument must be valid:

$$P(a) \Rightarrow Q(a)$$
$$\underline{Q(a)}$$
$$\therefore P(a)$$

This argument is now an argument with no quantifiers; $P(a)$ and $Q(a)$ are simply statements. We are claiming it is a valid argument, but this exhibits the converse error presented at the end of Section 1.3.

Before proceeding to the exercises, it is worth pointing out the differences in *presentation* of statements with multiple quantifiers. Consider the following methods for writing the same statement.

1. $\forall x \in D,\ \exists y \in E \mid x \leq y.$
2. $\forall x \in D\ \exists y \in E\ x \leq y.$
3. For every $x \in D$, there exists $y \in E$ such that $x \leq y.$

These statements have the same interpretation, and it is a stylistic preference as to which one an author would choose to use. From a readability perspective, however, the third one is much preferred to the first two. For the purposes of this section, though, our exercises use the more abbreviated versions (1) and (2), simply to emphasize the *concepts*. In the coming chapters, however, you are encouraged to incorporate the readability of (3) into your mathematical writing.

Exercises

1. Let $P(x, y)$ be the predicate, "$x = 2\sqrt{y} + 3$." Determine the truth value of the following.

   (a) $P(0, 3)$
   (b) $P(1, 5)$
   (c) $P(3(2\sqrt{2} + 1), \sqrt{18})$

2. Let $Q(x, y)$ be the predicate, "Letter x does not appear in word y." Determine the truth value of the following.

   (a) $Q(\text{e}, \text{three})$
   (b) $Q(\text{x}, \text{why})$
   (c) $Q(\text{s}, \text{quantifier})$

3. Let $P(x, y, z)$ be the predicate, "$x < y < z$." Determine the truth value of the following.

   (a) $P(1, 3, 2)$
   (b) $P(3.141, \pi, 3.142)$
   (c) $P(-\sqrt{2}, -\frac{3}{2}, -\frac{\sqrt{3}}{2})$

4. Let $Q(x, y)$ be the predicate, "x likes flavor y of ice cream," where the domain set for x is P, the set of all people in the world, and the domain set for y is I, the set of all ice cream flavors in the world. Write each of the following mathematically, using quantifiers.

   (a) All people in the world like some ice cream flavor.
   (b) Someone in the world enjoys every flavor of ice cream.
   (c) Everybody likes chocolate!
   (d) Someone does not like all flavors of ice cream.
   (e) There is somebody that does not like any flavor of ice cream.
   (f) Everybody likes every kind of ice cream.

5. Translate each statement to symbolic form using multiple quantifiers. Be sure to define all sets and predicates.

   (a) Every course has a grade policy.
   (b) Between any two real numbers there is a rational number.

(c) There is a person in the world who has longer hair than every other person.

(d) Every plan has a data limit.

(e) There is a maximum data limit that applies to all plans.

6. Translate the following statements into everyday language.

   (a) $\forall x \in \mathbb{Z}^*, \exists\, y \in \mathbb{R} \mid xy = 1$.
   (b) $\exists\, x \in \mathbb{Z} \mid \forall y \in \mathbb{Z},\ x + y = y$ and $xy = x$.
   (c) $\forall x \in \mathbb{R},\ \forall y \in \mathbb{R} - \{x\},\ \exists\, n \in \mathbb{Z}^+ \mid \frac{1}{n} < |x - y|$.

7. Negate the following statements.

   (a) $\forall x \in D, \exists\, y \in E \mid P(x, y)$.
   (b) $\exists\, x \in D \mid \forall y \in E,\ P(x) \wedge Q(y)$.
   (c) $\forall m \in \mathbb{R}, \forall n \in \mathbb{Q},\ P(m) \vee \sim Q(n)$.
   (d) $\exists\, x \in A \mid \exists\, y \in B,\ P(y) \Rightarrow Q(x)$.
   (e) $\forall x \in S,\ \exists\, y \in T \mid \forall z \in W,\ P(x, y, z) \Leftrightarrow Q(x, y, z)$.

8. For each of the following statements, (a) translate it to a mathematical statement involving multiple quantifiers (defining any sets you use), (b) negate the mathematical statement and (c) translate the negation back to a statement in everyday English.

   (a) Every professor has a favorite student.
   (b) There is a class that no professor likes to teach.
   (c) Every professor teaches every class that is offered.
   (d) Some student will complete all of the homework exercises.
   (e) Every student will complete at least one of the homework exercises.
   (f) Nobody can do every problem.

9. Find a counterexample that disproves each of the following statements, or explain why no such counterexample exists.

   (a) $\forall x \in \mathbb{R},\ \forall y \in \mathbb{R},\ 0 < x^2 + y^2$.
   (b) $\forall x \in \mathbb{Z},\ \forall y \in \mathbb{Z}$, if $x^2 - y^2 = 0$, then $x = y$.
   (c) $\forall x \in \mathbb{R},\ \exists\, y \in \mathbb{R} \mid$ if $xy = 0$, then $y = 0$.

10. Determine which quantified statements are true, given

$$L = \{a, b, c, d, e\}$$

and

$$W = \{\text{cat, dog, goat, frog, mouse}\}.$$

   (a) $\forall x \in L,\ \exists\, y \in W \mid x$ is a letter in y.

(b) $\forall x \in W, \exists\, y \in L \mid x$ contains letter y.

(c) $\exists\, x \in W \mid \forall y \in L, y$ is a letter in x.

(d) $\exists\, x \in L \mid \forall y \in W, x$ is a letter in y.

(e) $\exists\, x \in L \mid \exists\, y \in W \mid x$ is a letter in y.

11. Determine which quantified statements are true. For all of the statements,

$$D = \{x \in \mathbb{Z} \mid |x| \le 4\}.$$

(a) $\forall x \in D, \exists\, y \in D \mid x + y \in D$.

(b) $\exists\, x \in D, \forall y \in D, x + y \in D$.

(c) $\forall x \in D, \forall y \in D, x + y \in D$.

(d) $\exists\, x \in D, \exists\, y \in D \mid x + y \in D$.

12. Rewrite using only $\forall$ and $\exists$ quantifiers:

$$\exists! \, a \in \mathbb{R} \mid \forall x \in \mathbb{R}, ax = x.$$

13. Give two examples, in everyday English, exhibiting each of the following arguments or errors.

(a) Universal Modus Ponens

(b) Universal Modus Tollens

(c) Universal Transitivity

(d) Universal Inverse Error

(e) Universal Converse Error

14. Determine whether the following arguments exhibit Universal Modus Ponens, Universal Modus Tollens, Universal Transitivity, or either the Universal Inverse or Converse Error.

(a) If a customer intends to buy a car, then he or she is given a gift card.
Cooper received no gift card.
∴ Cooper must not have planned to purchase a car.

(b) I studied all day Saturday and Sunday for the exam.
Spending all weekend preparing for the test is sufficient for passing it.
∴ I will pass the test.

(c) The travel voucher is redeemable only if the owner filled out a survey.
Carolyn filled out a survey.
∴ Carolyn's travel voucher is redeemable.

(d) If the bark starts to peel, the tree is diseased.
A diseased tree must be cut down.
∴ A tree must be cut down if its bark starts to peel.

(e) A new phone comes free with every upgraded plan.
   Annie is still using her old phone.
   ∴ She mustn't have upgraded her plan.

(f) Professor Heller told us that the exam would begin at 8 AM if everyone arrived on time.
   As usual, Chase will probably arrive two minutes late.
   ∴ We won't have the full amount of time to work on the test because it won't start on time.

15. For nonempty sets D and E, suppose you know the statement

$$\forall x \in D, \exists y \in E \mid P(x, y)$$

is true. Explain why you can or cannot determine the truth value of the following.

(a) $\exists x \in D \mid \forall y \in E, P(x, y)$.

(b) $\exists x \in D, \exists y \in E, P(x, y)$.

16. In the previous question, can we determine the truth values of the statements if we do not assume D and E to be nonempty?

3

Introduction to Proofs

The concept of *proof* is fundamental to all of mathematics. It is not enough to simply demonstrate with an example that a certain general result holds. Similarly, one cannot argue that a result does *not* hold because a counterexample to it cannot be found. Two examples stemming from Pierre de Fermat (1601-1665), the great French mathematician, serve to highlight this reasoning. The first was mentioned in the introduction to Chapter 1, but we repeat it here to reiterate its importance.

Fermat conjectured that all numbers of the form $2^{2^n} + 1$, where n is a positive integer, are prime. Checking the first few cases ($n = 1, 2, 3, 4$) yielded primes: 5, 17, 257, and 65,537. The case $n = 5$ was much more difficult to investigate. Is 4,294,967,297 prime? Without a calculator or computer, this is no simple problem (and even with one it is not routine). It is easy to see that no small integers are factors of this beast of a number, but what about three-, four- and five-digit numbers? Fermat did not check (or at least did not check them thoroughly), and he believed that this number, and every remaining number of the form $2^{2^n} + 1$, was prime [32]. As one of mathematics' all-time all-stars, Fermat could not be wrong, could he?

In 1729, Swiss mathematician Leonhard Euler (1707-1783) set out to attack the problem [6]. Ultimately he showed that $2^{2^5} + 1$ was not prime and that it could be written as the product of 641 and 6,700,417 [17]. Euler's example did more than just prove that this single number was not prime, however. It demonstrated that Fermat's entire conjecture, that all numbers of the form $2^{2^n} + 1$ were prime, was incorrect. This is just one example of why proof is needed when making a claim about *all* things. Just because the result may hold for some or even most of the things, one cannot jump to the conclusion that it holds for all of the things.

Before proceeding to the second claim by Fermat, it is worth noting the impact Euler had on nearly *all* of mathematics; no introduction-to-higher-mathematics textbook would be complete without doing so. His mind was and is incomparable. He is known to have memorized not just the first 100 primes but also their squares through sixth powers! His discoveries were influential in the development of calculus and graph theory (which we shall discuss later), and he is amongst the "founding fathers" of topology and number theory. Analysis, musical theory, fluid dynamics, mechanics and optics all bear witness to his intellectual prowess. And quite interestingly, much of his work took place while he was partially blind (losing almost all sight in his right eye in 1738) and eventually totally blind (occurring in 1766) [17]. Upon going blind, he required not one but *two* scribes to write the mathematics that he spoke. Volume upon volume of work was developed post-blindness. Today, it is believed that a quarter to a third of *all* mathematics developed or published in the 18th century came from Euler alone, a simply astonishing thought.

The second mathematical claim by Fermat, which came to be known as "Fermat's Last Theorem," also highlights the necessity for proof. It had been known for at least 2,000 years that there were infinitely many solutions (consisting of positive integers x, y, and z) to the equation $x^2 + y^2 = z^2$ (or equivalently, that there are infinitely many non-congruent right triangles). But what happens if instead of squares one looks for solutions to $x^3 + y^3 = z^3$ or $x^4 + y^4 = z^4$? In 1637, Fermat conjectured that there were *no* positive integer solutions to the equation $x^n + y^n = z^n$ when n is a positive integer greater than two. Yet he supplied no argument as to why, instead famously jotting in the margin of a copy of Diophantus' *Arithmetica* that he had "discovered a truly marvellous proof of this, which this margin is too narrow to contain [38]."[1]

At first it was assumed that Fermat had indeed found a proof of the result; who would question one of the premier mathematical authorities of the time? None of his contemporaries could find such a proof, however. Years became decades which turned into centuries, with neither a counterexample nor a proof of the result. Some of the greatest mathematical minds to walk the planet either ignored the problem or made minimal advances on it: Gauss, Newton, Euler, and countless others. It was not until 1993 that Andrew Wiles announced his proof of a result that implied Fermat's Last Theorem was indeed true (that no solution to the equation exists).[2]

It took centuries to disprove the existence of a solution. This is simply another justification for the necessity of proof. Because a result could not be found did not mean a solution would never be found. It required a mathematical proof to show that no solution existed. Yet, since it has been proven, every mathematician now knows, without a doubt, that there are no positive integer solutions to $x^n + y^n = z^n$ for integers $n > 2$.

Thus is both the importance and the beauty of mathematical proof. Once proven, a result stands for all of eternity. It is not a hypothesis that is believed to be true. It *is* true and *always will be* true. Hence, the importance of proof. Perhaps it is best stated by Hermann Hankel [31]:

---

[1] Many college students have attempted the same technique on homework and exams. Such an approach is not encouraged.

[2] A flaw in Wiles' original proof was discovered, and eventually, the final result appeared in two papers: Wiles' original work and a piece co-authored with Richard Taylor addressing the flaw. Wiles worked in isolation for years, developing completely new areas of mathematics, in order to attack this one simple question [38].

> *In most sciences one generation tears down what another has built, and what one has established, another undoes. In mathematics alone each generation adds a new story to the old structure.*

Throughout this chapter we will investigate different types of proof techniques. Some will seem natural and others may take a bit of thinking to understand why they actually accomplish what they are claimed to do. Each, however, is grounded in the logical system we developed in the first two chapters. Our focus will not simply be on how to *write* proofs but rather how to approach the grunt-work required *before* writing a proof, and then, how to write *your* proofs.

3.1 What is Proof?

The discussion preceding this section addresses why proof is necessary, but it still leaves open the question: *what is proof?* Like certain geometric objects or a statement in Chapter 1, there is no way to define it without introducing further questions. Our definition of proof is similar in nature.

> **Definition 3.1.** A *proof* is a process by which a mathematical statement is shown to be true. Each step of the process follows logically from axioms, stated assumptions or previously established results until the desired mathematical result is reached.

It should have come as no surprise that certain words in Definition 3.1 come from Chapter 1: statement, axiom, logically, assumption, true. Moreover, the definition bears much resemblance to the definition of a logical deduction. Why? Because logical deductions were our first glimpse at the basic structure of a mathematical proof.

There is another word in the definition of proof that will be stressed throughout this text: "process." Processes have certain rules that they must follow, yet there is oftentimes room for personalization. Take writing, for example. When we first learned to write, we learned the requirements of forming sentences. In our early years, every sentence we wrote took the same standard shape. Yes, the subjects and the predicates changed, but the structure remained constant. *The wagon is red. My dog is funny.* From there, we were tasked with creating our first story. You were not aware of it at the time, but this was your first attempt at creative writing. Was it good? Here is the first story I ever wrote (sans the equally awkward illustrations).

> *This is me. I like to play ball. My favorite food is pizza. I want to be a racer. This is my mom.*

Grammatically the sentences are all correct. But I am sure all agree; it simply is not a good piece of writing (I was 6).

Fast forward a few years and think about how your writing process changed. In later elementary school or early middle school, how was every single paper structured? It was likely five paragraphs long. The first paragraph was your introduction and it included the main claim of the paper. Following the claim, there would be a few sentences laying out how you were going to support that claim. Each of those sentences would become or inspire the topic sentence for each of the following three paragraphs. Lastly, a concluding paragraph simply summarized the arguments, reiterating the main claim of the paper as well as what was emphasized in each of the middle three paragraphs.

This style of writing created papers that were *correct*, in that they accomplished the goal of making a claim and supporting it. A reader of the essay would fully comprehend the logic of the author. However, you learned that a *correct* paper need not be an *elegant* paper and that *good* writing made for *effective* writing. The reader would not just comprehend what you were saying but would engage with what was on the paper. You learned how transitions, flow, word choice, context and inclusiveness could persuade, influence and connect with a reader. Lastly, after exposing yourself to this plethora of writing tools, you developed *your style of writing*. Your papers became representative of you.

This is how we will approach proof throughout the remainder of this book. In this chapter and when new proof techniques are introduced, the basic structure of the process will be stressed. Once learned, it will be up to you to develop *your mathematical proof technique*. Before we get to our first technique, however, we must gain comfort with certain mathematical terminology.

The Language of Proofs

Flip through any advanced mathematics textbook and you are bound to stumble across certain words that indicate certain types of mathematical "things:" lemma, theorem, corollary, proof, conjecture. While we have defined a proof, these other terms require defining.

Definition 3.2. A *theorem* is a mathematical statement that is true and its truth is demonstrated via a proof.

Definition 3.3. A *lemma* is a mathematical statement that is true but often considered a subsidiary proposition used to prove another theorem.

Definition 3.4. A *corollary* is a mathematical statement that follows immediately from a previously proven result.

Definition 3.5. A *conjecture* or *proposition* is a mathematical statement that is believed to be true but has not been proven true.

Lemmas, theorems and corollaries are all true mathematical statements. Why not call them all theorems? Mathematicians have chosen these classifications to better organize their results. When reading a textbook or a paper, readers know that theorems are the main results. If some result is classified as a lemma, the reader may think, "This is a tool that leads to a bigger result." If that reader sees a corollary, they know that its proof depends heavily on a prior theorem or proof of a theorem.

These ideas can be better clarified through the following results. Do not get caught up in the details of the terms. The goal of stating these is not to impart mathematical wisdom about divisibility and primes but rather to exhibit each of the previous definitions.

Lemma: If an integer n divides both integers a and b, then n divides the difference $a - b$.

In some number theory textbooks this result may be listed as a theorem. In the context we are using it, however, we classify it as a lemma. It is "machinery" for proving the following main theorem.

Theorem: There are infinitely many prime numbers.

This is a main result and is consequently called a theorem. From it, the following corollary follows immediately.

Corollary: There are infinitely many prime numbers greater than 1,000,000,000,000.

The proof of this corollary is nearly immediate when considering the theorem (see Exercise 12a). Collectively, the order of presentation of these three results feels natural. It is the choice of words (lemma, theorem, corollary) that helps the reader recognize how they fit into this particular mathematical discussion. Additionally, from these results we may believe other mathematical results hold though they have never been proven; these are known as *conjectures*. The following is known as the *Twin Primes Conjecture*.

Conjecture: There are infinitely many pairs of twin prime numbers (that is, pairs of prime numbers that differ by two).

It should be noted that the choice of what to call certain results is an avenue for personalization in mathematics. Sometimes *result* or *proposition* may be used in place of *theorem*. Likewise, *lemma* may be replaced by *preliminary result*. The choice of words is not as important as their consistent and proper use. As you develop your mathematical voice and are exposed to a wider swath of mathematical literature, choose the organizational terms that best fit your style and stick with them. Develop *your* style.

We will often repeat this phrase: develop *your* style and find *your* voice. It is important to note, however, that this is not an open-ended invitation to introduce vivid language and fantastical metaphors into your mathematical writing. Choices to personalize one's writing should fall within the accepted norms appearing throughout mathematical literature. There are guidelines to writing clean and clear mathematics, of which many will be presented throughout the text as *Quality Tips*. All of these tips are summarized in Appendix B. Correct usage of terminology is the first tip.

> **Quality Tip 1.** *Use proper and uniform terminology for organizing results.*

Before proceeding to introduce *what* a mathematician uses, we step back and imagine *how* a mathematician works. Let us begin in the early days of a mathematician's studies. The basic tools are learned: arithmetic through algebra, trigonometry into calculus. Think of this as if it were a tree. The roots and base of the tree's trunk consist of these foundational things. To "do math," no matter what area of math, one must understand these concepts. However, upon completing the calculus sequence, the tree's trunk splits into two thick, hefty branches: pure mathematics and applied mathematics. Proceeding down the pure branch, mathematical theory and proof are introduced. From there, the branch forks further into smaller branches corresponding to areas of pure mathematics: analysis, algebra, number theory, topology, etc. Follow a branch and it too will continue to split into smaller and smaller branches. Point-set topology forms the foundation for more specific subareas: algebraic topology, differential topology, etc.[3]

To study mathematics means to progress down these branches. An undergraduate student explores various main stems and perhaps dabbles in some of the smaller extensions. In graduate school, students begin to move down branches to more specific areas. Eventually the branches become so specialized that there are no longer textbooks explaining the material; it is learned

---

[3]This is simply one visualization of the mathematical discipline. Some will argue that branches split earlier or later than calculus, or that labeling them "pure" and "applied" is faulty. Do not get caught up in these minute details and simply digest the entire metaphor. You become a mathematician by climbing the tree and working your way down branches. You specialize when you get far enough down the branch, and then "creating new mathematics" means to extend branches on this tree.

through dissecting journal papers and conference proceedings. Initially, the twig students find themselves on may be explained with papers from the 20th century (visualize this as the time this branch first appeared and began to grow). They read more papers, moving further down the smaller branches, as the papers become more recent. Students soon approach the tip of a tiny twig. It is here that they look for mathematics that is *currently* being done. Recent journal publications and the mathematics arXiv, an e-print repository housed through the Cornell University Library, present new scholarly work.[4]

The mathematician finds new math *to do* in the various literature sources. At the end of each branch on this tree there are countless conjectures and open questions. They are often so specialized that only those who have traversed from the trunk of the tree down every branch and through each fork to reach that single very tip of a twig will understand what the conjectures even ask. But it is those people who are suited, then, to begin their mathematical work. They aim to either prove the conjectures as true or to find counterexamples to them. And how do they do this? They look for patterns and consider specific subcases. They scratch out examples, draw pictures, consider related results and even change the conjectures to something easier to comprehend. They *think*, they *ask questions* and they *play*. Eventually, if they are lucky, they resolve the conjecture or some variant of it. And in doing so, the tree branch lengthens ever so slightly.

As discussed in the Preface, the pure mathematician is often asked, "What is that used for?" The pure mathematician knows better than to worry about this, however. She knows that someone will come along someday and extend the branch of knowledge a little bit further. From there, others may take the branch in multiple directions or fuse it to another branch. Someday, perhaps, an applied mathematician may be looking for the exact tool needed to address an issue arising in some real-world application he has been investigating. Lo and behold, there, on the tree of pure mathematics, is that tool. Follow the branch back and that small extension provided by you, the pure mathematician, is a necessary piece of the theoretic puzzle.

If this is *how* and *why* a pure mathematician does mathematics, *what* does the pure mathematician wield as his or her most powerful tool? Precise mathematical definitions.

Definitions

"Pick the right tool for the job and know how to use that tool." This is a response you may get if you ask a carpenter or a gardener for advice. Professional painters will not use small brushes to paint large walls and gardners will not dig a trench using a pitchfork. Having the right tools and knowing how to use them makes certain jobs significantly easier.

Mathematicians are no different. To successfully "do math" you must un-

---

[4]Visit https://arxiv.org/archive/math to see the latest "new math" being done.

derstand the tools of the trade and how to properly implement them. Perhaps no other tool is more important or essential to constructing proofs than *definitions*. To begin investigating their intricacies, we present three number theoretic definitions. They are terms you are undoubtedly familiar with.

Definition 3.6. An integer n is called *even* if $n = 2m$ for some $m \in \mathbb{Z}$.

Definition 3.7. An integer n is called *odd* if $n = 2m+1$ for some $m \in \mathbb{Z}$.

Definition 3.8. Let $n, d \in \mathbb{Z}$. We say d *divides* n, expressed as $d|n$, if $n = dm$ for some $m \in \mathbb{Z}$. We say d is a *divisor* or *factor* of n and that n is a *multiple* of d; d is called a *proper* divisor if $d < n$. If d does not divide n, we write $d \nmid n$.

Notice the precision in every one of these definitions. Ambiguity is eliminated by completely defining every variable. For example, it would be unclear to say, "An integer n is called *even* if $n = 2m$," or, "An integer n is called *even* if $n = 2m$ for some number m." In both situations, m is not explicitly defined. What type of number is it? By Definition 3.6, 6 is even because $6 = 2(3)$ and 3 is an integer. If we did not stipulate that m had to be an integer, then we might be able to claim that 7 is even, since $7 = 2(3.5)$.

Example 3.1. Both 86 and -112 are even:

$$86 = 2(43) \text{ and } -112 = 2(-56).$$

Both 55 and -1729 are odd

$$55 = 2(27) + 1 \text{ and } -1729 = 2(-865) + 1.$$

Is 0 even, odd, or neither? Because $0 = 2(0)$, it is even.

This example seems trivial. *Of course* 86 is even because we have known it is even for years. But this is not proper mathematical justification. Just like we cannot say, "A tiger is a mammal because it is a mammal," we cannot claim a number to be even because it is even. Similarly, we cannot claim 86 to be even because "2 divides 86." We must show that 86 precisely fulfills Definition 3.6.

Example 3.2. Why does the following equation not show that 21 is odd?

$$21 = 2(11) - 1$$

Definition 3.7 requires 21 be written as $2m+1$ for some integer m, not $2m - 1$. While it is indeed true that $21 = 2(11) - 1$, this does not satisfy the definition of being odd. We have that 21 is odd because

$$21 = 2(10) + 1.$$

The definition of divisibility is similar to that of even and odd in that it is an existence-type definition. To show a number n is even, we show there exists an integer m so that $n = 2m$. Likewise, to show that a divides b, we show there exists an integer c so that $b = ac$.

Example 3.3. Show that $8|40$ but 8 does not divide 41 (denoted $8 \nmid 41$).

Notice that

$$40 = 8(5) \text{ and } 41 = 8(5.125).$$

This shows, respectively, that $8|40$ (since $5 \in \mathbb{Z}$) but $8 \nmid 41$ (since $5.125 \notin \mathbb{Z}$).

Example 3.3 highlights other particularities of definitions. What is wrong with saying, "40 divided by 8 is 5, so 8 divides 40?" That claim is true, but again, to show that a definition holds you must precisely satisfy all of the requirements of that definition. Another peculiarity of Example 3.3 is the notation. Writing "$8|40$" is shorthand for the phrase, "Eight divides forty." It is not the fraction $\frac{8}{40}$ or $\frac{40}{8}$. It is incorrect to write

$$8|40 = 5.$$

This abuse of notation is as egregious as writing

$$\text{eight divides } 40 = 5.$$

Perhaps a more obscure detail is in showing $8 \nmid 41$. Our justification is that $41 = 8(5.125)$ and because 5.125 is not an integer, it must be that 8 does not divide 41. Here, we are assuming some sort of arithmetic uniqueness. We have not developed any theory telling us that there is not *another* way to write 41 as a product of 8 and something else. For the purposes of this section, this blind faith in the properties of the real number system is needed.[5]

---

[5] Assumed properties of the real number system appear in Appendix A titled "Properties of Real Number System.

112 *Introduction to Proofs*

In the next example, we start to present some of the thought processes involved with creating mathematical proofs. When we get to actually writing mathematical proofs, these thoughts will not be included in our final drafts. However, they are critical to the development of our first drafts. Why? Proofs are rarely ever straightforward. Work is involved in figuring out why the result actually holds.

One of the key methods for developing this is *asking questions*. This may seem counter-intuitive; you are looking to answer a question, so why would you are ask more questions? What you are doing is breaking the larger question down into different, easier-to-handle types of questions. This is the mathematical version of taking an open question (a question requiring an elaborate answer, in this case a proof) and turning it into numerous closed questions (questions with short, simple answers). Example 3.4 exhibits this questioning approach.

Example 3.4. Show that if 6 divides an integer m, then m must be even.

What do we know? The integer m is divisible by 6. Thus, $m = 6n$ for some $n \in \mathbb{Z}$.

What are we trying to show? That m is even.

What does this mean, according to the definition of being even? We must show $m = 2p$ for some $p \in \mathbb{Z}$.

How can we take what we know and turn it into what we must precisely show? Use algebra to rewrite what we know:

$$m = 6n$$
$$= 2(3n).$$

Can we claim that $3n$ is an integer? Because the integers are closed under multiplication, $3n$ is an integer. By Definition 3.6, m is necessarily even.

We finish our discussion of definitions with one last piece of advice. While we stress here the importance of *knowing* definitions, it should be noted that there are different levels of *understanding* them. Take the definition of even, for example. Mathematicians may claim they know the definition because they have simply memorized it. This memorization guarantees that they can show something is even because it fulfills parts of the definition. But how would they go about working through the mathematical derivation, mathematical intuition or scratch work leading up to writing up their final solution? For this, there is a deeper level of understanding required. It is that deep-down, in-your-gut understanding of what the definition is really saying. It is the ability to put the definition in one's own words without losing any of its integrity. To become a successful mathematician, taking your understanding of definitions beyond simple memorization is critical.

Existential and vacuously true statements

Definition 2.17 states that an existential statement of the form

$$\exists\, x \in D \mid P(x)$$

is true if and only if $P(x)$ is true for at least one x in the set D. This gives rise to our first proof technique.

> **Definition 3.9.** An *existential proof* is a proof of an existential statement of the form
>
> $$\exists x \in D \mid P(x).$$
>
> To prove the existential statement, at least one element $x \in D$ is shown to be in the truth set of the predicate $P(x)$.

This proof technique should feel completely natural as it coincides with exhibiting truth of certain statements in our everyday lives. Consider the following discussion:

Juan: The other soccer team has at least one player over six feet tall.

Erica: I don't believe you. Prove it to me.

Juan: Look here at their roster. It says that their goalie Manny is 6'4".

Erica: I guess you were right!

There may be no other tall player on the team, or, every single other player on the team could be taller than Manny. It does not matter, though. All it took to convince Erica about the truth of Juan's statement was the existence of a player over six feet tall.

> **Example 3.5.** Prove the following mathematical statement:
>
> $$\exists x \in \mathbb{Q} \mid \frac{98}{100} < x < \frac{99}{100}.$$
>
> **Proof** Notice that

$$\frac{98}{100} < \frac{985}{1000} < \frac{99}{100}.$$

Since this number is rational, we have shown that

$$\exists x \in \mathbb{Q} \mid \frac{98}{100} < x < \frac{99}{100},$$

we have shown the desired result.   □

Existential proofs, like all mathematical proofs, are not unique. In Example 3.5, the choice of the particular rational number satisfying the result was arbitrary; any rational number r satisfying

$$\frac{98}{100} < r < \frac{99}{100},$$

would have sufficed in proving the result.

There is a second proof technique that is related to existential proofs. Recall that a statement is either true or false. This means that for any given statement, either the statement is true or its negation is true. Also, we saw in Section 2.3 that the negation of the universal statement

$$\forall x \in D, \ P(x) \Rightarrow Q(x)$$

is the existential statement

$$\exists x \in D \mid P(x) \land \sim Q(x).$$

Thus, showing the universal statement

$$\forall x \in D, \ P(x) \Rightarrow Q(x)$$

is false equates to proving the existential statement

$$\exists x \in D \mid P(x) \land \sim Q(x)$$

is true (precisely the concept of finding a counterexample). But to do this, there must be some $x \in D$ that is in the truth set of both $P(x)$ and $\sim Q(x)$. What if the truth set of one of these (in particular, of $P(x)$) is empty? Then there could not possibly any such counterexample to the original statement, meaning the negation of the original statement must be false. Consequently the original universal statement must be true.

Definition 3.10. When the truth set of the predicate $P(x)$ (as a subset of a set D) of universal statement

$$\forall x \in D, \ P(x) \Rightarrow Q(x)$$

is empty, then the universal statement itself is said to be *vacuously true*. A proof of such a result is called a *vacuous proof*.

Example 3.6. Prove the following statement:

$$\forall x \in \mathbb{R}, \text{ if } x^2 + 2x + 1 < 0, \text{ then } x = \pi.$$

Proof Note that

$$x^2 + 2x + 1 = (x+1)^2,$$

and for any real number x,

$$0 \le (x+1)^2.$$

Thus,

$$x^2 + 2x + 1 < 0$$

is false for any $x \in \mathbb{R}$, meaning that the universal statement,

$$\forall x \in \mathbb{R}, \text{ if } x^2 + 2x + 1 < 0, \text{ then } x = \pi,$$

is vacuously true. $\square$

Note that in Examples 3.5 and 3.6 the word *Proof* signifies the beginning of each proof and a small square box $\square$ denotes the end of a proof. While not absolutely required, it is good practice to include them in your work. They are organizational in nature; the reader knows where the proof begins and where the proof ends.

Different mathematicians may choose to use alternative symbols or methods to accomplish the same goals that *Proof* and $\square$ do. Some will indent an entire proof, much like a lengthy quotation in expository writing. The open square signifying the end of the proof may be replaced with ∎. You may choose to end your proofs simply with the words *End of proof.* Others will use the letters *Q.E.D.* (or simply *QED*) in place of the $\square$ symbol. These letters stand for "quod erat demonstrandum," Latin for "that which was to be demonstrated."

These are simply stylistic options. No matter what style you develop, however, you should strive for consistency. IMAGINE READING AN INTRODUCTION TO MATHEMATICAL PROOFS TEXTBOOK AND MIDWAY THROUGH IT, THE AUTHOR CHANGES FONTS. WHY DID HE DO THAT? DOES IT SIGNIFY CHANGES IN THE MATERIAL? IS THERE EMPHASIS ON THIS MATERIAL THAT WAS NOT ON THE PREVIOUS MATERIAL? OR IS THE AUTHOR JUST DOING THIS TO MAKE A POINT? No matter the reason, you are distracted. This stylistic gaffe has taken away from the quality of the writing. Mathematical style is no different.

Quality Tip 2. *Be consistent with your style choices.*

Lastly, it is worth noting one other major difference in the proofs of the previous two examples. Note that the last sentence in the proof of Example 3.5 uses the pronoun "we" whereas throughout the proof of Example 3.6, there is no such first-person "action" taken.

Which approach is better? This is often a hotly debated topic, and ultimately, the answer is a personal decision. Choosing to include first-person pronouns such as 'we" conveys a sense of collaboration throughout the proof: "You, the reader, and I, the author, are working through this proof together." The proof writer is *the* author of the proof and he or she decides how readers should feel as they read the proof.

On the flip side, the author may *not* want the reader to have a connection to the proof. Perhaps the author wants to present the mathematics in a non-personal way. This approach is just as effective as the "collaborative" approach. As with style, determine what methodology fits *your* proof-writing approach, consistently use that approach and constantly work to improve how you craft your proofs.

Exercises

1. Provide an intuitive explanation for each of the following mathematical terms. That is, reword the definitions into loose, non-precise mathematical language.

    (a) even
    (b) odd
    (c) divides

2. Show that each of the following is even.

    (a) 42
    (b) $8m - 6$, for $m \in \mathbb{Z}$
    (c) $16n^4 + 12n^3 - 20n + 8$, for $n \in \mathbb{Z}$
    (d) $(p + 1)(p + 2)$, for $p \in \mathbb{Z}$

3. Show that each of the following is odd.

    (a) 129
    (b) -13
    (c) $72m - 5$ is odd for $m \in \mathbb{Z}$
    (d) n^2, if n is an odd integer

4. Determine if each of the following is even or odd.

    (a) $48p - 220q$, for $p, q \in \mathbb{Z}$
    (b) n^3, if n is even
    (c) $-12k^5 - 2$, if k is odd

    (d) $4m^2 - 7n$, if m and n are odd

5. List all the divisors of 24. Which of those divisors are proper?

6. Show that each of the following holds.

    (a) $6|1122$.

    (b) $9|0$.

    (c) $5|(125p + 50q)$ for $p, q \in \mathbb{Z}$.

    (d) Any multiple of 200 is divisible by 100.

    (e) $3|6^m$ for $m \in \mathbb{Z}^+$.

    (f) $4|(2^{n+1} - 8n^3)$, where $n \in \mathbb{Z}^+$.

7. Show that each of the following is false.

    (a) $8x$ is even for any $x \in \mathbb{R}$.

    (b) The sum of three consecutive integers is even.

    (c) For $p, q \in \mathbb{Z}$, if $p - q$ is odd, then p is even and q is odd.

    (d) For $m \in \mathbb{Z}$, if $3|m$, then m is odd.

    (e) Every integer divisible by both 2 and 6 is divisible by 12.

    (f) For $a, b, c \in \mathbb{Z}$,

        i. if $a|bc$, then $a|b$ or $a|c$.

        ii. if $a|(b + c)$, then $a|b$ or $a|c$.

8. Show the necessary scratch work to justify the following claims.

    (a) If p is even, then $p + 6$ is even.

    (b) If m is even, then $4|m^3$.

    (c) The sum of three odd integers is odd.

    (d) The product of three odd integers is odd.

    (e) If m and n are odd integers, then $m + n$ is even.

    (f) For $a, b, c \in \mathbb{Z}$, if $a|b$, then $a|bc$.

    Though we have yet to discuss mathematical proof in-depth, comment on the quality of the proofs in the following three exercises. What is odd, wrong or seems to be missing from them? Why?

9. *Claim*: If x is an odd integer, then $3x + 11$ is even.

    Proof Let x be an odd integer. x equals $2m + 1$. Take 3 times x. Substitute $2m + 1$. Distributing gives $6m + 3$. Now add 11 and it becomes $6m + 14$. This is of the form $2n$, so it's even.   □

10. *Claim*: If $p \in \mathbb{Z}$ and $6|p$, then p is even.

    Proof Let $p \in \mathbb{Z}$.

    Let $6|p$.

    Definition 3.8 gives $p = 6m$ for some $m \in \mathbb{Z}$.

    Algebra gives $p = 2(3m)$.

Because $m \in \mathbb{Z}$, properties of the integers give $3m \in \mathbb{Z}$.
Definition 3.6 gives that p is even.
This is what we wanted to show.   □

11. *Claim*: The sum of an odd integer and 7 is even.

    Proof Take an odd integer.

    $$n + 7 = (2m + 1) + 7$$
    $$n + 7 = 2m + 8$$
    $$n + 7 = 2(m + 4)$$

    $n + 7$ is even.   □

12. Prove the following existential results are true.

    (a) $\exists\, x \in \mathbb{R} \mid x^2 = 5$.
    (b) $\exists\, x \in \mathbb{R} \mid 14 - x = 2x^2 + 4$.
    (c) If W is the set of all words in the English language, then
        $\exists!\, x \in W$ such that x is 1 letter in length and x contains the
        vowel a.

13. Show that the following universal results are false.

    (a) $\forall x \in \mathbb{Z}$, if $x < 1$, then $x^2 + 1 > 1$.
    (b) $\forall a \in \mathbb{R}$, $ax = 3x^2 + 1$ has a solution.
    (c) $\forall x \in \mathbb{R}$, $x - 1 \in \mathbb{Z}$ if and only if $\dfrac{1}{x-1} \in \mathbb{Q}$.
    (d) If W is the set of all words in the English language, then
        $\forall x \in W$, x is spelled "eeee."

14. Explain why each of the following results is vacuously true.

    (a) $\exists\, x \in \varnothing$ such that $x = 4$.
    (b) $\forall x \in \mathbb{R}^-$, if $x > 0$, then $10x + 2x^2 = 3$.
    (c) $\forall x \in \mathbb{Z}^*$, $x = 0$ implies $x^2 = 5$.
    (d) $\forall x, y \in \{m \in \mathbb{Z} \mid m^2 = \pi^3\}$, if $x = y$, then $x < y$.

15. Three proofs of the following result are given. Comment on the
    quality of each. Which do you feel is stylistically the strongest?
    Why?

    Claim: For any $p \in \mathbb{Z}^+$, $p!$ is odd only if $p = 1$.

    Proof 1: Let $p \in \mathbb{Z}^+$. If $p = 1$, then $p! = 1$, which is odd. Otherwise,
    $p \geq 2$, meaning

    $$p! = p \cdot (p - 1) \cdots \cdots 2 \cdot 1$$
    $$= 2 \cdot (p \cdot (p - 1) \cdots \cdots 4 \cdot 3 \cdot 1).$$

Because $p \cdot (p-1) \cdots \cdots 4 \cdot 3 \cdot 1$ is an integer, $p!$ is even. Thus, $p!$ is odd only if $p = 1$.

Proof 2: If p is a positive integer, then $p!$ contains a factor of 2 only if $p \geq 2$. Thus, $p!$ is odd (namely, $p! = 1$) when, and only when, $p = 1$.

Proof 3: We know that $1! = 1$, which is odd. For $p \in \mathbb{Z}$, $p \geq 2$, we see that

$$p! = 2 \cdot \frac{p!}{2}.$$

Because

$$\frac{p!}{2} = \frac{p \cdot (p-1) \cdots \cdots 2 \cdot 1}{2},$$

it follows that

$$\frac{p!}{2} \in \mathbb{Z},$$

meaning, by definition, $p!$ is even, proving what we set out to prove.

16. Would it make sense, from a logical point of view, to say that a claim is *vacuously false*?

17. An integer n is said to be *perfect* if n equals the sum of its proper divisors. The integer n is *abundant* if it is less than the sum of its proper divisors and n is *deficient* if it is greater than the sum of its proper divisors.

    (a) Provide an intuitive explanation of the terms *perfect*, *abundant* and *deficient*.

    (b) Classify the first twelve positive integers as perfect, abundant or deficient.

    (c) Prove that 28 is perfect.

    (d) Prove that a prime number (Definition 3.18) is deficient.

18. Suppose the definition of divisibility were changed to be defined on all real numbers, as follows.

    For $x, y \in \mathbb{R}$, we say x *divides* y, written $x|y$, if there exists $d \in \mathbb{R}$ such that $y = xd$.

    (a) Show $\pi|4$ and $5|e$.

    (b) Is it true that $x|y$ for all $x, y \in \mathbb{R}$? Prove or disprove.

3.2 Direct Proofs

Logical deductions are an actual form of mathematical proof, yet their use is very limited. They can only be used to show that an argument is valid. Consider their structure, however. When we construct one, what is it that we are actually doing? We are assuming all of the hypotheses of the argument to be true. Then, we are showing that the conclusion must follow in some sort of logical way from those assumptions. In other words, we are showing that *if* the hypotheses are true, *then* the conclusion is true. It is a conditional statement that we are showing is true, and that is the focus of this section. It is the proof technique we refer to as *direct proof*, commonly known as an *if-then proof*.

> **Definition 3.11.** A *direct proof* is a proof of a conditional statement. The hypotheses of the conditional statement are assumed to be true. Then, using only these assumptions along with previously established facts (definitions, theorems, axioms, results, etc.), it is shown that the conclusion must necessarily be true.

The importance of definitions was stressed in Section 3.1. If the conclusion of a proposition is to show something is a particular *definition*, then to deduce this, one must have a firm understanding of what that *definition* is. In the proof itself, this understanding must be explicit. However, the proof is the end product of much hard work. The intuitive notion (the aforementioned in-your-gut understanding) of *definition* will guide you through your process of inquiry.

In some of our first proofs, and various proofs throughout the text, we will illustrate some of the scratch work necessary to actually develop the proof. Let us begin with a result you are probably familiar with: the sum of two odd integers is even.

> **Example 3.7.** Show that if x and y are both odd integers, then their sum is even.
>
> *Scratch work*: Let x and y be odd. Show $x + y$ is twice an integer. Since x and y are odd, we know $x = 2p + 1$ and $y = 2q + 1$.
>
> $$\begin{aligned} x + y &= (2p + 1) + (2q + 1) \\ &= 2p + 2q + 2 \\ &= 2(p + q + 1). \end{aligned}$$

Since $p + q + 1$ is an integer, we have figured out the structure of our proof and are ready to construct it.

Proof Suppose x and y are both odd integers. Then, by definition, $x = 2p + 1$ and $y = 2q + 1$ for some integers p and q. Then,

$$x + y = 2(p + q + 1).$$

By closure, $p + q + 1$ is an integer, so that, by definition, $x + y$ is even. □

How do the scratch work and the final proof differ in the above example? To begin with, the scratch work is not precise. Note that p and q are never defined as integers. Some may be aghast at this, but because this is scratch work and not our final draft, it is okay! As you strengthen your proof-writing techniques, you will also refine your approach to the inquiry process. Perhaps you will start to include phrases such as, "for p, $q \in \mathbb{Z}$." In scratch work, this is not necessary so long as you keep aware of what all of your assumptions are. Assuming something that is not actually an assumption is a fatal flaw to every proof!

Quality Tip 3. *Proofs must be self-contained. Precisely define each and every object appearing in your proof.*

Secondly, notice that the scratch work lacks cohesion. It jumps from one thought to the other, whereas the proof itself includes transitional language and has structural flow to it. It is a natural progression of thought. The reader understands exactly where each step of the proof comes from and where each idea is leading.

Quality Tip 4. *Give your proof cohesion and flow by having it follow a natural train of thought.*

Quality Tip 5. *Use writing techniques, such as transitions and word choice, to strengthen your proof and guide your audience.*

Just as proofs can be written in different styles, problems, too, may be written using various forms. Not every conditional statement is stated in "if-then" form, as we saw in Section 1.1. Identifying what it is you actually have to prove (that is, a statement's hypotheses and conclusion) is the first step towards creating a proof of the result.

Example 3.8. Show that the product of any two odd integers is odd.

Scratch work: What is this proposition really saying? Rewritten, it says, "If x and y are odd, then xy is odd." Thus, we must assume we have two odd integers and show their product is also odd.

Assume: $x = 2p + 1$, $y = 2q + 1$.

$$xy = (2p + 1)(2q + 1)$$
$$= 4pq + 2p + 2q + 1$$

Where to go from here? Often it is helpful to "keep your eyes on the prize." What is it we have to show? That xy is odd. Explicitly, this means we must show $xy = 2m+1$, for some integer m. A "looser" way of thinking of this: show $xy = 2(\ ) + 1$. A little bit of algebra leads us in the right direction:

$$4pq + 2p + 2q + 1 = 2(2pq + p + q) + 1.$$

With this, we are ready to prove the result.

Proof Let x and y be odd integers, so that $x = 2p+1$, $y = 2q+1$ for some $p, q \in \mathbb{Z}$. Then,

$$xy = (2p + 1)(2q + 1)$$
$$= 2(2pq + p + q) + 1.$$

By closure, $2pq + p + q \in \mathbb{Z}$, so that xy is odd. $\square$

Notice that in the previous example, one of the algebra steps from the scratch work was not included in the final proof. Because we assume our readers are astute mathematicians, there is no harm in eliminating basic steps such as these.

Quality Tip 6. *Know your audience.*

In lieu of this Quality Tip, it is worth noting the importance of knowing not only your audience but the *expectations* of your audience. If you are using this textbook in a classroom setting, you are likely writing proofs *for* your instructor. This does not mean that you must write the proof in his or her style, trying to mimic what he or she does does. Rather, it means you need to be aware of what they expect in terms of showing or justifying "minor" steps of the proof, such as basic algebra. If your instructor expects every step be shown and every step be fully justified, then the proof of the claim from Example 3.8 may look something like the following.

Proof Let x and y be odd integers, so that $x = 2p + 1$, $y = 2q + 1$ for some $p, q \in \mathbb{Z}$. Then,

$$
\begin{aligned}
xy &= (2p+1)(2q+1) \text{ (Substitution)} \\
&= (2p+1)(2q) + (2p+1)(1) \text{ (Multiplicative distribution)} \\
&= (2p)(2q) + (1)(2q) + (2p)(1) + (1)(1) \text{ (Multiplicative distribution)} \\
&= 4pq + 2q + 2p + 1 \text{ (Multiplicative commutativity)} \\
&= 2(2pq + q + p) + 1 \text{ (Multiplicative distribution)}.
\end{aligned}
$$

By multiplicative closure, $2pq \in \mathbb{Z}$, and by additive closure, $2pq + p + q \in \mathbb{Z}$, so that by Definition 3.7, xy is odd. □

Prior to constructing a proof, you should identify your intended audience and make sure the proof addresses the needs of that audience. But what is the range of possible audiences? As you begin your proof-writing journey, you are writing for an audience of beginners: those at your level. Your proofs will include more algebra (making sure it is easy for the reader to follow precisely what you are doing) and more complete justifications (perhaps even justifying steps of algebra by citing axioms of the real numbers, as we did in the proof above). Introductions and conclusions will be thorough and explanations will be in-depth (perhaps even repetitive). Writing for beginners means that you explain *everything*; you want nobody to get lost in your work.

But even writing so-as not to lose a reader can be broken down into two subcases: internal and external audiences. If you are writing a proof for someone intimately related to the material you are working with, such as a classmate or your instructor, then you assume their familiarity with the material. They know what you have been studying and they know what tools you have at your disposal.

Writing for an *external* audience means that you are writing for someone who does not know what your background is and is not familiar with what level of mathematics you understand. More importantly, you are not familiar with his or her background or current mathematical level. Such situations would include submitting solutions in various mathematics competitions (such as the William Lowell Putnam Mathematical Competition) or submissions to the problem solving sections of undergraduate mathematics journals (*Math Horizons*, *Mathematics Magazine* and *The College Mathematics Journal* are three such journals).

As you progress in mathematics, your audience progresses with you. Proofs in upper-level undergraduate mathematics courses will have a very different style than those in a second-year college-level course. The importance of style and voice increases as level increases, while the necessity for showing each and every step decreasing. You assume the reader can connect the dots more easily, even when those dots are further apart[6].

---

[6] An example of this might be in a proof requiring completing the square for a quadratic

Lastly, the *style* of proof may change based upon the area of its intended audience. Certain areas of mathematics use computational-style or constructive-style proofs, while others rely heavily on visual or graphical arguments. Those in applied areas, such as computer science, may focus heavily on the establishment of *correctness* (such as an algorithm or recursive process) over the elegance of the argument.

For now, we will work on strengthening the fundamentals of proper proof writing and periodically move towards developing voice and personalization. In that sense, then, note the stylistic differences in the proof of Example 3.7 and that of Example 3.8. The latter proof utilizes symbols more than that of the first. "Let x be an integer" and, "Let $x \in \mathbb{Z}$," are two ways of saying the same thing. One way is not better than the other; it is simply the proof writer's style that determines which approach is used.

Why reinvent the wheel? What does this mean in the context of mathematical proofs? It means to stay conscious of previously shown results. This can save you a lot of work. If something has already been proven, there is no need to prove it again. Expedite the process by citing the results!

Example 3.9. Prove that if x, y, and z are odd integers, then $x(y + z)$ is even.

Scratch work: Considering how we approached the previous two examples, it might make sense to assume $x = 2p + 1$, $y = 2q + 1$ and $z = 2r + 1$ and proceed as before. What would we show first? That $y + z$ is even. But this step of the proof would be a replica of the proof of Example 3.7. If we simply call on that result, it will make for a much more efficient proof.

We assume x, y, and z are odd integers. By Example 3.7, $y + z$ is an even integer. Thus, if we show that the product of an even and an odd integer is even, we will have proven our result.

Proof Let x, y, and z be odd integers. The sum $y + z$ is even, by Example 3.7. Thus, it suffices to show that the product of two integers, one even and one odd, is even.

Let m be an even integer and n an odd integer. By definition, $m = 2p$ for some integer p. Then

$$mn = 2pn,$$

and since pn is an integer, we have by definition that mn is even. It follows then that $x(y + z)$ is even, as was to be shown.   □

Note the use of variables in Example 3.9, particularly in the second half

---

expression. In the lower-level course, every step of the process might be exhibited. In an upper-level course, simply showing the first and last steps of the process would suffice.

of the proof. Why was the use of the variable x discontinued? It is because the progression of thought in the proof changed direction. The proof turned into a proof that the product of an even and an odd integer is even. Once we said, "Let x, y, and z be odd integers," those became *arbitrary but specific* odd integers. We were not allowed to redefine x. Hence, we introduced m and n to be new even and odd integers, respectively. While it would have been just fine to continue the use of x as an odd integer and introduce m, an even integer, into the problem, this could have confused the reader. As the goal of the proof changed, the use of all new variables to guide the reader changed as well.

While seemingly quite simple and straightforward, divisibility is a very important aspect of basic number theory (the study of the integers and their properties). It is a key component to the definition of prime numbers. While we address primes and their properties in the coming sections, we consider here basic properties of divisibility. We begin with two lemmas about divisibility and its relation to the integers -1, 0, and 1.

Lemma 3.12. *Every nonzero integer divides* 0.

Proof Let n be a nonzero integer. Then,

$$0 = n \cdot 0,$$

proving that every nonzero integer divides 0. □

Lemma 3.13. *Every integer is divisible by both* 1 *and* -1.

Proof Let $n \in \mathbb{Z}$. Then,

$$n = 1 \cdot n$$
$$= (-1) \cdot (-n),$$

proving that $1 | n$ and $-1 | n$, respectively. □

Lemma 3.13 shows that 1 (and -1) divides every integer. But what integers divide each of them? Clearly 1 and -1 are divisors of both themselves and each other. But are there other divisors? Our familiarity with arithmetic tells us surely there are no other integers dividing them. But how can we prove it? The following lemma is both familiar and necessary for proving our conjecture. It says that a positive integer's divisors can be no larger than the integer itself.

Lemma 3.14. *If* $d, n \in \mathbb{Z}^+$ *with* $d | n$, *then* $d \leq n$.

Proof Assume that $d, n \in \mathbb{Z}^+$ with $d | n$. By Definition 3.8, there exists $m \in \mathbb{Z}$ such that $n = dm$. Because $n > 0$, d and m must have the same sign. Thus, m is positive, or equivalently, $1 \leq m$. Then, $d \leq md$ (since $d \in \mathbb{Z}^+$), so that $d \leq n$, as desired. □

Having these lemmas in our arsenal, we are prepared to attack Theorem 3.15.

Theorem 3.15. *For $a \in \mathbb{Z}$, if $a|1$ or $a|(-1)$, then $a = \pm 1$.*

Proof Suppose $a \in \mathbb{Z}$ with $a|1$. Then, $1 = am$ for some $m \in \mathbb{Z}$. Moreover, because 1 is positive, a and m must have the same sign. If $a > 0$, then by Lemma 3.14,

$$0 < a \leq 1,$$

so that $a = 1$. The case where $a < 0$ is left as an exercise. $\square$

The following two results are properties of divisibility that are used when further developing its theory. In particular, Lemma 3.16 is a key result in proving Theorem 3.31.

Lemma 3.16. *If an integer n divides both integers a and b, then n divides both the sum $a + b$ and the difference $a - b$.*

Proof Let $a, b, n \in \mathbb{Z}$ with $n|a$ and $n|b$. By definition, there exist $x, y \in \mathbb{Z}$ so that $a = nx$ and $b = ny$. Then,

$$a + b = nx + ny$$
$$= n(x + y),$$

and similarly,

$$a - b = n(x - y),$$

proving n divides both $a + b$ and $a - b$ since, by closure, $x \pm y \in \mathbb{Z}$. $\square$

Theorem 3.17. *Divisibility is transitive: for integers a, b, and c, if $a|b$ and $b|c$, then $a|c$.*

Proof Let a, b, and c be integers and suppose $a|b$ and $b|c$. By Definition 3.8, there are integers m and n such that $b = am$ and $c = bn$. Then,

$$c = (am)n$$
$$= a(mn).$$

Since $mn \in \mathbb{Z}$, we have shown that $a|c$. $\square$

This is a natural point to discuss what is often deemed as the scope (within a proof) of a variable. It can be thought of as the range in a proof over which the variable is defined. If we were to begin speaking of x here, it would not make sense. The variable x is not defined in this context; it was defined in the proof of Example 3.9, and once that proof has ended, x is no longer defined as an odd integer. The scope of the variable x defined in the proof of Example 3.9 is solely the proof of Example 3.9.

The scope of a variable in proofs is identical to the scope of variables in computer programming. Consider the following piece of Java code.

```
1
2   float x = 1;
3
4   void function1(){
5   float y = 2;
6   println(x);
7   println(y);
8   }
9
10  void function2(){
11  float z = 3;
12  println(x + z);
13  }
14
```

There are two functions defined in this code, function1 and function2. Notice that x is defined prior to both of them; x is then understood within each function (it is sometimes called a *global* variable). Each function is within the scope of x. When y is defined to equal 2 in function1, it is only within function1 that y has this definition. If line 12 were to be println($x + y$), the code would have an error. The variable y is not defined within function2(); it is outside the scope of y.

Thus, keep track of variables and when they are or are not still defined. Also, it is good practice to be consistent with your choice of variables. Something would look awkward with the proof in Example 3.8 had we began with, "Let x and G be odd integers." Is it incorrect? No. But consistency in variable choices aids the reader in following the thought progression of the proof. Upper- versus lowercase letters, those at the beginning of the alphabet versus those at the end, Greek versus Roman ... use the same types of variable to represent similar sorts of mathematical objects.

Quality Tip 7. *Do not reuse a specific variable within its scope in a proof.*

Quality Tip 8. *Be consistent with what style of variables represents what type of mathematical object.*

The following example demonstrates the idea of variable scope.

Example 3.10. Prove that the square of an integer that is one or three more than a multiple of 4 is itself one more than a multiple of 4.

Scratch work: Our assumption here is that we begin with an integer either 1 or 3 more than a multiple of 4. What does this mean? It means if n is

---

such an integer, then $n = 4m + 1$ or $n = 4m + 3$ for some integer m. We want to show that $n^2 = 4(\ ) + 1$. A little algebra, similar to the proofs of our previous even and odd results, is the key. The difference here is that we know n takes one of *two* possible forms. We simply need to repeat the process, showing the result holds, for either form.

Proof Suppose that n is an integer that is 1 or 3 more than a multiple of 4. This means that $n = 4m + 1$ or $n = 4m + 3$ for some integer m.

If $n = 4m + 1$, then

$$n^2 = 16m^2 + 8m + 1$$
$$= 4(4m^2 + 2m) + 1,$$

and since $4m^2 + 2m \in \mathbb{Z}$, we have that n^2 is one more than a multiple of 4.

Now, if $n = 4m + 3$, then

$$n^2 = 4(4m^2 + 6m + 2) + 1,$$

which is also one more than a multiple of 4. In either case, the result holds. □

Notice that the variable m appeared in two different places in the proof of Example 3.10.[7] In saying, "If $n = 4m + 1$," m became defined as long as the proof was operating under that particular hypothesis. When the proof moved to a new hypothesis, m was no longer defined and could be redefined under the new hypothesis. This choice of reusing m in the second hypothesis ("if $n = 4m + 3$") was actually a wise choice. The consistency in using m did not distract the reader. She never asked herself, "Why did we change variables? Does that represent another significant change?" The variable m was representing the same thing but simply in a new context.

Related to the scope of a variable is the concept of *when* a variable is defined inside a proof. Once a phrase such as, "Let $n \in \mathbb{Z}$," is used, n becomes an arbitrary but *fixed* integer. To exhibit this idea, consider the following two uses of the integer n.

1. Let $x \in \mathbb{Z}$ be even. Then, by definition, $x = 2n$ for some $n \in \mathbb{Z}$.

2. Let $x, n \in \mathbb{Z}$ with x even. Then, by definition, $x = 2n$.

[7]The proof in Example 3.10 is technically a proof by cases, which we investigate fully in Section 3.5.

In (1), x is defined to be an arbitrary even integer. The definition of being even says there must be another integer n, dependent upon x, so that $x = 2n$.

In (2), x and n are first defined to be arbitrary even integers. At this point, they both become fixed. It need not be the case that $x = 2n$; the initial choice of n was not dependent upon x in any way. This approach is incorrect.

> **Quality Tip 9.** *When variables within a proof are dependent upon one another, define them in an appropriate order.*

We conclude our introduction to direct proofs by investigating a few key properties of prime numbers. Even though the concept of a prime number is simple (they are introduced shortly after learning about division), the study of properties of prime numbers has been and continues to be a dynamic area of mathematics research. In fact, there are numerous famous unanswered questions (called *open questions*) involving primes that we will state at the end of this section. First we supply a precise definition of what it means to be *prime*.

> **Definition 3.18.** A positive integer $n > 1$ is called *prime* if its only divisors are 1 and itself, and it is called *composite* if it is not prime.

The definition of being composite may seem a bit awkward: n is composite if it is *not* prime. Why define it this way? It mirrors concepts from symbolic logic. Just like a statement must be true or false, one or the other but not both, positive integers greater than 1 must be prime or composite (not prime).

If we are to assume within a proof that an integer n is composite, we do not have much to work with. All we know is that n is not prime. From there, we can derive further information by dissecting the definition of being (or not being) prime. But there is an alternative approach. Definitions such as that of *composite* posed in Definition 3.18 can often be restated into an equivalent, usable definition. We cannot simply say that the two definitions are equivalent, however. Such a claim requires proof. Lemma 3.19 provides such an alternative definition to being composite.[8]

Lemma 3.19. *Let $n \in \mathbb{Z}$, $n > 1$. Then n is composite if and only if $n = ab$, with $a, b \in \mathbb{Z}$ and $1 < a < n$, $1 < b < n$.*

Scratch work: We are asked to prove an "if and only if" statement. This equates to proving *two* conditional statements (by Theorem 1.17): (1) if n is composite, then $n = ab$ where $1 < a < n$, $1 < b < n$, and (2) if $n = ab$ where $1 < a < n$, $1 < b < n$, then n is composite.

Proof Let $n \in \mathbb{Z}$, $n > 1$. To prove the biconditional, let us first assume n is composite. Definition 3.18 give that n is not prime, meaning that there is a divisor a of n that is neither 1 nor itself. By Definition 3.8,

---

[8]Note that some books will actually define *composite* in the form of Lemma 3.19 and then proceed to show that that definition is equivalent to not being prime.

$$n = ab$$

for some $b \in \mathbb{Z}$. The inequality $1 < a$ implies

$$b < ab,$$

or equivalently,

$$b < n.$$

Likewise, $a < n$ is equivalent to

$$a < ab,$$

giving

$$1 < b.$$

Thus, $1 < b < n$, as required.

Now assume $n = ab$, with $a, b \in \mathbb{Z}$ and $1 < a < n$, $1 < b < n$. This says precisely that n has a proper divisor greater than 1, showing that n is not prime and consequently proving our result. □

Lemma 3.19 provides an equivalent and more useful definition for being composite. The next lemma, though, expands upon this result. While Lemma 3.19 says that a composite number has some proper divisor greater than 1, Theorem 3.20 says that a composite number must have a *prime* divisor.

Where would we even begin in trying to construct a proof of such a result? One question that often helps is, "Can I think about specific cases and generalize?" For example, how would you go about determining if 72 had a prime factor? If we do not know if it has a prime divisor, we at least know it factors: $72 = 8 \cdot 9$. Are either of these prime? No, but they too factor. We continue this process. What are we doing? In basic arithmetic terms, we are find the *prime factorization* of 72. When first learning about prime factorization of integers, we are often taught to draw trees, such as Figure 3.1.

How can we generalize what happened specifically in Figure 3.1 to a general proof? In constructing the factorization tree of Figure 3.1, we started with 72. Is it prime? No; we factored it. We then repeated this question for each of the factors. Is 9 prime? Is 8 prime? If the integer is not prime, we extend the tree. But if it is prime, the branch of the tree stops. Why do we know each branch must ultimately stop? As we progress down the tree, the integers get smaller, but since they are positive, there is a limit to how small they can get.

This idea is *exactly* the inspiration (and scratch work) needed to prove Theorem 3.20.

Theorem 3.20. *Every positive integer greater than 1 has a prime divisor.*

Proof Let $n \in \mathbb{Z}$, $n > 1$. We know that n is either prime or composite. If n is prime, then the result holds (since $n|n$; see Exercise 3a). Otherwise, n is composite, and by Lemma 3.19,

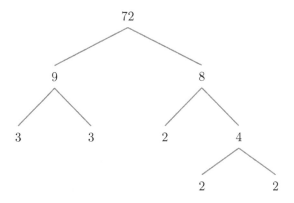

FIGURE 3.1
Finding prime factors

$$n = a_1 b_1$$

where $a_1, b_1 \in \mathbb{Z}$ with $1 < a_1 < n$ and $1 < b_1 < n$.

Consider a_1. It divides n, and it is either prime or composite. Should it be prime, then the result holds. Otherwise, it is composite, and by Lemma 3.19, there are $a_2, b_2 \in \mathbb{Z}$ with $1 < a_2 < a_1$ and $1 < b_2 < a_1$ with

$$a_1 = a_2 b_2.$$

Now consider a_2. It divides a_1 (and consequently n by Theorem 3.17) and is either prime or composite. As before, if it is prime, the result holds. Otherwise, it is composite, and by Lemma 3.19, we can find $a_3, b_3 \in \mathbb{Z}$ with $1 < a_3 < a_2$ and $1 < b_3 < a_2$ with

$$a_2 = a_3 b_3.$$

Because there are only finitely many integers between 1 and n, we can continue this process, knowing that it must terminate with some integer a_r $(r \in \mathbb{Z}^+)$, $1 < a_r$ and $a_r | n$. For the process to terminate, it must be that a_r is prime. Thus, a_r is the desired prime divisor of n, as was to be shown. $\square$

Primes can take many forms, and in the open questions presented at the end of this section, we can see that it is unknown if there are infinitely many primes (in Section 3.4 we show that there are actually infinitely many primes) that take various forms. However, some forms have unique representatives.

Lemma 3.21. *The only prime number that is one more than a perfect cube is 2.*

Proof Since $2 = 1^3 + 1$, it is one more than a perfect cube. Now, suppose $n^3 + 1$ is prime. We know

$$n^3 + 1 = (n+1)(n^2 - n + 1).$$

To be prime, it must be the case that either $n + 1 = 1$ (in which case $n = 0$, and $n^3 + 1 = 1$ is not prime) or

$$n^2 - n + 1 = 1$$

(in which case $n = 1$, which yields $n^3 + 1 = 2$). Thus, the result holds. □

Primes form the building blocks of our natural number system. Because of this, researchers long to understand as much about them as possible. Some of the greatest minds in mathematics, past and present, have attempted to answer the following questions, yet they remain open [18].

1. **Goldbach Conjecture**: Every even integer greater than 2 is the sum of two primes.

2. **Odd Goldbach Conjecture**: Every odd integer greater than 5 is the sum of three primes.

3. **Twin Prime Conjecture**: Are there infinitely many *twin primes* (i.e., primes that differ by 2, such as 5 and 7, or, 29 and 31)?

4. Do there exist infinitely many prime numbers of the form $n^2 + 1$?

5. Must a prime number exist between n^2 and $(n+1)^2$ for $n \in \mathbb{Z}^+$?

Throughout this section we have highlighted some of the key properties of direct proofs. We continue our investigation of them in the next section by looking at a particular type of direct proof involving sets.

Exercises

1. If you were to prove the following results, state what the primary assumptions are and what the desired conclusion to be reached is.

    (a) The sum of an odd integer and an even integer is odd.

    (b) The cube of any odd integer is odd.

    (c) The product of a composite number and a prime number is a composite number.

    (d) Every nonzero integer divides itself.

2. Prove the following results about even and odd integers.

    (a) The product of an even integer and any integer is even.

    (b) For all $x \in \mathbb{Z}^+$, $x^2 + 5x + 1$ is odd.

    (c) The difference of an even and an odd integer is odd.

    (d) The product of two consecutive integers is even.

3. Prove the following results about divisibility.

    (a) Every nonzero integer divides itself.

    (b) Knowing that 100 divides an integer is sufficient for knowing that 10 divides the integer.

    (c) Let $x, y \in \mathbb{Z}$. Then, $x|y$ only if $x^2|y^2$.

    (d) If $d|a$, $d|b$, and $d|c$, where $a, b, c, d \in \mathbb{Z}$, then $d|(2a^2 + 3b^3 + 4c^4)$.

4. Write a second proof for each of the results in the previous problem whose style differs significantly from your original proof.

5. Prove the following biconditional results.

    (a) If $n \in \mathbb{Z}$, then n is even if and only if $n + 1$ is odd.

    (b) An integer x is odd if and only if $-x$ is odd.

    (c) Suppose $a, b, q, r \in \mathbb{Z}$ with $a = bq + r$. Then $d \in \mathbb{Z}$ divides both a and b if and only if d divides both b and r.

    (d) For $a, b, c \in \mathbb{Z}$ with $c \neq 0$, then $a|b$ if and only if $ca|cb$.

6. Let $a, b, c \in \mathbb{Z}$ with $c|a$ and $c|b$ (c is called a *common divisor* of a and b). Prove that for any $x, y \in \mathbb{Z}$, $c|(ax + by)$.

7. Disprove each of the following:

    (a) For $p \in \mathbb{Z}$ with p prime, $(-1)^p = -1$.

    (b) An integer a is said to be a *perfect square* if $a = x^2$ for some $x \in \mathbb{Z}$. Then, for any $m, n \in \mathbb{Z}$, if mn is a perfect square, then m and n are both perfect squares.

    (c) For all $a \in \mathbb{Z}$ with $a > 1$, $a^2 - 1$ is composite.

    (d) The product of two consecutive integers is positive.

    (e) For all $a, b \in \mathbb{Z}$, $(a + b)^2 = a^2 + b^2$.

8. Prove or disprove each of the following.

    (a) For $a, b, c \in \mathbb{Z}$, if $a|b$, then $a|bc$.

    (b) Let $n \in \mathbb{Z}^*$. Then, $n|n^p$ for any $p \in \mathbb{Z}^+$.

    (c) For integers m and n, if $n|m$ and $m|n$, then $m = n$.

    (d) If integers a and b are composite, then $a + b$ is composite.

9. Finish the proof of Theorem 3.15.

10. Find any errors, either mathematical or stylistic, in the following proof.

Claim: For integers a, b and c, if $a|b$ and $a|c$, then $a|(b + c^2)$.

Proof Let $a, b, c \in \mathbb{Z}$ with $a|b$ and $a|c$. By definition, $b = am$ and $c = am$ for some $m \in \mathbb{Z}$. Then,

$$b + c^2 = am + (am)^2$$
$$= am + a^2 m^2$$
$$= a(m + am^2).$$

By properties of the integers, $m + am^2 \in \mathbb{Z}$. Consequently, by definition, we have shown $a|(b + c^2)$, as was to be shown.   □

11. Comment on the style of the following proof.

    Claim: If the difference of two integers is even, then the difference of their cubes is also even.

    Proof Let A and x be integers with $A - x$ even. By Definition 3.6, there exists an integer θ such that

    $$A - x = 2\theta.$$

    By algebra and substitution we see

    $$
    \begin{aligned}
    A^3 - x^3 &= (A - x)(A^2 + Ax + x^2) \\
    &= (2\theta)(A^2 + Ax + x^2) \\
    &= 2(\theta(A^2 + Ax + x^2)).
    \end{aligned}
    $$

    Closure properties of the integers give that $\theta(A^2 + Ax + x^2)$ is an integer, which proves that $A^3 - x^3$ is even.   □

12. Comment on the style of the following proof.

    Claim: For every $n \in \mathbb{Z}$, $3(n^2 + 3) - 2(n^2 + n + 4)$ is a perfect square.

    Proof We will prove that $3(n^2 + 3) - 2(n^2 + n + 4)$ is a perfect square for any $n \in \mathbb{Z}$. Then, begin by taking an integer $n \in \mathbb{Z}$. Then,

    $$(n - 1)^2$$

    is a perfect square (as it is the square of an integer). Then,

    $$n^2 - 2n + 1 = (n - 1)^2.$$

    Then,

    $$3(n^2 + 3) - 2(n^2 + n + 4) = n^2 - 2n + 1.$$

    Then,

    $$3(n^2 + 3) - 2(n^2 + n + 4) = (n - 1)^2.$$

    Then, because we assumed n to be an integer, we have that $n - 1$ is an integer, and then, we deduce systematically by the definition of a perfect square then that $3(n^2 + 3) - 2(n^2 + n + 4)$ must be a perfect square.   □

13. Find the error in the following proof.

    Claim: For $n \in \mathbb{Z}$, the quantity $n^2 + 5n - 14$ is composite when $n > 2$.

    Proof Let $n \in \mathbb{Z}$ with $n > 2$. Then,

    $$n^2 + 5n - 14 = (n - 2)(n + 7).$$

    By closure, both $n - 2 \in \mathbb{Z}$ and $n + 7 \in \mathbb{Z}$. So, $n^2 + 5n - 14$ is the product of two integers. Thus, it is, by definition, composite. □

14. Find the error in the following proof.

    Claim: If $m, n \in \mathbb{Z}$ have a common factor greater than 1, then $m^2 + n^2$ is composite.

    Proof $m^2 + n^2 = c^2(x^2 + y^2)$ if we let $m = cx$ and $n = cy$ be the integers with a common factor $c > 1$. This number is composite since both factors are greater than 1. The result holds. □

15. List three tips for improving the stylistic quality of a proof. Alternatively, list three stylistic errors that detract from a proof's overall quality.

16. Suppose you are asked to prove the following claims. For each, write different two proofs, with the first being written for an audience of calculus-level mathematics students and the latter being written for a collection of college mathematics professors.

    (a) If $m, n, p \in \mathbb{Z}$ are odd, then $mn + p$ is even.
    (b) If d is a common divisor of integers a, b and c, then $d^2 | (ab - ac + bc)$.

3.3   Direct Proofs: Set Element Method

We direct our focus now on sets. Recall that the intuitive notion of a subset is that it is a subcollection of a larger collection; set S is a subset of set T if everything that is in S is also in T. How would one go about proving that one set is a subset of another? We use the precise definition: S is a subset of T if every element of the set S is also an element of the set T. That is, to show $S \subseteq T$, we prove that an implication holds: *if $s \in S$, then $s \in T$.*

Definition 3.22. The *set element method of proof* is a proof technique for showing one set S is a subset of another set T. An arbitrary but particular element s of S is defined, and s is then shown to be an element of T.

It is important to note that the set element method of proof proves that one set *is* a subset of another set. It is not a method for showing one set is *not* a subset of another set. Our intuitive notion of a subset yields the method for showing this latter scenario. Consider a particular calculus class, for example. In that class, a number of students are athletes and a number of students are majoring in economics. How would one determine if the group of economics majors is a subcollection of the group of athletes? You would ask all the economics majors if they were athletes. If any one single student said, "no," then you would not need to query any more of the future economists. That single student proves that the collection of economics majors is not a subcollection of the collection of athletes.

Mathematically speaking, that economics major who is not an athlete is playing the role of the *counterexample* to the claim that the set of economics majors is a subset of the set of athletes. It disproves the universal statement,

$$\forall x \in \text{Calculus, if } x \in \text{Economics, then } x \in \text{Athletes.}$$

The following is a mathematical example of finding such a counterexample.

Example 3.11. Let

$$S = \{x \in \mathbb{Z} \mid x = 3n + 1, \text{ for some } n \in \mathbb{Z}\}$$

and let T be the set of all odd integers. Prove that neither $S \subseteq T$ nor $T \subseteq S$.

Before jumping into the proof of the result, let's walk through the thought process behind the proof's construction. We search for a single counterexample to show $S \nsubseteq T$. A counterexample is not a symbolic argument such as, "Elements of S are of the form $3n + 1$ while elements of T are of the form $2m + 1$." Rather, in terms of the example above, the economist is not an athlete: a specific element of S that we can point to and say, "This is in S but it is not in T."

Sometimes it helps to simply list the elements in the sets. We have here that

$$S = \{\ldots, -2, 1, 4, 7, \ldots\}$$

and

$$T = \{\ldots, -3, -1, 0, 1, 3, \ldots\}.$$

Right away we can find the two counterexamples that will help us prove neither set is a subset of the other.

Proof Suppose S is defined as

$$S = \{x \in \mathbb{Z} \,|\, x = 3n + 1, \text{ for some } n \in \mathbb{Z}\}$$

and T is the set of all odd integers. We show that neither set is a subset of the other. To show S is not a subset of T, we must find one specific element of S that is not an element of T. Observe that $4 \in S$ since $4 = 3(1) + 1$; 4 is not odd, so that $4 \notin T$. Thus, S is not a subset of T.

Likewise, we see that $3 \in T$, but $3 \notin S$, for if it where, there would exist $n \in \mathbb{Z}$ so that $3 = 3n + 1$. Algebra shows that the only such value of n is $n = \frac{2}{3}$, which is not an integer. □

The main takeaway from Example 3.11 should be how one set is shown to not be a subset of another: the use of specific counterexamples. Next we present an important lemma that will be used in various upcoming results.

Lemma 3.23. *If S is any set, then $\varnothing \subseteq S$.*

Proof Let S be any set. To prove $\varnothing \subseteq S$, we must show that every element of $\varnothing$ is also an element of S. Since $\varnothing$ has no elements, this claim is vacuously true and the result holds. □

Lemma 3.23 is a critical result that underlies many set element method proofs. For some sets A and B, to prove $A \subseteq B$, we begin by assuming $x \in A$. In doing so, we assume that A is nonempty. However, if A is empty, Lemma 3.23 proves the desired result. Because of this, in proofs where we aim to prove one set A is a subset of another, we will simply take a general element of A and show it is an element of the other set. We will not explicitly examine the case when $A = \varnothing$.

We next prove a few of the set identities of Theorem 2.8, with the remaining ones left as exercise

Example 3.12. Prove the Inclusion of Intersection laws of Theorem 2.8.

Proof Let A and B be sets. To prove

$$A \cap B \subseteq A,$$

let $x \in A \cap B$. By Definition 2.6, $x \in A$ and $x \in B$. Therefore the two Inclusion of Intersection laws hold:

$$A \cap B \subseteq A, \text{ and, } A \cap B \subseteq B.$$

$\square$

Maybe you are thinking, "Wow, that proof was short. Was it really necessary to prove that? Clearly, an element of $A \cap B$ is an element of A *and* an element of B." The latter thought here, that an element of $A \cap B$ is an element of both A and B, is the crux of the actual proof. But as we have discussed previously, no matter how simple a result seems, it is not assumed to be true until mathematically proven. Even within proofs, we must provide valid justification.

Quality Tip 10. *Appropriately justify your steps.*

The proof of the next result assumes a previous part of Theorem 2.8 has been shown true.

Example 3.13. Prove the first Identity law of Theorem 2.8.

Proof Let A be a set. To prove

$$A \cup \varnothing = A,$$

we must prove that both

$$A \cup \varnothing \subseteq A$$

and

$$A \subseteq A \cup \varnothing.$$

The latter follows from the Inclusion of Union identity law.

To prove the first, let $x \in A \cup \varnothing$. The definition of set union gives $x \in A$ or $x \in \varnothing$. Since $\varnothing$ has no elements, it must be that $x \in A$. This shows that

$$A \cup \varnothing \subseteq A,$$

and consequently, that the desired result holds. $\square$

Notice in Example 3.13 that previous results were referenced by name rather than number ("definition of set union" versus "Definition 2.6"). Either method is acceptable; this is another way you can begin to develop *your* style of proof writing.

Example 3.14. Prove the Double Complement law of Theorem 2.8.

Proof Let A be a set in some universal set U. To prove

$$(A^C)^C = A,$$

we begin by showing

$$(A^C)^C \subseteq A.$$

To this end, take $a \in (A^C)^C$. By Definition 2.6,

$$a \notin \{x \in U \mid x \notin A\},$$

which implies that $a \in A$ (since an element of U is either an element of or not an element of A). Hence,

$$(A^C)^C \subseteq A.$$

To show

$$A \subseteq (A^C)^C,$$

let $a \in A$. So,

$$a \notin \{x \in U \mid x \notin A^C\},$$

meaning by Definition 2.6 $a \in (A^C)^C$.

Since the sets $(A^C)^C$ and A are subsets of one another, we have that

$$(A^C)^C = A,$$

as was to be shown. □

Here, the proof is succinct and complete. Even though it is short, it introduces and defines the necessary parts and concludes with a summary of what was to be shown. It outlines what it is going to accomplish by saying, "First, we will do this, and then we will do this." It lets the reader know what direction the author intends to take. You do not want your reader to have to pause and think, "Wait ... why did this step just happen? Why is the proof taking this direction?"

Quality Tip 11. *Where necessary, describe what the goal of an upcoming section is or what organizational approach the proof will take.*

Example 3.15. Prove the Set Difference law of Theorem 2.8.

Proof Let A and B be sets in a universal set U. In order to prove that

$$A - B = A \cap B^C,$$

we show that the sets $A - B$ and $A \cap B^C$ are subsets of one another. First, let $x \in A - B$. Definition 2.6 gives that $x \in A$ and $x \notin B$, meaning $x \in B^C$. Hence, $x \in A \cap B^C$, by the same definition. Consequently,

$$A - B \subseteq A \cap B^C.$$

Now, let $x \in A \cap B^C$. It follows from Definition 2.6 that both $x \in A$ and $x \in B^C$, or equivalently, $x \notin B$. This means $x \in A - B$, showing that

$$A \cap B^C \subseteq A - B.$$

Because of these two results, we have proven the desired set equality.
$\square$

The ultimate goal of writing a mathematical proof is not to prove a result as quickly as possible. It is to create both a mathematically correct *and* an elegant display of why a result must be true. Elegance means writing complete and correct sentences, not simply connecting ideas symbolically. This is one of the main differences between being asked to show that some results holds versus being asked to "prove" that result holds. Hence, the following tip, though obvious by this point in the text, is emphasized.

Quality Tip 12. *Use correct grammar.*

The next example illustrates that the set element method of proof applies to results involving a general finite number of sets as well. Note the necessity to prove $A \subseteq A_i$ for just a single general integer i.

Theorem 3.24. *For sets A and A_i ($i = 1, 2, \ldots, n$, where $n \in \mathbb{Z}^+$), if*

$$A \subseteq \bigcap_{i=1}^{n} A_i,$$

then $A \subseteq A_i$ for $i = 1, 2, \ldots, n$.

Proof Let A and A_i ($i = 1, 2, \ldots, n$, where $n \in \mathbb{Z}^+$) be sets with

$$A \subseteq \bigcap_{i=1}^{n} A_i.$$

Consider one of the sets A_i, for some $i \in \{1, 2, \ldots, n\}$. To prove $A \subseteq A_i$, let $x \in A$. Since

$$A \subseteq \bigcap_{i=1}^{n} A_i,$$

by Definition 2.4 we know that

$$x \in \bigcap_{i=1}^{n} A_i.$$

By the definition of an indexed intersection, then, $x \in A_i$. Thus, $A \subseteq A_i$, as was to be shown. $\square$

When working through *how* to prove a result, it sometimes helps to step back from the problem and analyze the structure of the objects present. In the following example, it is crucial to note that $A \times B$ is itself a set, but it is very different from sets A and B. The sets A and B can be thought of as consisting of elements; the set $A \times B$ consists of *ordered pairs* of those same elements.[9] Structurally, then, the sets simply *look* different.

Example 3.16. If A and B are nonempty sets, prove that

$$A \times B = B \times A$$

if and only if $A = B$.

Proof Let A and B be nonempty sets. We wish to prove the biconditional statement

$$A \times B = B \times A$$

if and only if $A = B$. Such a proof requires proving two implications. We begin with assuming that $A = B$ and show that

$$A \times B = B \times A.$$

If $A = B$, then

$$A \times B = A \times A = B \times A,$$

as was to be shown.

To prove the other conditional statement, assume

---

[9]It is a bit of an oxymoron to say that A is a set consisting of elements; *every* set consists of elements. What we mean here is that if we arbitrarily write out the set A, we might begin by writing $A = \{a_i \mid i \in I$, where I is some index set$\}$.

$$A \times B = B \times A.$$

To prove $A = B$, we prove that A and B are subsets of one another. Proceed by taking $x \in A$. Since B is nonempty, choose $b \in B$. Then, $(x, b) \in A \times B$. Since

$$A \times B = B \times A,$$

we have that $(x, b) \in B \times A$. Definition 2.14 gives then that $x \in B$, proving that $A \subseteq B$. The proof that $B \subseteq A$ is similar. Hence, $A = B$ and consequently the desired biconditional statement holds.   □

In Example 3.16, it appears that we waved our hands near the end of the proof, stating that the additional claim is proven similarly. This is an allowable technique when part of a proof is identical, with the exception of variables, to a previously proven part of the proof. It would be redundant to repeat the same argument, simply swapping the roles of sets A and B. Nothing new would be learned by the reader. If, however, there would be an additional hypothesis, step or method in the second part of the proof, it would be necessary to then prove the additional claim.

The technique for notating this tool varies. Some will use language as seen in Example 3.16. Another common choice of language is the use of the Medieval Latin phrase *mutatis mutandis*, meaning "the necessary changes having been made"; "the proof that $B \subseteq A$ follows mutatis mutandis." Choose the phrasing that fits your style, or, develop an equivalent approach to tell your reader how your proof is proceeding.

> **Quality Tip 13.** *Do not reinvent the wheel within a proof. If additional cases of a proof are proven identically to something already proven, use that previous part of the proof to streamline your work.*

We conclude this section with one last example exhibiting the importance of notation. In particular, note the choice of variables in the following proof. All sets within the proof are denoted using upper-case Roman letters, and elements of sets are denoted with lower-case Roman letters. This consistency is nothing new, mathematically speaking. In geometry, angles are typically assigned lower-case Greek letters (α, β, θ). Variables in equations tended to be lower-case Roman letters from the end of the alphabet (x, y, z), while real-valued functions often are named using the letter f, g, or h. This uniformity eases readers' potential confusion.[10]

---

[10]This idea is highlighted by simply considering the real-valued function $x(f) = f^2$.

Quality Tip 14. *Be consistent with notational norms.*

Example 3.17. For sets A and B, if $A \subseteq B$, then

$$\mathcal{P}(A) \subseteq \mathcal{P}(B).$$

Before presenting a clean final draft of the proof, let us discuss the thought process that goes into *how* this result is proven. First note what we can assume: A is a subset of B. Thus, we know that every single element of A is also an element of B.

What is it we are trying to show? We want to show that the *power set* of A is a subset of the *power set* of B. We cannot simply say, "Since every element of A is an element of B, the result holds." Doing that equates to saying, "Assume the hypothesis. Then, the conclusion must hold." We did not actually show *why* the conclusion holds. The work is missing.

To show

$$\mathcal{P}(A) \subseteq \mathcal{P}(B),$$

we take an arbitrary element of $\mathcal{P}(A)$ and show it is in $\mathcal{P}(B)$. What is the structure of such objects? Elements of $\mathcal{P}(A)$ are themselves sets; namely, they are subsets of A. If we take an arbitrary element of $\mathcal{P}(A)$, then, it would not be notationally correct to say, "Let $a \in \mathcal{P}(A)$." Why? The reader is trained to think, "The symbol a represents an element of A." That is not what we want. We want the reader to initially recognize that our arbitrarily chosen element is actually a *set*. Thus, a better choice of variable would be S (or some other capital letter, which typically denote sets).

Now, we then have $S \in \mathcal{P}(A)$. By definition of the power set, the set S is a subset of A; that is, every element of S is an element of A. But how does that help us? Keeping in mind what we want to show (that $S \in \mathcal{P}(B)$) and what we have assumed (that $A \subseteq B$), the result is ours for the taking. Example 2.8 tells us that S must also be a subset of B, yielding the desired result: $S \in \mathcal{P}(B)$.

We take all of this thinking and scratch work and after numerous drafts, the following efficient proof results.

Proof Let A and B be sets and assume $A \subseteq B$. Let $S \in \mathcal{P}(A)$. By definition of the power set, $S \subseteq A$ and by Example 2.8, $S \subseteq B$. Hence, $S \in \mathcal{P}(B)$, proving that

$$\mathcal{P}(A) \subseteq \mathcal{P}(B).$$

$\square$

Our discussion of direct proofs, including the set element method, leads naturally towards the two proof techniques of the next section, which we classify as *indirect* proof techniques.

Exercises

1. State the initial assumptions and desired conclusion for a proof of each of the following results. For example, if asked to prove $A \subseteq B$, the initial assumption would be, "Assume $a \in A$," and the conclusion would be, "Show $a \in B$." Assume A, B and C are sets.

    (a) $A - B \subseteq A$.
    (b) $A \cap B \cap C \subseteq A \cap C$.
    (c) $(A - B) - C \subseteq A - C$.
    (d) $(A - C) - (B - C) \subseteq (A - B) - C$.

2. Disprove the following claims by providing an explicit counterexample. Assume A, B and C are sets.

    (a) If $A \subset B$ and $A \subset C$, then $A \subset (B \cap C)$.
    (b) If A is a set, then $A \subseteq \mathcal{P}(A)$.
    (c) $A \times (B \cap C) = (A \times B) \cap C$.
    (d) $\mathcal{P}(A \times B) = \mathcal{P}(A) \times \mathcal{P}(B)$.

3. Prove the following set identities from Theorem 2.8:

    (a) Inclusion of Union
    (b) Commutativity
    (c) Distribution
    (d) Universal Complement

4. Prove each of the following claims that that one set is a subset of another. Assume that A, B and C are sets.

    (a) If $A \subseteq B$ and $A \subseteq C$, then $A \subseteq B \cap C$.
    (b) If $A \subseteq B$, then C, $A \cup C \subseteq B \cup C$.
    (c) If $A \subseteq B$, then $A \cap C \subseteq B \cap C$.
    (d) If $A \subseteq B$, then $A - C \subseteq B - C$.
    (e) If $A \times B \subseteq A \times C$ (assuming A, B and C are nonempty), prove that $B \subseteq C$.
    (f) If $A, B \subseteq C$ and $A \subseteq B$, then $C - B \subseteq C - A$.

5. Prove each of the following claims that one set is equal to another set. Assume that A, B, C and D are sets.

    (a) $A \cup (A \cap B) \cup (A \cap C) = A$.
    (b) $(A \cap B) \cup (A \cap B^C) = A$.
    (c) $(A - C) - (B - C) = (A - B) - C$.

    (d) $(A \cap B) \times (C \cap D) = (A \times C) \cap (B \cap D)$.

6. Prove each of the following. Assume that A, B and C are sets.

    (a) $A \times (B \cap C) = (A \times B) \cap (A \times C)$.
    (b) If $A \subseteq B$, then $A \times A \subseteq B \times B$.
    (c) $A \subseteq B$ if and only if $A - B = \varnothing$.
    (d) If $A \subseteq B$, then $\mathcal{P}(A) \subseteq \mathcal{P}(B)$.
    (e) If A is a set, then $A \in \mathcal{P}(A)$.
    (f) If $\mathcal{P}(A) \subseteq \mathcal{P}(B)$, then $A \subseteq B$.

7. Comment on the style of the following proof of the first Complement identity from Theorem 2.8. Identify changes that can be made to make the proof more stylistically consistent.

Proof Let A be in U.

$A \cup A^C \subseteq U$: Let $x \in A \cup A^C$. According to Definition 2.6, $x \in A$ or $x \in A^C$. But, because U is the universal set, both $A \subseteq U$ and $A^C \subseteq U$. Thus, in either case, by Definition 2.4, we have that $x \in U$, showing that $A \cup A^C \subseteq U$.

$U \subseteq A \cup A^C$: Take $u \in U$. Then, either $u \in A$ or $u \notin A$, or equivalently, $u \in A$ or $u \in A^C$, by the definition of set complement. Thus, $u \in A \cup A^C$, as was to be shown. $\square$

8. Comment on the style of the following proof of the second Universal Bound identity from Theorem 2.8. Identify changes that can be made to strengthen the proof.

Proof Let A any set in universal set U. Then $A \cap \varnothing = \varnothing$.

First, vacuously true $\varnothing \subseteq A \cap \varnothing$, as $\varnothing$ equals no elements.

Second, consider arbitrary element of $A \cap \varnothing$ that element must reside in $\varnothing$ but no such elements so $A \cap \varnothing \subseteq \varnothing$.

Combining what was shown in the previous two paragraphs, shown $A \cap \varnothing = \varnothing$. $\square$

9. Comment on the proof of the Associativity set identity of Theorem 2.8.

Claim: For sets A, B, and D, $(A \cap B) \cap D = A \cap (B \cap D)$.

Proof Let A, B, and D be nonempty sets (for if they were empty, the intersection of the empty set with any set is itself empty, and the result would hold).

To show $(A \cap B) \cap D \subseteq A \cap (B \cap D)$, begin with $x \in (A \cap B) \cap D$. Then, $x \in A \cap B$ and $x \in D$, and consequently $x \in A$ and $x \in B \cap D$, showing $x \in A \cap (B \cap D)$.

Likewise, to show $A \cap (B \cap D) \subseteq (A \cap B) \cap D$, take $x \in A \cap (B \cap D)$. This means $x \in A$ and $x \in B \cap D$, so that $x \in A \cap B$ and $x \in D$, making $x \in (A \cap B) \cap D$, as was to be shown.

Since both inclusions hold, we have that $(A \cap B) \cap D = A \cap (B \cap D)$. □

10. Comment on the quality of the following disproof of the claim that for sets A, B, and C, if $A \subseteq (B \cup C)$, then $A \subseteq B$ or $A \subseteq C$.

    Disproof: $\{1, 2\}$, $\{1\}$, and $\{2\}$ disproves the result.

11. Comment on the quality of the following disproof of the claim that for sets A, B, and C, $A \cup (B \cap C) \subseteq (A \cup B) \cap C$.

    Proof Let A, B and C be sets such that $A \cup (B \cap C) \subseteq (A \cup B) \cap C$. Then, if $x \in A \cup (B \cap C)$, then $x \in A$ or $x \in B \cap C$, meaning that $x \in A$, or both $x \in B$ and $x \in C$. Because we assumed the result holds, we also have that $x \in (A \cup B) \cap C$. This tells us that $x \in (A \cup B)$ and $x \in C$, or equivalently, $x \in A$ or $x \in B$, and, $x \in C$. In the latter situation, x must be an element of C. In the first case it need not be, thus disproving the claim.   □

12. Does Theorem 3.24 hold if the intersection is replaced by a union? Prove or disprove.

13. For $i \in \mathbb{Z}^+$, let $B_i = \{in \mid n \in \mathbb{Z}\}$.

    (a) Prove that $B_6 \subseteq B_2$.
    (b) Prove that if $n|m$ $(n, m \in \mathbb{Z}^+)$, then $B_m \subseteq B_n$.

14. Prove the following.

    (a) $\{4^n \mid n \in \mathbb{Z}\} \subseteq \{2^n \mid n \in \mathbb{Z}\}$.
    (b) $\{4^n \mid n \in \mathbb{Q}\} = \{2^n \mid n \in \mathbb{Q}\}$.

15. Prove: $(\mathbb{Z} \times \mathbb{Z}^+) \cap (\mathbb{Z}^+ \times \mathbb{Z}) = \mathbb{Z}^+ \times \mathbb{Z}^+$.

16. If S is any set and A_i are sets, $i \in I$, where I is some index set, prove that

$$S - \bigcap_{i \in I} A_i = \bigcup_{i \in I} (S - A_i).$$

17. Prove the following results about the disjunctive union (see Exercise 2.2.23). Assume that A, B and C are sets.

    (a) $A \oplus B = (A - B) \cup (B - A)$.
    (b) $(A \oplus B) \oplus (B \oplus C) = A \oplus C$.
    (c) $A \cap (B \oplus C) = (A \cap B) \oplus (A \cap C)$.
    (d) $A \oplus B = A^C \oplus B^C$.

3.4 Proof by Contrapositive and Contradiction

In Section 3.2, statements of the form "if P, then Q" were proven directly. That is, the hypothesis P was assumed to be true and after a series of steps, each following logically from prior steps or known facts, the conclusion Q was reached. There are, however, alternative ways to prove these types of conditional statements. We investigate two in this section, called *proof by contrapositive* and *proof by contradiction*. These proof techniques are sometimes called *indirect* proof techniques because they approach a conditional statement in a way that does not seem straightforward or direct.

Recall that Theorem 1.14 proved a conditional statement and its contrapositive are logically equivalent. That is, for statements P and Q,

$$(P \Rightarrow Q) \equiv (\sim Q \Rightarrow \sim P).$$

What does this mean? A conditional statement is true if and only if its contrapositive is true. Thus, an alternative method for proving a conditional statement true is to prove its contrapositive true.

Definition 3.25. The conditional statement "if P, then Q" is *proven by contrapositive* by directly proving the statement "if $\sim Q$, then $\sim P$". That is, the negation of Q is assumed to be true and it is shown that the negation of P must be true.

There are no hard rules for when to try proving an if-then statement directly versus by contrapositive. One hint that may lead you towards using the contrapositive is that in your attempts to prove the statement directly you simply get stuck or lack any possible direction for trying to reach the desired conclusion. For instance, suppose we are to attempting to prove the following claim and our scratch work is similar to the work that follows:

Claim: If the square of an integer is even, then the integer itself must be even.

Scratch work: Suppose n^2 is even. We know then that $n^2 = 2k$ for some k. We want to show $n = 2m$.

Right away our scratch work comes to a halt. We could attempt to take square roots, but that introduces $\sqrt{2}$ into our work. If we assume more about n, we risk introducing faulty hypotheses or over-complicating our work. So we back up and try looking at the problem from a different perspective. It is an implication. Maybe a proof by contrapositive will be fruitful, and indeed it is!

Theorem 3.26. *Let $n \in \mathbb{Z}$. If n^2 is even, then n is even.*

Proof Suppose $n \in \mathbb{Z}$. We proceed by proof by contrapositive, assuming n to be odd and showing that n^2 is odd. By definition, $n = 2m + 1$ for some $m \in \mathbb{Z}$. Then,

$$n^2 = (2m + 1)^2$$
$$= 4m^2 + 4m + 1$$
$$= 2(2m^2 + 2m) + 1.$$

By closure, $2m^2 + 2m \in \mathbb{Z}$, so that we have shown that n^2 is odd. Consequently, the contrapositive of what we have proven must also be true: if n^2 is even, then n is even.   □

Theorem 3.26 could also have been proven by calling upon Example 3.8. Having assumed n to be odd, we know that $n^2 = n \cdot n$, the product of two odds. Either proof "does the trick" and proves the desired result. However, if one is building a catalog of results, it is often preferred to use previous work to expedite later proofs. Such an approach finds its roots in ancient Greece in one of the most widely read texts of all time: Euclid's *Elements* [27, 28, 26].

Though attributed to Euclid (approx. 300 B.C.), the *Elements* is a collection of 13 books, with parts authored by numerous famous Greek mathematicians, such as Pythagoras, Hippocrates of Chios (not of the Hippocratic oath), and Eudoxus [17]. It is as famous for its breadth (covering geometry (known today as Euclidean geometry) and elementary number theory) as it is for its axiomatic approach to mathematics. Beginning with just 23 definitions, 5 postulates[11], and 5 general axioms, Euclid systematically developed the remainder of the work. No result was proven unless it could rest upon only the previously proven results and stated assumptions. The *Elements* became the foundational gold-standard for deductive, logical mathematics.

Looking back at the proof of Theorem 3.26, note that the proof technique is introduced early on in the proof. While not a required part of a proof, it is good practice when using an indirect proof to inform your readers how you will proceed. It is natural to believe a statement of the form "if P, then Q" will be proven by first assuming P to be true. If it begins with some other assumption, your reader may have to pause and try to figure out from what angle you are attacking the problem.

> **Quality Tip 15.** *State when you are using an indirect proof technique.*

There is another situation when a proof by contrapositive may be a more obvious option to use. Mathematical objects are often defined as to *have* a certain property, yet a claim may involve an object *not* having that property. Employing the definition becomes cumbersome, if not impossible. Consider the definition of divisibility, for example. Saying an integer n is divisible by 12 provides a tool to work with:

---

[11]Euclid's famous 5th postulate became the source of much investigation, and ultimately it was found that not assuming it gave rise to non-Euclidean geometries [34].

$$n = 12m$$

for some integer m. But what if we knew or had to show that n was *not* divisible by 12? We would only be able to say that there is no such integer m so that $n = 12m$. How would we go about showing that no such m exists? In terms of being a mathematical tool, this information gives us nothing to either work with or work towards. Its negation would, however.

Example 3.18. Prove that an integer not being divisible by 12 is a necessary condition for knowing that integer is not divisible by 4.

Before proving the result, we rewrite the statement in standard if-then form.

If an integer is not divisible by 4, then it is not divisible by 12.

We see that a proof by contrapositive may be helpful here and proceed to prove the below statement.

If an integer is divisible by 12, then it is divisible by 4.

Proof We prove that an integer not being divisible by 12 is a necessary condition for knowing that integer is not divisible by 4 by proving its contrapositive: if an integer is divisible by 12, then it is divisible by 4.

Assume an integer n is divisible by 12. Since $4|12$ and divisibility is transitive (Theorem 3.17), we know $4|n$. Consequently, the result holds. $\square$

There is no need to investigate proof by contrapositive any further, since the actual proofs themselves are simply direct proofs. It is only necessary to learn to recognize *when* to use a proof by contrapositive.

The second indirect proof technique, called *proof by contradiction*, relies on the fact that a statement is either true or false. It must have exactly one of these truth values; it cannot have both. Thus, if you suppose a statement has a particular truth value and arrive at some inconsistency, then that original assumption could not possibly have been true and consequently the statement must have the opposite truth value from what was originally assumed.

Definition 3.27. To perform a *proof by contradiction*, assume the negation of the desired result. Use this assumption, along with previously determined results, to arrive at a logical contradiction. Consequently, the original assumption of the negation of the conclusion could not have held.

Why would one even want to use a proof by contradiction?[12] Perhaps proof by contradiction may be the only method available if all other proof techniques have failed. Additionally, when the negation of the desired conclusion is assumed, it often becomes a critical tool that can be used throughout the proof. If you find yourself stuck in your scratch work and you have yet to use this assumption, see if it is somehow a benefit to you. Since you are assuming the negation of what actually is true, the assumption should be a pivotal piece leading to a contradiction.

Be aware that whatever contradiction may arise may not be obvious. It is only through scratch work that you determine what logical fallacy you will arrive at. It may be that you originally assumed a statement P to be true and you deduced $\sim P$ to be true. Or it may turn out that you prove $1 < 1$ or that there is a real number whose square is negative. Contradictions come in all shapes and sizes.

Note that a proof by contradiction is not just a technique for proving conditional statements, as we will see in our first example, Theorem 3.28. However, when proving a conditional statement $P \Rightarrow Q$ via contradiction, we assume the *entire statement* to be false. Recall, then, that this means we assume P to be true and Q to be false, or equivalently, P to be true and $\sim Q$ to be true.

Proof by contradiction traces its history to over 2,000 years ago. Archimedes, considered by most as one of the greatest mathematicians of all time, was able to determine the area of a circle using *two* proofs by contradiction (*double reductio ad absurdum* (reduction to absurdity)). He let the area of a particular circle be A and area of a corresponding triangle be T. By showing the assumptions $A < T$ and $T < A$ led to absurdities, it must have been the case that $A = T$ [17].

Two millenia later, we find ourselves investigating this proof technique. Our first example of such a proof is a fairly obvious result.

Theorem 3.28. *There is no greatest integer.*

If given this result to prove, without any suggestion to use a particular proof technique, what might lead us to try contradiction? Here, there really is no assumption to work with. In proceeding with a proof by contradiction, we at least have *something* to use: the existence of a greatest integer. What would be a contradiction to that assumption? If we could find an integer that were larger than the assumed greatest one. We know exactly how to do this: any integer plus a positive integer yields a greater integer. This is the crux of the proof.

Proof To prove there is no greatest integer, suppose for contradiction that

---

[12]The 20th century English mathematician G. H. Hardy says of proof by contradiction: "It is a far finer gambit than any chess gambit: a chess player may offer the sacrifice of a pawn or even a piece, but a mathematician offers the game [24]."

n is the greatest integer. By closure, $n+1 \in \mathbb{Z}$, and $n < n+1$, a contradiction. Hence, the result holds. $\qquad\square$

As in a proof by contrapositive, notice that the proof technique is named within the proof itself. This is particularly important in a proof by contradiction. Before we see other results proven by contradiction, we note an important notational part of the previous proof.

Example 3.19. What is incorrect with the following alternative proof to Theorem 3.28?

Proof To prove there is no greatest $\mathbb{Z}$, suppose for contradiction that n is the greatest $\mathbb{Z}$. By closure, $n+1 \in \mathbb{Z}$, and $n < n+1$, a contradiction. Hence, the result holds. $\qquad\square$

There is only one simple difference between this proof and the original: the word "integer" is replaced by the symbol "$\mathbb{Z}$." Why is this incorrect? The symbol $\mathbb{Z}$ represents the set of all integers; it is not shorthand for the word "integer." Likewise, it would be incorrect to write, "the values x and y sum to a positive number" as "the values x and y + to a positive number."

Quality Tip 16. *Do not incorrectly use mathematical notation in place of written words.*

We continue our exploration of indirect proofs with two of the most famous proofs in mathematics history. They are famous for their elegance, simplicity and efficiency. The results themselves stand out as major milestones in the development of the real number system, as well. The first proves that there are real numbers that are *not* rational (called *irrational numbers*). The second result deals with the building blocks of the positive integers: primes.

Theorem 3.30 requires a definition. Despite referencing rational numbers previously, we define them explicitly.

Definition 3.29. A real number n is called a *rational number* if it can be written in the form

$$n = \frac{a}{b},$$

where $a, b \in \mathbb{Z}$ and $b \neq 0$. The set of all rational numbers is denoted $\mathbb{Q}$. A real number that is not rational is called *irrational*.

Theorem 3.30. (The Irrationality of $\sqrt{2}$) *The real number $\sqrt{2}$ is irrational.*

Before diving into the proof, notice that this result is just *begging* to be proven using a proof by contradiction. The definition of being *irrational* is simply *not being rational*. There is nothing to work with there, except that such a number cannot be written in the form described in Definition 3.29. In assuming that a number *is* rational, we obtain a tool to work with.

Proof To prove $\sqrt{2}$ is irrational, proceed by contradiction and assume $\sqrt{2} \in \mathbb{Q}$. Then,

$$\sqrt{2} = \frac{a}{b},$$

where $a, b \in \mathbb{Z}$ and $b \neq 0$. Moreover, we assume a and b have no common factors greater than 1. We have

$$b\sqrt{2} = a$$

or equivalently,

$$2b^2 = a^2.$$

Since $b^2 \in \mathbb{Z}$, then a^2 by definition is even. Theorem 3.26 tells us that a is necessarily even. Hence, $a = 2k$ for some $k \in \mathbb{Z}$. Substituting, we see:

$$2b^2 = (2k)^2$$
$$= 4k^2,$$

or equivalently,

$$b^2 = 2k^2.$$

As before, since $k^2 \in \mathbb{Z}$, we have that b^2, and as a consequence b, must be even. Thus, $b = 2j$ for some $j \in \mathbb{Z}$.

We have shown that 2 is a factor of both a and b, a contradiction to the assumption that they had no common factors. In turn, $\sqrt{2} \notin \mathbb{Q}$, as was to be shown. $\square$

In the previous proof and many of the proofs preceding it, note the use of white-space, linebreaks and alignment to make the proof readable and easy to follow. Again, while not a requirement of a correct proof, it is a small technique that can improve the aesthetic quality of a proof.

Quality Tip 17. *Use white-space, linebreaks and alignment to improve a proof's presentation.*

The second of our famous results is widely considered one of the most beautiful mathematical proofs of all time, and it is a classic proof by contradiction. It comes to us from Euclid's *Elements*, appearing as the 20th proposition in the 4th of its 13 books [28]. Before proceeding to the proof, it is worthwhile to discuss this phrase: "most beautiful mathematical proofs."

What makes mathematics *beautiful*? The 20th century Hungarian mathematician Paul Erdős (1913 - 1996) explained it quite intimately [29].

> *Why are numbers beautiful? It's like asking why is Beethoven's Ninth Symphony beautiful. If you don't see why, someone can't tell you. I know numbers are beautiful. If they aren't beautiful, nothing is.*

If any recent mathematician had the authority to discuss *beautiful* mathematics, it was Erdős. In terms of quantity, he is the most published mathematician in history, authoring or co-authoring over 1500 items from a wide array of mathematical areas, including graph theory, number theory, probability and set theory, just to name a few [37]. Erdős believed some of his results were more elegant than others and that in all of mathematics, some results stood above the rest. Those, Erdős claimed, belonged in "The Book," a text kept by the Supreme Fascist (Erdős' reference to God).

After Erdős' death, some of his closest friends and colleagues published their take on "The Book," titled "Proofs from The Book" [2]. Amongst the proofs appearing in it is that of Theorem 3.31 (six proofs of the result, to be precise, are included). Moreover, the text includes a number of the results from Chapter 7 (Theorems 7.29, 7.31, 7.35 and 7.36).

Theorem 3.31. (The Infinitude of Primes) *There are infinitely many prime numbers.*

Proof To prove there are infinitely many prime numbers, assume for contradiction that there are only finitely many of them:

$$p_1, p_2, \ldots, p_n,$$

for some $n \in \mathbb{Z}^+$. Construct a new number

$$N = p_1 p_2 \ldots . p_n + 1.$$

Since $p_i \geq 2$ for all i, we have that $N > p_i$ for all i. If N is prime, we have reached a contradiction, having found a prime not in the original list (since it is greater than every number in the list). Else, N must be composite.

Since N is composite, Theorem 3.20 says that N has a prime divisor. Having assumed that $p_1, p_2, \ldots, p_n$ constitute all the prime numbers, we can conclude p_i is a divisor of N for some particular $i \in \{1, 2, \ldots, n\}$.

Now, note that $p_i | (N - 1)$, since

$$\begin{aligned} N - 1 &= p_1 p_2 \ldots p_n + 1 - 1 \\ &= p_1 p_2 \ldots p_n \\ &= p_i (p_1 p_2 \ldots p_{i-1} p_{i+1} \ldots p_n). \end{aligned}$$

Thus, $p_i|N$ and $p_i|(N-1)$. By Lemma 3.16, $p_i|[N-(N-1)]$, or equivalently, $p_i|1$, a contradiction to Lemma 3.15. Consequently, our original assumption that there are only finitely many primes must be false, proving the desired result. $\qquad\qquad\square$

Some of the notation in the proof of Theorem 3.31 may come across as overly thorough, and you may be wondering if certain steps of the proof could be presented in a different manner. Indeed they can be. For example, in showing that $p_i|(N-1)$, it is not necessary to include the last line of the argument ("$= p_i(p_1p_2 \ldots p_{i-1}p_{i+1} \ldots p_n)$"). This was included simply to verify for the reader that the precise definition of divisibility was satisfied. You will find alternative presentations of the proof that will simply say, "Because p_i is among the primes p_1, p_2, $\ldots$, p_n, it follows that

$$N - 1 = p_1 p_2 \ldots p_n$$

is divisible by p_i."

Another common technique that often appears in this proof (and others) would be assigning the divisor of N to be p_1, rather than some general p_i. In cases like this, a phrase would likely be used: "without loss of generality." Thus, the proof may read,

"Since N is composite, Theorem 3.20 says that N has a prime divisor. Having assumed $p_1, p_2, \ldots, p_n$ constitute all the prime numbers, we can conclude, without loss of generality, that p_1 is that divisor of N."

What does this mean? By saying "without loss of generality" (or "WLOG," for short), we are making an assumption in our proof that narrows the proof to a special case without introducing additional hypotheses to the problem. It is used when the proof technique can be applied trivially to all additional cases.

Often you will see "WLOG" used when choice of variable appears within a proof. In the proof above, it was true that either p_1 or p_2 or $\ldots$ or p_n divides N. Why introduce a new variable, i, to help represent that divisor when we could just as easily choose one of them and tell the reader, "It doesn't really matter what we call it; the result will hold for any of the primes, so let's just assume it to be the first one in the list." In doing this, we would no longer have to reference the variable i, cleaning up the proof a bit near its end.

> **Quality Tip 18.** *"Without loss of generality" can be used in a proof when particular cases of the proof can be applied and generalized to all others.*

Next, we combine the technique of last section (the set element method) and proof by contradiction in showing that a particular set equals the empty set. Why is such a combination of techniques necessary? In showing two sets are equal, we show that the two sets are subsets of one another (the set element method), but no set that is a subset of the empty set has any elements! To show that a set has *no* elements, assume it has an element and arrive at some absurdity (proof by contradiction).

Theorem 3.32. *If A is any set, then $A \times \varnothing = \varnothing$.*

Proof Let A be any set. Since $\varnothing$ has no elements, we vacuously have

$$\varnothing \subseteq A \times \varnothing.$$

To show

$$A \times \varnothing \subseteq \varnothing,$$

assume for contradiction that $A \times \varnothing$ is nonempty. Thus, there is an element $(a, x) \in A \times \varnothing$, where $a \in A$ and $x \in \varnothing$. However, $\varnothing$ has no elements. Thus, we have reached a contradiction. Consequently, $A \times \varnothing$ must be empty, proving that

$$A \times \varnothing = \varnothing.$$

$\square$

In Chapter 2, we discussed that there is only one empty set. How do we demonstrate that there is indeed only one of these sets? More generally, how do we go about showing something is unique? You have encountered facts about uniqueness throughout your mathematical upbringing. For example, you know that every positive integer can be factored one and only one way as a product of primes (see Theorem 4.29).

To show that a certain object is *not* unique is simple. Find two of those objects. In doing so, you have found a counterexample to the claim that that object is unique or has some unique property. But what if it actually is unique? How can we go about showing that everywhere, in all of creation, there is not one other of these objects? We perform a proof by contradiction.

To prove uniqueness, we make an assumption that we know is false (that there are more than one of the thing we wish to show is unique). But in doing so, we introduce a tool to the problem that helps us reach some absurd result. Even though we know there is only one of the thing we are trying to show is unique, if we assume there are two of those things, we suddenly have something to work with.

Definition 3.33. A *proof of uniqueness* is used to show an there is only one of a particular object, method or form of something. To prove that that object is unique, assume that there are two of them and derive that those two things must actually be equal.

Our first example of a proving uniqueness is in showing something we have already claimed: that there is only one empty set. We defined the empty set as *the* empty set. In proving the uniqueness of the empty set, we become justified in saying precisely this (rather than *an* empty set). We begin with a lemma.

Lemma 3.34. *If E is a set with no elements and A is any set, then $E \subseteq A$.*

Proof Assume for contradiction that E is a set with no elements, A is any set, and that E is not a subset of A. By definition, then, there is an element of E that is not an element of A. But E is taken to have no elements. Thus, we have a set that both has no elements and has an element. This is a contradiction, and hence, we must have $E \subseteq A$. $\square$

This lemma helps prove the uniqueness of the empty set. In uniqueness proofs such as this, choice of notation is vital. Oftentimes subscripts denote the two objects that ultimately will be shown to be equal and therefore proving uniqueness. That approach is taken in Theorem 3.35.

Theorem 3.35. *The empty set is unique.*

Proof Assume $\varnothing_1$ and $\varnothing_2$ are two empty sets. Because they are sets, in order to prove that they are equal we show that they are subsets of one another.

Since $\varnothing_1$ is a set with no elements and $\varnothing_2$ is some set, Lemma 3.34 implies that

$$\varnothing_1 \subseteq \varnothing_2.$$

An identical argument shows that

$$\varnothing_2 \subseteq \varnothing_1.$$

Thus,

$$\varnothing_1 = \varnothing_2,$$

proving that the empty set is unique. $\square$

We added various methods into our toolbox of proof techniques in this section: proofs by contrapositive and contradiction, as well as proofs of uniqueness. Our last proof technique of this chapter is powerful. The logic behind is comes directly from a rule of inference in Chapter 1, and the implementation of the method will not be new. It is a method we have actually already used (see Example 3.10): *proof by cases*.

Exercises

1. Suppose the following are to be proven using a proof by contrapositive. For each, state both what the assumptions of such a proof would be and what would need to shown.

(a) Let $a, b \in \mathbb{Z}$. If $3 \nmid ab$, then $3 \nmid a$ and $3 \nmid b$.

(b) For $x, y \in \mathbb{Z}$, if $x + y$ is even, then x and y must both be odd or both be even.

(c) For $p \in \mathbb{Z}$, $p > 2$, if p is prime, then p is odd.

(d) Let $x \in \mathbb{Z}$. Then, $x^4 + 2x + 13$ being even implies that x is odd.

2. Suppose the following were to be proven using a proof by contradiction. State all assumptions of such a proof.

(a) Let $m \in \mathbb{Z}$. If $5m + 4$ is odd, then m is odd.

(b) The negation of an irrational number is irrational.

(c) There do not exist positive integers x and y such that $x^2 - y^2 = 1$.

(d) For all $x, y \in \mathbb{Z}$, $12x + 16y \neq 1$.

3. Prove each of the statements in (1) using a proof by contrapositive.

4. Prove each of the statements in (2) using a proof by contradiction.

5. Using a proof by contrapositive, show the following results hold.

(a) If x^2 is odd ($x \in \mathbb{Z}$), prove that x is odd.

(b) For $x \in \mathbb{Z}$, if $(x + 1)(x + 3)(x + 5)$ is odd, then x is even.

(c) Let $a \in \mathbb{R}$ with $a^3 - a^2 \geq 0$. Prove that $a \geq 0$.

(d) If $x^2(y^2 + 4y)$ is odd, for $x, y \in \mathbb{Z}$, prove that both x and y are odd.

(e) Two integers are said to have the same *parity* if they are both even or both odd. If $x, y \in \mathbb{Z}$ and $x + y$ is even, then x and y have the same parity.

6. Prove each of the following using a proof by contradiction.

(a) There exists no smallest odd integer.

(b) If $n, m \in \mathbb{Z}$ with $n \geq 2$, then $n \nmid m$ or $n \nmid (m + 1)$.

(c) There do not exist integers x and y such that $12x + 18y = 1$.

(d) For all $a, b, c \in \mathbb{Z}$, if $a \nmid b$ but $a | c$, then $a \nmid (b + c)$.

(e) There does not exist $x, y \in \mathbb{Z}^+$ such that $x^2 - y^2 = 10$.

7. Critique the following proof by contrapositive. How could it be changed to make it stronger?

Claim: For integers x, if $x^2 + 8x + 7$ is even, then x is odd.

Proof Let x be an even integer. Assume $x = 2k$ for some integer k. Then,

$$x^2 + 8x + 7 = 2(2k^2 + 8k + 3) + 1,$$

making $x^2 + 8x + 7$ by definition odd, proving the desired result. $\square$

8. Critique the following proof by contradiction. How could it be changed to make it stronger?

Claim: Every prime number greater than 2 is odd.

Proof Every prime number greater than 2 is odd, because if it were even, then $p = 2k$, for some integer k, so that p has at least three distinct factors (1, 2, and p), but p is prime, so this cannot happen. $\square$

9. Let A, B and C be sets. Prove the following.
   (a) $A \cap (B - A) = \varnothing$.
   (b) If $(A \cap B) \subseteq C$, then $(A - C) \cap (B - C) = \varnothing$.
   (c) The second Universal Bound Law for Sets (Theorem 2.8).

10. Prove each of the following.
    (a) If a^3 is even, then a is even.
    (b) $\sqrt[3]{2}$ is irrational.

11. Prove the following pair of statements.
    (a) Suppose p is prime and $n \in \mathbb{Z}$. Then, $p \nmid n^2$ is sufficient for knowing $p \nmid n$.
    (b) $\sqrt{p}$ is irrational when p is prime.

12. Prove the following claims.
    (a) There are infinitely many prime numbers greater than 1,000,000,000,000.
    (b) For $a, b \in \mathbb{R}$, if $a^3 + ab^2 \leq b^3 + a^2b$, then $a \leq b$.
    (c) For all $x \in \mathbb{R}^+$, prove that $\dfrac{x}{x+1} < \dfrac{x+1}{x+2}$.
    (d) There do not exist primes a, b, and c such that $a^2 - b^2 = c^2$.

13. Using any proof technique, prove the following results about rational numbers.
    (a) The sum of two rational numbers is rational.
    (b) If $a, b \in \mathbb{Q}$, then $a^2 b^3 \in \mathbb{Q}$.
    (c) $3\sqrt{2}$ is irrational.
    (d) $\dfrac{1 + \sqrt{2}}{2}$ is irrational.
    (e) If $x \in \mathbb{Q}$ and $y \notin \mathbb{Q}$, then $xy \notin \mathbb{Q}$.
    (f) Every rational number x has an additive inverse. That is, if $x \in \mathbb{Q}$, then there exists $y \in \mathbb{Q}$ such that $x + y = 0$.
    (g) Every nonzero rational number x has a multiplicative inverse. That is, if $x \in \mathbb{Q}^*$, then there exists $y \in \mathbb{Q}$ such that $xy = 1$.

14. Disprove the following claims about irrational numbers.

    (a) The sum of two irrational numbers is irrational.
    (b) The product of two irrational numbers is irrational.
    (c) The quotient of two nonzero irrational numbers is irrational.

15. Prove that $\sqrt{6}$ is irrational.

16. Prove that every nonzero rational number can be expressed as the product of two irrational numbers.

3.5 Proof by Cases

Suppose set in front of you is a map of Iowa. Its 99 counties are outlined, but to make it easier to distinguish them, you want to color each of them a single color, but with one rule: no two adjacent counties are the same color. How many colors would we need? Clearly 99 colors would do the trick. Without knowing the geography of Iowa, you are probably reasoning that 98 colors would suffice. What about 97? 96? 50? What is the *fewest* number of colors we would need? More generally, what is the fewest number of colors required to color *any* contiguous region that is subdivided into contiguous sub-regions (such as states broken up into counties)?[13]

This is the basic premise of the famous Four-Color Theorem. Francis Guthrie, a student at University College London, first conjectured in 1852 that it would take only four colors to color any such map. It was not until 1976, however, that a valid proof of the result was formulated. The authors of the solution, Kenneth Appel and Wolfgang Haken, used a computer to break the problem down; they showed that there were only 1936 map types that must be considered. Because they were then able to show all of these maps could be colored with just four colors, it stood that the result holds [40].

This is precisely the idea behind our next proof technique, called *proof by cases* or *proof by exhaustion*.

> **Definition 3.36.** A *proof by cases* or *exhaustion* proves that a particular proposition holds by breaking the proposition into a finite number of cases and showing the result holds for each and every one of the cases. Every occurrence of the proposition must fall under the guise of at least one of the cases.

What is the logic behind proof by cases? It should come as no surprise that

---

[13]Contiguous regions are those that are not broken up into distinct "parts," such as the upper- and lower-peninsulas of Michigan.

it stems from the logical rule of inference known as Cases. Recall the argument:

$$P \vee Q$$
$$P \Rightarrow R$$
$$\underline{Q \Rightarrow R}$$
$$\therefore R$$

Rather than just two statements, P and Q, we see that this rule generalizes to any finite number of initial statements $P_1, P_2, \ldots, P_n$:

$$P_1 \vee P_2 \vee \cdots \vee P_n$$
$$P_1 \Rightarrow R$$
$$P_2 \Rightarrow R$$
$$\vdots$$
$$\underline{P_n \Rightarrow R}$$
$$\therefore R$$

How does this correspond to how a proof by cases actually *works*? Let us suppose we are trying to prove a statement of the form, "if P, then R," using a proof by cases. As in any direct proof, the hypothesis P is assumed to be true. However, we now proceed in breaking up P into cases. If we call them $P_1, P_2, \ldots, P_n$, we have that

$$P \equiv P_1 \vee P_2 \vee \cdots \vee P_n.$$

Thus, in assuming P to be true, the above logical equivalence implies that we are assuming $P_1 \vee P_2 \vee \cdots \vee P_n$ to be true. Then, the "work" of our proof is in showing, for every i, that $P_i \Rightarrow R$. Once we have done this, the generalization of Cases shows that the conclusion R must hold true.

In the real world, proof by cases occurs regularly. Populations are many times naturally broken into subpopulations. If we can show a certain thing is true for each of those subpopulations, then it must be true for the entire population.

Example 3.20. At a particular liberal arts college, the following conversation is overheard.

Max: My advisor told me that every student must take a lab course. I don't believe her!

Minnie: I can prove it to you.

Max: How?

Minnie: Let me list every major for you and show you which lab course they take. All the humanities majors take this physics course called "Light, Sound, and Waves." Math and computer science majors take "Quantitative Modeling." Social science majors take this biology course ... etc.

> *Max*: I understand now. Thank you for that exhaustive list!
>
> How does this exhibit a proof by cases? Minnie creates her cases by stratifying the student population into groups. Then, she proceeds to show that in each and every case, the result holds.

Our first mathematical proof by cases is a proof of a familiar result from algebra. Recall the following definition.

Definition 3.37. For $x \in \mathbb{R}$, define the *absolute value of* x as

$$|x| = \begin{cases} x & \text{if } x \geq 0 \\ -x & \text{if } x < 0 \end{cases}.$$

Example 3.21. Let $x, y \in \mathbb{R}$ with $y \neq 0$. Prove that

$$\left| \frac{x}{y} \right| = \frac{|x|}{|y|}.$$

Like many of our first proofs we have seen thus far, this result may seem obvious. Perhaps you are tempted to argue, "The absolute value of a real number is either the number itself or its opposite, if the number is negative. If we take the absolute value of a fraction, we can make the numerator positive in this way, as well as the denominator. Or, we can simply just do this with the whole fraction."

While this thought process is intuitively correct, there are flaws to it. It uses imprecise language ("make the numerator positive"). Additionally, it reads argumentatively. Nowhere does it call upon strict mathematical definitions or results. It simply is *not* a proof. But it *is* good initial scratch work and thinking.

If we compare it to the definition of the absolute value, we see that a proof by cases falls naturally into our laps. Why? Because in computing the absolute value of a number, we consider if the number falls into one of two cases: is the number non-negative or is the number negative? In this case, we begin with two numbers, x and y, and consider four possible scenarios, whether each of them is non-negative or negative.

Proof Let $x, y \in \mathbb{R}$ with $y \neq 0$. To prove

$$\left| \frac{x}{y} \right| = \frac{|x|}{|y|},$$

we consider four cases, with Definition 3.37 used in each:

Case 1 ($x \geq 0$, $y \geq 0$): Notice that

$$\left|\frac{x}{y}\right| = \frac{x}{y} = \frac{|x|}{|y|}.$$

Case 2 ($x \geq 0$, $y < 0$): In this case,

$$\left|\frac{x}{y}\right| = -\frac{x}{y} = \frac{x}{-y} = \frac{|x|}{|y|}.$$

Case 3 ($x < 0$, $y \geq 0$): We have

$$\left|\frac{x}{y}\right| = -\frac{x}{y} = \frac{-x}{y} = \frac{|x|}{|y|}.$$

Case 4 ($x < 0$, $y < 0$): Here,

$$\left|\frac{x}{y}\right| = \frac{x}{y} = \frac{-x}{-y} = \frac{|x|}{|y|}.$$

Since the result holds for all possible cases, we have shown the claim to be true. $\square$

Notice the structure of Example 3.21. Cases are clearly labeled and line-breaks help underscore the notion of a new case. Take these away and notice how difficult it becomes to read and follow the proof:

Proof Let $x, y \in \mathbb{R}$ with $y \neq 0$. To prove $\left|\frac{x}{y}\right| = \frac{|x|}{|y|}$, we consider four cases. If $x \geq 0$, $y \geq 0$, notice that $\left|\frac{x}{y}\right| = \frac{x}{y} = \frac{|x|}{|y|}$. If $x \geq 0$, $y < 0$, then $\left|\frac{x}{y}\right| = -\frac{x}{y} = \frac{x}{-y} = \frac{|x|}{|y|}$. If $x < 0$, $y \geq 0$, we have $\left|\frac{x}{y}\right| = -\frac{x}{y} = \frac{-x}{y} = \frac{|x|}{|y|}$. If $x < 0$, $y < 0$, then $\left|\frac{x}{y}\right| = \frac{x}{y} = \frac{-x}{-y} = \frac{|x|}{|y|}$. Since the result holds for all possible cases, we have shown the claim to be true. $\square$

Quality Tip 19. *Indicate the cases in a proof by exhaustion and present them clearly.*

Always keep in mind that clarity and readability are important. How you label your cases and present them is a personal decision. It is one more way to develop *your* mathematical voice.

The definition of absolute value lends itself to proofs of other results using cases. One of the more famous of these is the *triangle inequality*. Two lemmas are necessary for its proof.

Lemma 3.38. *If $x \in \mathbb{R}$, then $-|x| \le x \le |x|$.*

Proof Let $x \in \mathbb{R}$. To prove $-|x| \le x \le |x|$, we consider the following two cases.

Case 1 ($x \ge 0$): By Definition 3.37, $|x| = x$, so that $x \le |x|$. Moreover, $-|x| \le 0 \le x$, yielding

$$-|x| \le x \le |x|.$$

Case 2 ($x < 0$): In this case, Definition 3.37 gives that $|x| = -x$. Consequently, $-|x| \le x$. Since $x < 0 < |x|$, we have that

$$-|x| \le x \le |x|.$$

Since the result holds in all cases, it follows that

$$-|x| \le x \le |x|.$$

$\square$

Notice that the work in each case is very similar. Though you may be tempted to say that Case 2 follows mutatis mutandis from Case 1, you cannot. It uses a different part of Definition 3.37 and the algebraic steps are not identical.

Lemma 3.39. *Let $x \in \mathbb{R}$. Then, $|-x| = |x|$.*

In trying to develop the proof for this result, we may initially find ourselves working with two cases: $x < 0$ and $x \ge 0$. When working with the latter of these, $x \ge 0$, chances are that we would break that case into two *subcases*: $x > 0$ and $x = 0$. Using subcases in a proof by cases is completely legitimate. When considering the real numbers, having two cases, $x \ge 0$ and $x < 0$, with the first then having two subcases, $x > 0$ and $x = 0$, is awkward. Why not simply consider three cases: $x > 0$, $x < 0$, and $x = 0$?

Proof Let $x \in \mathbb{R}$. There are three cases to consider: $x > 0$, $x < 0$, or $x = 0$. We employ Definition 3.37 to justify the steps in each case.

Case 1 ($x > 0$): Here,

$$\begin{aligned} |x| &= x \\ &= -(-x) \\ &= |-x| \text{ (since } -x < 0). \end{aligned}$$

Case 2 ($x < 0$): Then,

$$\begin{aligned} |x| &= -x \\ &= |-x| \text{ (since } -x > 0). \end{aligned}$$

Case 3 ($x = 0$): Since $-x = 0$,

$$|x| = 0$$
$$= |-x|.$$

With the result holding for all possible values of x, it must be that

$$|-x| = |x|$$

for every $x \in \mathbb{R}$. ☐

Equipped with these two lemmas, we are prepared to prove the Triangle Inequality.

Theorem 3.40. (Triangle Inequality) *Let* $x, y \in \mathbb{R}$. *Then,*

$$|x + y| \leq |x| + |y|.$$

Proof Let $x, y \in \mathbb{R}$. We proceed by considering two cases: $x + y$ is either negative or non-negative.

Case 1 ($x + y < 0$): Since $x + y < 0$,

$$\begin{aligned}
|x + y| &= -(x + y) \quad \text{(Definition 3.37)}\\
&= -x + -y \\
&\leq |-x| + |-y| \text{ (Lemma 3.38)}\\
&= |x| + |y| \quad \text{(Lemma 3.39)}.
\end{aligned}$$

Case 2 ($x + y \geq 0$): Because $x + y \geq 0$,

$$\begin{aligned}
|x + y| &= x + y \quad \text{(Definition 3.37)}\\
&\leq |x| + |y| \text{ (Lemma 3.38)}.
\end{aligned}$$

Because

$$|x + y| \leq |x| + |y|$$

in both cases, the result holds for all $x, y \in \mathbb{R}$. ☐

Claims about piecewise functions are often proven using proof by cases, as the cases are naturally determined by the definition of the function. These are not the only times when cases are used, however. Any proposition where the hypothesis can be stratified via some property is a good candidate for trying a proof by exhaustion. The property that determines the cases often provides another tool that can be used in the proof. Some claim about the set of all integers can be proven true if it is shown to be true for both even integers and odd integers. When considering these cases, then, we obtain the additional hypothesis that we are working with an even integer or an odd integer, respectively. Lemma 3.41 exhibits this.

Lemma 3.41. *The product of any two consecutive integers is even.*

In thinking about how such a proof might proceed, we realize that we have little to work with other than starting with two consecutive integers. What do these quantities look like? Yes, we could give all of them variable names, say, n and m, and simply say that $m - n = 1$, but that complicates matters. We aim to use as few of variables as possible in proofs. Thus, we can take our consecutive integers to be n and $n + 1$.

Proof Let $n \in \mathbb{Z}$. We wish to show $n(n + 1)$ is even. To do so, consider that n is either even or odd.

Consider n to be even, so that $n = 2k$ for some $k \in \mathbb{Z}$. Then,

$$n(n + 1) = 2[(k)(k + 1)],$$

which is even because $k(k + 1) \in \mathbb{Z}$.

Next, consider the case when n is odd. Then, there is some $k \in \mathbb{Z}$ with $n = 2k + 1$, and,

$$\begin{aligned} n(n + 1) &= (2k + 1)(2k + 2) \\ &= 2((2k + 1)(k + 1)), \end{aligned}$$

which, by closure properties, is even.

Since $n(n + 1)$ is even for all possible occurrences of n, the result holds. $\square$

In the proof above, the cases were not explicitly labeled as they were in our first few results. However, the reader understands that this is a proof by cases and the cases are clearly noted in an expository fashion. Determine *your* style.

Additionally, notice in the proof of Lemma 3.41 the repeated use of the variable k in each case. The scope of k in the case when n is even is *only* this case. We are allowed to, and should, reuse k in second case. In both cases, k plays the same role. Because of this, the reader is more likely to easily follow the logic of the proof.

It is natural to partition the integers into evens and odds. But this is just one specific instance of a more general way to break up the integers. Whole number division provides the tool for this, as stated in Theorem 3.42. It is called the Quotient-Remainder Theorem, the proof of which appears in Section 4.3.

Theorem 3.42. (Quotient-Remainder Theorem)*: Let $a, b \in \mathbb{Z}$ with $b > 0$. Then there exist unique integers q and r such that*

$$a = bq + r$$

with $0 \leq r < b$.

> **Definition 3.43.** In the statement of the Quotient-Remainder Theorem, we call q the *quotient* and r the *remainder* when a is divided by b, with b called the *divisor*.

Note that the cases of being even or odd are actually applications of the Quotient-Remainder Theorem with a quotient of 2.

How does Theorem 3.42 partition the integers? It does so by the possible remainders for a particular fixed divisor. For example, every integer takes exactly one of the following forms: $5m$, $5m + 1$, $5m + 2$, $5m + 3$ or $5m + 4$, for some $m \in \mathbb{Z}$. Suppose then we make some claim about the entirety of the integers. If we can prove the proposition that an arbitrary integer fulfills the claim, regardless of which of these five forms the arbitrary integer takes, then the proposition would hold for every integer. That is, we have to show the result holds *for every case*. As mentioned previously, within each of these specific cases, we obtain an additional tool to work with: the form of the general integer (for example, assuming it to be of the form $5m + 2$).

Theorem 3.44. *The square of any integer is of the form* $3m$ *or* $3m + 1$ *for some* $m \in \mathbb{Z}$.

Let us consider how to determine the nuts-and-bolts of the proof. Our results in the past have broken the integers into two cases: being even or being odd. We can anticipate that this result may not work, however, as the conclusion of the theorem involves being a multiple of 3. When this is the case, it is often natural to consider the Quotient-Remainder Theorem with a divisor of 3. With a little algebra, that is indeed the tool needed to construct this proof.

Proof Let $n \in \mathbb{Z}$. By Theorem 3.42, $n = 3m$, $n = 3m + 1$ or $n = 3m + 2$ for some $m \in \mathbb{Z}$. Consider each case.

Case 1: Suppose $n = 3m$ for some $m \in \mathbb{Z}$. Then,

$$n^2 = 9m^2$$
$$= 3(3m^2).$$

Case 2: Suppose $n = 3m + 1$ for some $m \in \mathbb{Z}$. Then,

$$n^2 = 9m^2 + 6m + 1$$
$$= 3(3m^2 + 2m) + 1.$$

Case 3: Suppose $n = 3m + 2$ for some $m \in \mathbb{Z}$. Then,

$$n^2 = 9m^2 + 12m + 4$$
$$= 3(3m^2 + 4m + 1) + 1.$$

In all possible cases, by closure properties of $\mathbb{Z}$, we see that n^2 is of the form $3m$ or $3m + 1$ for some $m \in \mathbb{Z}$, as was to be shown. □

In the next example, we show how the Quotient-Remainder Theorem can be used to show certain properties do *not* hold.

Example 3.22. If $n \in \mathbb{Z}$, show that $n^2 + 1$ is not divisible by 4.

The key observation to make is in the statement of the Quotient-Remainder Theorem itself. When an integer is divided by 4, it leaves a *unique* remainder of 0, 1, 2 or 3. Because of the definition of divisibility, only when the remainder is 0 is the integer divisible by 4.

In Theorem 3.44, we proceed with using the Quotient-Remainder Theorem with a quotient of 3, yielding three cases. Likewise here we might jump into this proof with four cases. While the result can be shown by considering our integer n to be of the form $4m$, $4m+1$, $4m+2$ or $4m+3$, we will see that the result can be shown with only two cases: n being either even or odd. Both methods will indeed prove the theorem, but succinct proofs carry a bit more elegance. Using an exorbitant number of cases can cause redundancy, and consequently, your reader questioning your methods. As you progress in your proof writing, you will find yourself attempting to craft not just correct proofs but concise proofs as well.

Proof Let $n \in \mathbb{Z}$. We know n is either even or odd. Let us consider both cases.

Case 1: Suppose $n = 2m$ for some $m \in \mathbb{Z}$. Then,

$$n^2 = 4m^2 + 1;$$

by Theorem 3.42, since $m^2 \in \mathbb{Z}$, n^2 is not divisible by 4.

Case 2: Suppose $n = 2m + 1$ for some $m \in \mathbb{Z}$. We have that

$$n^2 = 4(m^2 + m) + 1,$$

and as in Case 1, n^2 is not divisible by 4.
Thus, the result holds. □

A proof by exhaustion is nothing more than multiple proofs with additional assumptions. Particular cases may themselves require direct proofs while others use a proof by contradiction. Some may even be vacuously true. When you are working through your scratch work and trying to figure out how to actually prove a specific case, approach it as if it were a brand new claim. List all your assumptions, consider what you have to show and think about how you proceed. Whatever techniques you used in previous cases may be helpful in a new case, but they need not force the new case to proceed a certain way.

Exercises

1. Find the quotient and remainder guaranteed by the Quotient-Remainder Theorem when

   (a) 17 is divided by 6.
   (b) 4 divides 1960.
   (c) -214 is divided by 11.
   (d) An odd integer $n > 1$ divides $8n^3 + 5n + 2$.

2. Explain why the following cases would *not* suffice in proving the stated claims.

   (a) *Claim*: For all integers m and n, $m + n$ and $m - n$ have the same parity.

      Cases: Consider the cases where m and n are both even and where m and n are both odd.

   (b) *Claim*: If $n \in \mathbb{Z}^+$, then $n^7 - n$ is divisible by 7.

      Cases: Because we are trying to prove divisibility by 7, it is enough to consider only when the remainder upon division by 7 is odd: $n = 7k + 1$, $n = 7k + 3$, and $n = 7k + 5$, for some $k \in \mathbb{Z}$.

   (c) *Claim*: The product of four consecutive integers is divisible by 8.

      Cases: Since an arbitrary list of four consecutive integers is of the form m, $m + 1$, $m + 2$, and $m + 3$, there is only the one case to consider: the product of $8k$, $8k + 1$, $8k + 2$, and $8k + 3$.

3. For each of the claims below, list the cases that seem most natural to consider when constructing a proof of the result.

   (a) For all $n \in \mathbb{Z}$ with $0 \le n < 3$, $(n + 1)^2 > n^3$.
   (b) In any string of three consecutive integers, there is one integer that is divisible by 3.
   (c) For any $n \in \mathbb{Z}$, $n^2 + n$ is even.
   (d) If $3 \nmid n$ for $n \in \mathbb{Z}$, then $3 \mid (n^2 - 1)$.
   (e) For all $n \in \mathbb{Z}^+$, $n^3 + n$ is even.
   (f) For all $n \in \mathbb{Z}^+$, $4 \nmid (n^2 + 1)$.

(g) Every perfect cube is a multiple of 9, one more than a multiple of 9, or one less than a multiple of 9.

4. Prove each of the results in the previous problem.

5. Prove the following claims about the absolute value function. Assume all variables represent real numbers.

   (a) $|xy| = |x||y|$
   (b) $|x - 4| + 4 \leq |x| + 8$
   (c) If $n \in \mathbb{R}^+$, then $|x| \leq n$ if and only if $-n \leq x \leq n$.
   (d) If $n \in \mathbb{R}^+$, then $n \leq |x|$ if and only if $n \leq x$ or $x \leq -n$.

6. Find the flaw in the following proof:

   Claim: If n is an integer not divisible by 3, then $3|(n^2 - 1)$.

   Proof Let n be an integer not divisible by 3. By the Quotient-Remainder Theorem, $n = 3k + 1$ or $n = 3k + 2$ for some integer k.

   Let us assume that $n = 3k + 1$. We have that

   $$n^2 - 1 = (3k + 1)^2 - 1$$
   $$= 9k^2 + 6k$$
   $$= 3(3k^2 + 2k).$$

   By closure, $3k^2 + 2k$ is an integer, thus proving that $3|(n^2 - 1)$. □

7. Find the flaw in the following proof:

   Claim: The square of any integer is of the form $5n$, $5n+1$, or $5n+4$ for some integer n.

   Proof By the Quotient-Remainder Theorem, an integer n is of the form $5k$, $5k+1$, $5k+2$, $5k+3$, or $5k+4$, for some integer k. Consider the case when $n = 5k$. Then,

   $$n^2 = (5k)^2$$
   $$= 5(5k^2),$$

   and since $5k^2 \in \mathbb{Z}$ (by closure), as was to be shown in this case. Because all other cases follow mutatis mutandis, the claim holds. □

8. Prove that if n is even, then $n = 4k$ or $n = 4k + 2$ for some $k \in \mathbb{Z}$ using

   (a) A proof by cases.

(b) A proof by contrapositive.

9. For all $x, y \in \mathbb{R}$, prove that $\big||x| - |y|\big| \leq |x - y|$.

10. Prove the following.

   (a) The square of any odd integer is of the form $8k + 1$ for some integer k.

   (b) Every odd integer is of the form $4k \pm 1$ for some integer k.

   (c) The product of any two consecutive integers has the form $3m$ or $3m + 2$ for some $m \in \mathbb{Z}$.

   (d) Let $a, b, c \in \mathbb{Z}$ with $a^2 + b^2 = c^2$. Then, at least one of a or b is even.

   (e) For $n \in \mathbb{Z}^+$, $7|(n^7 - n)$.

11. (a) Prove that the sum of four consecutive integers is of the form $4m + 2$ for some integer m.

   (b) Prove the claim that 8 divides the product of four consecutive integers.

12. Prove the following results involving primes.

   (a) If $p > 3$ is prime, then p is of the form $6k + 1$ or $6k - 1$ for some $k \in \mathbb{Z}$.

   (b) If $n \in \mathbb{Z}^+$, $n \geq 4$, then n, $n + 2$ and $n + 4$ cannot all be prime.

13. Assuming the Quotient-Remainder Theorem to be true, prove the following:

   If $a, b \in \mathbb{Z}$ with $b > 0$, then there exist unique $q, r \in \mathbb{Z}$ satisfying $a = qb + r$ with $b \leq r < 2b$.

4

Mathematical Induction

The focus of this chapter is a proof technique, taken as a mathematical axiom, known as *mathematical induction* or, more simply, *induction*.[1] The first explicit appearance of a proof by induction is attributed to Blaise Pascal (1623-1662) in his *Traité du Triangle Arithmetiqué* (1654). Though he lived only to the age of 39, Pascal, a Frenchman, had a profound impact on mathematics (as well as philosophy, physics and a trove of other areas). While his work in projective geometry stands as one of his major achievements, Pascal, alongside Fermat, is credited with the development of probability theory [11]. We use one of its fundamental concepts as the vehicle for introducing induction.

Anyone familiar with basic probability is aware of the importance of binomial coefficients. In the aforementioned text, Pascal explored them, giving rise to what we know today as Pascal's Triangle. One particular result is pertinent to our conversation here. Pascal claimed that for *all* natural numbers n and non-negative integers r less than n,

$$\frac{\binom{n}{r+1}}{\binom{n}{r}} = \frac{n-r}{r+1}$$

(where this notation for the binomial coefficient is defined in Definition 4.18). How would one go about proving this? It is a claim involving infinitely many values of n; such a proof required a technique beyond a simple direct proof. The technique used is what we are introducing here, mathematical induction, and the proof of the above is the first appearance of mathematical induction in print. Since that appearance, induction has become one of the most important tools a mathematician can wield.

It should be noted that Pascal was not the first mathematician to actually use induction, however. Pascal was simply the first to explicitly lay out an inductive proof close to the form we commonly use today. Maurolico (16th century), al-Karaji (10th and 11th centuries), and even Euclid presented proofs that contained pieces of the inductive process [18].

---

[1]It should be noted that this technique is unrelated to inductive reasoning. Rather, mathematical induction is another form of deductive reasoning, as discussed in the previous chapter.

The applicability of induction touched upon in this chapter is only the tip of the iceberg. A perusal of Gunderson's *Handbook of Mathematical Induction* [22] provides a glimpse at just how powerful of a tool mathematical induction truly is. In addition to those number theoretic results mentioned below, certain major theorems in set theory, logic, graph theory, game theory, functional analysis, abstract algebra, geometry, algorithmic complexity and probability are proven inductively. The power of the tool is obvious.

In this chapter we introduce both mathematical induction and strong mathematical induction and see how they can be used to prove a variety of number theoretic results. In particular, we will use induction to prove various major theorems: the Binomial Theorem, the Fundamental Theorem of Arithmetic and Euclid's Lemma. Moreover, armed with this new tool, we will be able to address deeper concepts in the chapters to come.

4.1   Basics of Mathematical Induction

Imagine young children lining up dominoes around a room. Their intention is to push over the first domino and watch the entire collection fall. Before pushing over that first domino, they say to you, "We want to see *every* domino fall over, not just some of them. What do we have to do to make sure each and every one falls over?" What would be required of the collection in order guarantee this?

First, the children would need to make sure the first domino is is initially pushed over. If they were to start the sequence of falling dominoes by pushing over the second, third or tenth domino, those that come before whichever is first pushed would be left standing. Next, they need to verify that every domino would then cause the following domino to fall. If the dominoes are two inches tall, having a three inch gap between two adjacent dominoes would not be good. In general terms, they would need to check that the nth domino, upon falling, causes the $(n+1)$st domino to fall (though you probably would not phrase it to the group of children this way).

If the sequence of dominoes was infinitely long, would knowing these two pieces of information be enough to guarantee that, eventually, every domino in the sequence would fall? Even though dominoes in the sequence would continue falling indefinitely and that we could never say, "Every domino has fallen over," we know that sooner or later, the millionth, billionth or trillionth domino would tip, all from knowing these two small pieces of information.

This is precisely the concept of *mathematical induction*. It is a form of direct proof used to show that some property holds for an infinite subset of the positive integers[2]. Once mastered, it becomes one of the mathematician's most powerful tools.

Definition 4.1. A predicate $P(n)$ defined for $n \in \mathbb{Z}^+$ is proven true using the *Principle of Mathematical Induction* (or simply *mathematical induction*) by showing the following.

    1. $P(1)$ is true.

    2. For integers $k \geq 1$, $P(k)$ being assumed true implies $P(k+1)$ is true.

Before proceeding, it is worth noting that we are assuming the Principle of Mathematical Induction as an *axiom*. It is not a result that must be proven true; rather, it is simply assumed to hold.

Notice that Definition 4.1 is a mathematical formulation of the falling dominoes metaphor, with the statement $P(n)$ representing "the nth domino will fall over." The first step, in showing $P(1)$ is true, equates to showing that it is possible to tip over the first domino. Then, exhibiting that $P(k)$ is true implies that $P(k+1)$ is true is the mathematical way of saying that a general kth domino does indeed knock over the $(k+1)$st domino.

It should be noted that the Principle of Mathematical Induction is applicable not just to predicates whose domain is the set of all positive integers. Rather, if the claim is about all integers $n \geq a$ for some $a \in \mathbb{Z}$, then the first step of the mathematical induction process would be to show $P(a)$ is true, and then the second step would be to show for all integers $k \geq a$, $P(k)$ being true implies $P(k + 1)$ is true.

Before seeing our first example of an inductive proof[3], let us present the process of mathematical induction a bit more casually than it appears in the formal definition. Think of this as the how-to guide for writing an inductive proof. It is simply the outline of the proof; the in-betweens, the actual mathematics and the overall style are up to you. First, suppose you are asked to prove some claim:

Claim: $P(n)$ is true for all positive integers $n \geq a$, where $a \in \mathbb{Z}^+$.

Then, the proof proceeds as follows:

(1) Prove that the statement holds for the first integer it claims to be true for (the integer a).

This is called proving the *base case*. It is typically routine and shown by

---

[2]Induction can be used to show a property holds for a finite set as well, though simply exhibiting the property holds is often more direct when the set is small.

[3]The phrase "proof by induction" is often shortened to "inductive proof."

exhibiting the claim to be true. As trivial as it likely may seem, failing to show the base case holds true is a fatal error in any inductive proof.

(2) Assume that the statement is true for a general kth integer.

The assumption that $P(k)$ is true is called the *inductive hypothesis* of the proof. Clearly state and label the inductive hypothesis so that when it gets used later on in your proof, there are no questions about where it came from or why we are allowed to assume it to be true.

Some proof writers will initially choose to use the $P(k)$ notation, stating, "Let $P(k)$ be the predicate ... and assume it to be true for a general $k \in \mathbb{Z}^+$." While this is correct, as you advance in your proof writing you will find it adds an unnecessary layer into your proof. You will learn to be completely direct, stating, "Assume ... to be true for a general $k \in \mathbb{Z}^+$," eliminating the use of another variable in $P(k)$.

(3) Proceed in showing that the statement is true for the $(k+1)$st integer.

This is called the *inductive step* of the proof and is typically where most of the work goes into your proof. It will require using your inductive hypothesis at some point. When you do, state explicitly that that step follows from the inductive hypothesis.

When seeing mathematical induction for the first time, questions about the validity of the method are often raised. In particular, in assuming the inductive hypothesis (that the statement is true for a general integer k), are we not assuming the *exact thing we are trying to prove*? While it may seem so, step back and analyze the goal of this part of the inductive proof. The inductive hypothesis and inductive step are ultimately proving, "*If* the result holds for a general integer k, *then* the result holds for the integer $k+1$." To prove a conditional statement holds, it is necessary to assume the premise of the if-then statement. Thus, inductive proofs do not assume the result holds generally; they are simply necessary assumptions in proving that one instance of the proposition being true forces the subsequent instance of the proposition to be true.

Let us proceed to our first proof by mathematical induction. It is a famous result that appears in most calculus textbooks. When calculating integrals via the *definition* of the integral (rather than the Fundamental Theorem of Calculus), we are often told to assume three certain sum formulas

1. $\displaystyle\sum_{i=1}^{n} i = 1 + 2 + \cdots + n = \frac{n(n+1)}{2}$

2. $\displaystyle\sum_{i=1}^{n} i^2 = 1^2 + 2^2 + \cdots + n^2 = \frac{n(n+1)(2n+1)}{6}$

3. $$\sum_{i=1}^{n} i^3 = 1^3 + 2^3 + \cdots + n^3 = \left[\frac{n(n+1)}{2}\right]^2$$

These three results provide quick formulas for summing the first n positive integers, their squares or their cubes, respectively. In calculus textbooks, you are simply handed them (or referred to an appendix where they are proven or roughly justified). But besides crunching out a few particular examples, say, $n = 2$, $n = 4$ and $n = 10$, how can we prove that for *any* general $n \in \mathbb{Z}^+$, these results hold? Mathematical induction, of course!

Theorem 4.2. *For any $n \in \mathbb{Z}^+$,*

$$\sum_{i=1}^{n} i = \frac{n(n+1)}{2}.$$

Before presenting a final draft of a proof of this result, let us work through the scratch work. Just as the proof itself will have three major parts, we break our scratch work up into those same pieces.

Base case: Because the claim is for $n \geq 1$, the base case is when $n = 1$. We need to see if the left-hand side of the equality equals the right-hand side of the result. Indeed, since

$$\sum_{i=1}^{1} 1 = 1 = \frac{1(1+1)}{2},$$

the base case holds. As mentioned before, it is absolutely necessary to include this step in our actual proof. Though you may be thinking, "Why do I need to include $1 = 1$?" We are not saying, "The base case is $1 = 1$." Rather, we are showing the claim,

$$\sum_{i=1}^{n} i = \frac{n(n+1)}{2}.$$

is true when $n = 1$. It just so happens in this case that the left-hand side and the right-hand side of the equation both equal 1.

Inductive hypothesis: The inductive hypothesis is an assumption that the result holds for a general $n = k$. Simply put, we rewrite the result with k substituted for n and we assume this to be true. That is, we assume

$$1 + 2 + \cdots + k = \frac{k(k+1)}{2}.$$

Note here that we are writing out the sum long-hand rather than using the condensed summation form. This is a personal choice, but as you will notice in the coming portion of our scratch work, it is sometimes easier to see where the inductive hypothesis comes into play during the inductive step when written out this way. In scratch work, there is no harm in changing notations: long-hand versus summation notation, in this case. Inside a proof, however, choose one approach and be consistent. Additionally, we should note (because the final draft of our proof will require it) that we are assuming this hypothesis to be true for a general integer $k \geq 1$.

Inductive step: This is the main work of our proof. We must show the result holds when we substitute $n = k + 1$. That is, we must show

$$1 + 2 + \cdots + (k + 1) = \frac{(k + 1)((k + 1) + 1)}{2}.$$

What we cannot do at this point is simply cite the inductive hypothesis and claim, "By letting $k = k + 1$, the result holds." This is not an allowable variable substitution. Saying, "Let $k = k + 1$" is equivalent to saying, "Let $0 = 1$," which is clearly absurd!

If we find ourselves stuck in trying to work through the inductive step, keep in mind that we want to use the inductive hypothesis at some point. What can we do to introduce $1 + 2 + \cdots + k$ to the problem? We need not do much; it is actually already hiding there in plain sight. We just need to rewrite the left-hand side of our "to show" statement as

$$1 + 2 + \cdots + (k + 1) = 1 + 2 + \cdots + k + (k + 1).$$

There it is! Now call upon the inductive hypothesis, substituting it for the first k terms of the sum:

$$1 + 2 + \cdots + (k + 1) = 1 + 2 + \cdots + k + (k + 1)$$
$$= \frac{k(k + 1)}{2} + (k + 1).$$

From here, we simply add the resulting terms, proceed with a little algebra, and the desired expression falls in our laps:

$$\frac{k(k + 1)}{2} + (k + 1) = \frac{k(k + 1)}{2} + \frac{2(k + 1)}{2}$$
$$= \frac{k(k + 1) + 2(k + 1)}{2}$$
$$= \frac{(k + 1)(k + 2)}{2}$$
$$= \frac{(k + 1)((k + 1) + 1)}{2}.$$

With this, we can construct the final draft of our proof.

Proof We proceed by mathematical induction to prove, for $n \in \mathbb{Z}^+$, that

$$\sum_{i=1}^{n} i = \frac{n(n+1)}{2}.$$

Base case ($n = 1$): Because

$$\sum_{i=1}^{1} i = 1 = \frac{1(1+1)}{2},$$

the base case holds.

Inductive hypothesis (IH): Assume for a general integer $k \geq 1$ that

$$\sum_{i=1}^{k} i = \frac{k(k+1)}{2}.$$

Inductive step: We must show that

$$\sum_{i=1}^{k+1} i = \frac{(k+1)(k+2)}{2}.$$

Then,

$$\begin{aligned}
\sum_{i=1}^{k+1} i &= \sum_{i=1}^{k} i + (k+1) \\
&= \frac{k(k+1)}{2} + (k+1), \text{ by the IH} \\
&= \frac{k(k+1)}{2} + \frac{2(k+1)}{2} \\
&= \frac{(k+1)(k+2)}{2},
\end{aligned}$$

as required. Thus, the result holds. $\square$

Though we chose to work through our scratch work using long-hand notation for our sums, notice that in the proof we consistently used the short-hand summation notation. Neither notation is a better choice than the other. Choose which style matches *your* style. Consistency, as usual, is key.

Before discussing further tips for inductive proofs, let us prove the second of the aforementioned common summations. The scratch work is left to the reader. The proof of the sum of the first n cubes is left as an exercise.

Theorem 4.3. *For any $n \in \mathbb{Z}^+$,*

$$\sum_{i=1}^{n} i^2 = \frac{n(n+1)(2n+1)}{6}.$$

Proof Using the Principle of Mathematical Induction, we will prove that

$$\sum_{i=1}^{n} i^2 = \frac{n(n+1)(2n+1)}{6}$$

for any $n \in \mathbb{Z}^+$. First, note that the base case, when $n = 1$, is true:

$$1^2 = \frac{1(1+1)(2+1)}{6}.$$

Next, for a general $n \in \mathbb{Z}^+$, assume the inductive hypothesis to be true:

$$\sum_{i=1}^{n} i^2 = \frac{n(n+1)(2n+1)}{6}.$$

Then,

$$\sum_{i=1}^{n+1} i^2 = \sum_{i=1}^{n} i^2 + (n+1)^2$$

$$= \frac{n(n+1)(2n+1)}{6} + (n+1)^2 \quad \text{(inductive hypothesis)}$$

$$= \frac{n(n+1)(2n+1)}{6} + \frac{6(n+1)^2}{6}$$

$$= \frac{(n+1)[n(2n+1) + 6(n+1)]}{6}$$

$$= \frac{(n+1)(n+2)[2(n+1)+1]}{6}.$$

Thus, the inductive step holds, and consequently proving the formula true.
□

The proofs of Theorems 4.2 and 4.3 are different but share particular aspects that make for a quality inductive proof. First, the reader is told at the outset of the proof that it is going to proceed by induction. As with proofs by contradiction or contrapositive, this ensures the reader does not wander aimlessly through the proof and knows what to expect.

Note that in the proof of each theorem the inductive hypothesis is fully presented. It is clear and there are no questions about what it is saying. Just saying, "Assume the inductive hypothesis to be true" (the reader is not told what the inductive hypothesis actually says), or, "Assume for a general k that ..." (k is not explicitly defined), are poor presentations of the inductive hypothesis. State that what you are assuming *is actually the inductive hypothesis*, state it precisely and when the inductive hypothesis is employed, make certain to state so. The reader knows exactly where it works its way into the proof.

Why this emphasis on the inductive hypothesis? It is *the* tool that makes an induction proof an actual induction proof! It is not enough to just assume the inductive hypothesis. Likewise, we do not simply show a general $(k+1)$st case holds. Neither step can stand alone; they require each other. We must use the inductive hypothesis assumption to show that the inductive step holds.

> **Quality Tip 20.** *When proving via induction, be clear when you state and use the inductive hypothesis.*

There are two major differences in the proofs of the previous two theorems. First is the choice of variable in the inductive step. While both theorems are stated in terms of a variable n, the proof of Theorem 4.2 utilizes k in the inductive step ("Assume for a general $n = k$ ...") while Theorem 4.3 is in terms of n ("Assume for a general n that ..."). The choice here is up to the author. For some, changing to a new variable, such as k, differentiates the statement of the result from the work of the inductive step. For others, sticking with the same variable eliminates the introduction of one more variable into the proof. The difference is simply stylistic.

The second major difference in the two proofs is in the algebraic presentation of the work. The proof of Theorem 4.2 writes out the sum in long-hand notation, whereas the proof of Theorem 4.3 keeps the work in sum-notation. Just as with choice of inductive variable, decide which you, as the author of the proof, prefer.

One thing that *cannot* be done is using different variables for the inductive hypothesis and the inductive step, however. It is incorrect to assume for a general $n = k$ that the inductive hypothesis holds, and then to proceed and show the inductive step holds for $n + 1$, without reference to the variable k. In terms of the domino metaphor, this is analogous to assuming a general kth domino falls over and showing the $(n + 1)$st domino falls over.[4]

Mathematical induction is a proof technique that can be used in a variety of contexts. The next few results serve to show its versatility, with the first being used to prove an inequality. The proofs are presented in a variety of styles, providing you numerous examples from which to formulate *your* style.

Theorem 4.4. *For non-negative integers n, $1 + n \leq 2^n$.*

Proof We proceed by induction to prove that

$$1 + n \leq 2^n$$

for every non-negative integer n. The base case $(n = 0)$ is clearly true: $1 \leq 1$. Then, assume for the inductive hypothesis that

$$1 + n \leq 2^n$$

for a general $n \in \mathbb{Z}$, $n \geq 0$. Then,

$$
\begin{aligned}
1 + (n + 1) &= (1 + n) + 1 \\
&\leq 2^n + 1 \text{ (IH)} \\
&\leq 2^n + 2^n \text{ (since } n \geq 0) \\
&= 2^{n+1},
\end{aligned}
$$

---

[4]Explain it this way to the child asking about his long line of dominoes and he will be *really* confused.

as was required to be shown in the inductive step. Consequently, it follows that

$$1 + n \leq 2^n$$

for every non-negative integer n.   □

Notice that the proof of Theorem 4.4 is written in a much more condensed style than the previous inductive proofs we have seen. The proof is not sectioned off into its three distinct parts (base case, inductive hypothesis, inductive step). However, the reader is never lost; transitional and directional language appropriately guides the reader.

Previously we have seen how to prove divisibility results using direct proofs, but sometimes general claims about divisibility require mathematical induction.

Theorem 4.5. *For all integers $n \geq 0$,*

$$3 | (4^n - 1).$$

Scratch work: Let us focus on the scratch work of the inductive step, having assumed that the result holds for a general n. We must show that

$$3 | (4^{n+1} - 1),$$

or equivalently,

$$4^{n+1} - 1 = 3(\ ).$$

The tool that we have to work with is that

$$4^n - 1 = 3d.$$

There are two possible directions to proceed with our scratch work. We can work *from* the inductive hypothesis and attempt to *show* the inductive step, or we can start *with* part of the inductive step and work towards *introducing* the inductive hypothesis into the process. Let's show both approaches.

(1) If we start with

$$4^n - 1 = 3d$$

we can multiply both sides of the equation by 4. This will "get" 4^{n+1} into the equation:

$$4^{n+1} - 4 = 12d.$$

Keeping our eyes on the prize, adding three to both sides of the equation and a little factoring gets us to our goal:

$$4^{n+1} - 1 = 3(4d + 1).$$

(2) For this approach, let us simply work from one side of our "to show" statement. The left-hand side, $4^{n+1} - 1$, has potential for introducing the inductive hypothesis into it. We must manipulate the expression until $4^n - 1$ appears. That is not overly difficult:

$$4^{n+1} - 1 = 4 \cdot 4^n - 1$$
$$= 3 \cdot 4^n + (4^n - 1).$$

Introduce the inductive hypothesis and the result falls into place.

With the scratch work done, we are ready to proceed in constructing our proof. We choose the approach (2) for presenting the inductive step.

Proof Induction is used to prove that $3|(4^n - 1)$ for integers $n \geq 0$. First, $3|0$, so that the base case holds. Then, assume for any integer $n \geq 0$ that the result holds: $3|(4^n - 1)$. Note that

$$4^{n+1} - 1 = 4 \cdot 4^n - 1$$
$$= 3 \cdot 4^n + (4^n - 1).$$

Because, by definition

$$3|(3 \cdot 4^n)$$

and, by the inductive hypothesis

$$3|(4^n - 1),$$

it follows from Lemma 3.16 that

$$3|(4^{n+1} - 1).$$

Hence, by the Principle of Mathematical Induction, the claim holds. $\square$

Though we have previously used the factorial operator, we never explicitly defined it. We do so in Definition 4.6 and see that results using it are natural candidates for inductive proofs.

Definition 4.6. Let n be a non-negative integer. Then, the *factorial* operation on n, denoted $n!$, is defined as

$$n! = \begin{cases} n \cdot (n-1)! & \text{if } n > 0 \\ 1 & \text{if } n = 0. \end{cases}$$

Note that in Definition 4.6, factorial is defined in terms of itself. Such a definition is referred to as being *recursively defined*. To compute 3!, we need to know 2!, for which 1! must be known, and ultimately, 0! also. But this last quantity is known (and defined to be 1). Thus, recursive definitions are similar to mathematical induction in that a base definition is stated, and then subsequent expressions are defined in terms of previous ones. Hence, it is no surprise that induction is the proof method used in Theorem 4.7.

Theorem 4.7. *For non-negative integers n,*

$$2^n < (n+2)!.$$

Proof The claim that

$$2^n < (n+2)!,$$

when n is a non-negative integer, can be proven via mathematical induction.

Base ($n = 0$): Because

$$2^0 = 1$$
$$< 2$$
$$= (0+2)!,$$

the base case holds.
Induction hypothesis: Suppose for a non-negative integer k that

$$2^k < (k+2)!.$$

Now,

$$2^{k+1} = 2 \cdot 2^k$$
$$< 2 \cdot (k+2)! \text{ (IH)}$$
$$< (k+3) \cdot (k+2)! \text{ (since } k \geq 0)$$
$$= (k+3)!.$$

Having shown $2^{k+1} < (k+3)!$, the result follows.   □

We conclude our introduction to mathematical induction by proving a set-theoretic result. If A is a set with n elements, how many elements are in the power set $\mathcal{P}(A)$? A few quick computations show that $\mathcal{P}(\{\,\})$ has 1 element, $\mathcal{P}(\{1\})$ has 2 elements, $\mathcal{P}(\{1,2\})$ has 4 elements, and $\mathcal{P}(\{1,2,3\})$ has 8 elements. It appears that the power set of a set with n elements has 2^n elements, and that is indeed the case.

Note that our intuition agrees with this, as well. If we were to construct an element S of $\mathcal{P}(A)$ (that is, construct a subset S of A), for each $a \in A$ there are two options: $a \in S$ or $a \notin S$. Since A has n elements, there are two options for each of those elements of A: it is either in the subset S of A or not. A counting rule (the multiplication principle) says that there would be 2^n such subsets.

This argument is in no ways a proof. How would we prove such a claim? Because we are making a claim about a set with n elements, where n can be any non-negative integer, induction is a natural candidate for the proof method. And the intuition behind our counting (taking a specific element of A and considering whether or not it is in or not in a constructed subset) is the key idea behind the induction step.

The variable n represents the number of elements in the set. Thus, our inductive hypothesis will be a claim about the power set of any set of n elements. The inductive step will require us to consider a general set containing $n + 1$ elements. To call upon the inductive hypothesis, then, we will need to revert to a set containing just n elements. How can we do this? By taking out one element in the set of size $n + 1$, we are left with a set of size n.

Theorem 4.8. *If A is a set with n elements, where $n \in \mathbb{Z}^+$, then $\mathcal{P}(A)$ has 2^n elements.*

Proof Consider an arbitrary a set with n elements:

$$A = \{a_1, a_2, \ldots, a_n\}.$$

We induct on n to show that $\mathcal{P}(A)$ has 2^n elements.

The base case, $n = 1$, occurs when $A = \{a_1\}$. Then,

$$\mathcal{P}(A) = \{\varnothing, \{a_1\}\},$$

a set with 2^1 elements, showing that the base case holds true.

Now suppose the inductive hypothesis to be true, that the power set of any set with k elements contains 2^k elements, where k is a non-negative integer k. Consider, then, if A has $k + 1$ elements:

$$A = \{a_1, a_2, \ldots, a_{k+1}\}.$$

Because $k \geq 0$, we know that A, with $k + 1$ elements, is nonempty. Then, we can write $\mathcal{P}(A)$ as a union of two sets: those subsets of A containing a_1 and those that do not. Specifically, $A = A_1 \cup A_2$, where,

$$A_1 = \{S \subseteq A \mid a_1 \in S\},$$

and

$$A_2 = \{S \subseteq A \mid a_1 \notin S\}.$$

Note that this is a partition of $\mathcal{P}(A)$ since $A_1 \cap A_2 = \varnothing$ and both sets are nonempty ($A \in A_1$, $\varnothing \in A_2$). Because of this, the number of elements of $\mathcal{P}(A)$ is the sum of the number of elements A_1 and the number of elements of A_2. Consider A_1 and A_2 separately:

A_1: A set S is an element of A_1 if and only if $S = \{a_1\} \cup S_1$, where

$$S_1 \subseteq A - \{a_1\}.$$

The number of such sets S_1 is the number of subsets of

$$\{a_2, a_3, \ldots a_{k+1}\},$$

a set with k elements. By the inductive hypothesis, there are 2^k such sets.

A_2: A set S is an element of A_2 if and only if

$$S \subseteq A - \{a_1\}.$$

As in the previous case, there are 2^k such sets. Consequently, there are $2^k + 2^k = 2^{k+1}$ elements of $\mathcal{P}(A)$, proving the inductive step and as a result the entire claim.   □

We have seen various types of mathematical claims provable via induction. It is a versatile and powerful proof technique. While its basic structure is predetermined, a proof by induction has room for personalization. Practicing will help you develop your unique style. As you do, mathematical induction will become a routine tool that you are not afraid to employ; when to call upon it will become second nature to you.

In the next section, we expand upon these basic ideas and introduce a variant of induction called the *Principle of Strong Mathematical Induction*.

Exercises

1. Prove the base case and state the inductive hypothesis for a proof by induction of each of the following claims.

    (a) For $n \in \mathbb{Z}, n \geq 0$,

    $$1 + 3 + 3^2 + \cdots + 3^n = \frac{1}{2}(3^{n+1} + 1).$$

    (b) For $n \in \mathbb{Z}^+$,

    $$1^2 + 2^2 + \cdots + (2n)^2 = \frac{n(2n+1)(4n+1)}{3}.$$

    (c) If $n \in \mathbb{Z}^+$, then

    $$\sum_{i=1}^{n} i^4 = \frac{n(n+1)(2n+1)(3n^2 + 3n - 1)}{30}.$$

    (d) For $n \in \mathbb{Z}, n \geq 7$,

    $$3^n < n!.$$

    (e) For $n \in \mathbb{Z}^+$,

    $$\frac{1}{1 \cdot 3} + \frac{1}{3 \cdot 5} + \frac{1}{5 \cdot 7} + \cdots + \frac{1}{(2n-1) \cdot (2n+1)} = \frac{n}{2n+1}.$$

    (f) The quantity $n^3 + 2n$ is divisible by 6 for all integers n.

2. Prove the following summation results using mathematical induction.

   (a) The sum of the first n odd integers equals n^2. That is,
   $$1 + 3 + 5 + \cdots + (2n - 1) = n^2.$$

   (b) If $k \in \mathbb{Z}^+$, then
   $$2 + 5 + 8 + \cdots + (3k - 1) = \frac{k(3k + 1)}{2}.$$

   (c) For $k \in \mathbb{Z}^+$,
   $$1 + 4 + 7 + \cdots + (3k - 2) = \frac{k(3k - 1)}{2}.$$

   (d) For $n \in \mathbb{Z}^+$,
   $$1 + 4 + 4^2 + \cdots + 4^{n-1} = \frac{4^n - 1}{3}.$$

   (e) For $n \in \mathbb{Z}^+$,
   $$\sum_{i=1}^{n} i^3 = \left[\frac{n(n+1)}{2}\right]^2.$$

   (f) For positive integers n,
   $$\sum_{i=1}^{n} i(i+1) = \frac{n(n+1)(n+2)}{3}.$$

   (g) When $n \in \mathbb{Z}^+$,
   $$\sum_{i=1}^{n} \frac{1}{i(i+1)} = \frac{n}{n+1}.$$

3. Prove the following result about the sum of a finite geometric series:
   $$\sum_{i=0}^{n} r^i = \frac{r^{n+1} - 1}{r - 1},$$
   where $n \geq 0$ and $r \in \mathbb{R}$ with $r \neq 1$.

4. Use mathematical induction to prove the following claims about divisibility.

   (a) For all $n \in \mathbb{Z}^+$,
   $$5 | 6^n - 1.$$

   (b) The quantity $n^3 - n$ is divisible by 6 for all integers $n \geq 2$.

   (c) One less than every positive integer power of 4 is divisible by 3.

   (d) For $n \in \mathbb{Z}^+$,
   $$3 | (n^3 + 5n + 6).$$

   (e) When n is any non-negative integer,

$$5 \mid (2^{3n} - 3^n).$$

(f) For $m \in \mathbb{Z}^+$, 21 divides $4^{m+1} + 5^{2m-1}$.

(g) For all odd integers m and all $n \in \mathbb{Z}$, $m^{2^n} - 1$ is divisible by $4 \cdot 2^n$.

5. Use induction to validate the following results involving inequalities.

   (a) For $n \in \mathbb{Z}^+$,
   $$3^n > n^2.$$

   (b) For any integer $m \geq 4$,
   $$m! > m^2.$$

   (c) When $n \geq 2$,
   $$\sum_{i=1}^{n} i^3 < n^4.$$

   (d) For $n \in \mathbb{Z}$, $n \geq 6$,
   $$6n + 6 < 2^n.$$

   (e) For integers $m > 1$,
   $$4^m + 7 < 5^m.$$

   (f) For all integers $n \geq 2$ and positive real numbers x,
   $$(1 + x)^n > 1 + nx.$$

6. Prove the following results using mathematical induction.

   (a) For $n \geq 2$,
   $$\left(1 - \frac{1}{2}\right)\left(1 - \frac{1}{3}\right) \cdots \left(1 - \frac{1}{n}\right) = \frac{1}{n}.$$

   (b) For $n \in \mathbb{Z}^+$, prove that
   $$1 \cdot 3 \cdot 5 \cdots \cdot (2n - 1) = \frac{(2n)!}{n!2^n}.$$

   (c) For $n \in \mathbb{Z}^+$,
   $$\sum_{i=1}^{n} i \cdot 3^i = \frac{(2n - 1)3^{n+1} + 3}{4}.$$

   (d) When $n \in \mathbb{Z}^+$,
   $$\frac{1}{1 \cdot 4} + \frac{1}{4 \cdot 7} + \frac{1}{7 \cdot 10} + \cdots + \frac{1}{(3n - 2) \cdot (3n + 1)} = \frac{n}{3n + 1}.$$

   (e) The squares of first n odd positive integers sums to
   $$\frac{4n^3 - n}{3}.$$

7. Use mathematical induction to prove the generalized distribution rule, where $n \in \mathbb{Z}^+$ and $m, a_i \in \mathbb{R}$.

$$m(a_1 + a_2 + \cdots + a_n) = ma_1 + ma_2 + \cdots + ma_n$$

8. Prove the *Generalized Triangle Inequality*: for $x_i \in \mathbb{R}$ and $n \in \mathbb{Z}^+$,

$$|x_1 + x_2 + \cdots + x_n| \leq |x_1| + |x_2| + \cdots + |x_n|.$$

9. Prove the basic power rule for differentiation:

$$\frac{d}{dx}x^n = nx^{n-1},$$

when $n \in \mathbb{Z}^+$, assuming that $\frac{d}{dx}x = 1$ and that the product rule holds:

$$\frac{d}{dx}(f(x) \cdot g(x)) = \frac{d}{dx}f(x) \cdot g(x) + f(x) \cdot \frac{d}{dx}g(x).$$

10. If 3 distinct points on a circle are connected with straight lines, the interior angles of the resulting polygon sum to 180 degrees. Prove, using induction, that if n distinct points on a circle $(n \geq 3)$ are connected consecutively with straight lines, then the interior angles of the resultant polygon sum to $(n - 2) \cdot 180$ degrees.

4.2   Strong Mathematical Induction

Sequences are common mathematical objects, appearing in areas ranging from abstract algebra and topology to computational theory and mathematical biology. It is assumed that the reader is familiar with them, having encountered them in calculus when defining the convergence of a series (recall that the convergence of a series is dependent upon its sequence of partial sums). The mathematician yearns to understand *all aspects* of this infinitely long string of terms. Does one term impact the behavior of any others? Do the terms converge to a specific value, or do they diverge? Do the terms behave in unforeseen yet beautiful ways? Are there patterns amongst the relationships between the terms?

Thinking this way leads us to believe that mathematical induction is a helpful tool in understanding the entire sequence. Indeed that is the case. There is more than that, however. Sequences, when defined in particular ways, inspire the Principle of Strong Mathematical Induction, a variation on the Principle of Mathematical Induction introduced in the last section. We first define the *Fibonacci sequence* to serve as our vehicle for introducing this new mathematical tool.

Definition 4.9. The infinite sequence $\{F_n\}$, $n \in \mathbb{Z}$, $n \geq 0$, defined recursively by $F_0 = 1$, $F_1 = 1$, and

$$F_n = F_{n-1} + F_{n-2}$$

for $n \geq 2$ is called the *Fibonacci sequence*. The terms of the sequence are called *Fibonacci numbers*.

The recursive definition of the Fibonacci sequence can be summarized using everyday language. To find a term in the sequence, sum the two previous terms. For example, if we wanted to determine F_{10}, we would need to know the values of F_9 and F_8, which in turn would require us to know the values of F_7 and F_6, and so on. The term F_2 equals the sum $F_1 + F_0$, which are explicitly defined in Definition 4.9.

$$F_0 = 1$$
$$F_1 = 1$$
$$F_2 = F_1 + F_0 = 1 + 1 = 2$$
$$F_3 = F_2 + F_1 = 2 + 1 = 3$$
$$F_4 = F_3 + F_2 = 3 + 2 = 5$$

Suppose we aim to prove some claim $P(n)$ about the general nth term of the Fibonacci sequence. Mathematical induction seems like a natural choice for a proof technique; the sequence is defined recursively, so assuming something is true about a previous term may lead to information about the next term. The inductive hypothesis would be an assumption that $P(n)$ is true for some general integer n; that is, the claim holds for the nth term of the sequence. We would then proceed to attempting to prove that the claim holds for the $(n+1)$st term of the sequence (showing $P(n+1)$ is true). Because the $(n+1)$st Fibonacci number is the sum of the nth and $(n-1)$st, if proving that $P(n+1)$ is true requires knowing that the claim is true for *both* of the previous Fibonacci numbers, we find ourselves in a bind. The inductive hypothesis would only provide information about the nth Fibonacci number. A remedy for such situations is a tool called *strong mathematical induction*.

Definition 4.10. Let $a, b \in \mathbb{Z}^+$ with $a \leq b$. A predicate $P(n)$ defined for $n \in \mathbb{Z}$, $n \geq a$, is proven true using the *Principle of Strong Mathematical Induction* (or simply *strong induction*) by showing the following.

1. $P(a), P(a+1), \ldots, P(b)$ are true.

2. For $k \geq b$, $k \in \mathbb{Z}$, if $P(n)$ is true for all $n \in \mathbb{Z}$, $a \leq n \leq k$, then $P(k+1)$ is true.

There are two major differences between mathematical induction and strong mathematical induction. First, the base case for mathematical induction requires showing only one instance of $P(n)$ is true. In a proof by strong mathematical induction, the base case may require two, three or more such justifications. Why? Suppose a sequence is defined similarly to the Fibonacci sequence, where a general term is dependent upon the two previous terms, and let us revisit our domino metaphor, where the terms in this sequence are thought of as dominoes. In our base case, suppose we only show the first domino falls. We proceed to show that knowing two successive dominoes fall force the next domino to fall. Have we proven that *all* dominoes must fall? Not quite. How do we know that the second domino falls? It could not follow from the inductive step, as this part of the proof requires two dominoes to precede it although the second domino only has one previous domino. Nothing we have done proves that this second domino must fall, meaning that we have not shown that *every* domino must eventually fall.

The other major difference is in what the inductive hypothesis assumes. Unlike in mathematical induction, the inductive hypothesis in a strong inductive proof assumes that not only $P(n)$ is true for an arbitrary integer n but also that $P(i)$ is true for *every* integer i from the base case through n. In terms of the domino metaphor, we are assuming not just that the previous domino falls but that every domino up to and including the previous domino falls. Then, we must show the next domino falls.

Example 4.1 exhibits our first proof using strong mathematical induction. Notice that its overall structure is similar to the inductive proofs we saw in Section 4.1.

Example 4.1. Let $\{a_n\}$ be the sequence defined by $a_0 = 1$, $a_1 = 2$, $a_2 = 3$ and

$$a_n = a_{n-1} + a_{n-2} + a_{n-3},$$

for integers $n \geq 3$. Prove that $a_n \leq 2^n$ for all non-negative integers n.

Before proceeding with the proof of this result, let us investigate the sequence. First, it is recursively defined, much like the Fibonacci sequence. Here, however, the term a_n, for $n \geq 3$, is defined to be the sum of the *three* previous terms. You may be saying to yourself, "But a_2 is not the sum of the three previous terms! It only has two terms before it!" You are correct, and your latter thought, that a_2 has only two terms appearing in the sequence before it, is why a_2, along with a_0 and a_1, is *explicitly* defined.

What we are asked to prove is that

$$a_n \leq 2^n.$$

The fact that a_n is dependent upon more than one previous term leads us to think strong mathematical induction is necessary, and the first step in proceeding with it is to determine what must be shown in the base case.

A natural question that arises when first learning about strong mathematical induction deals with this particular part of the proof: how many specific cases of the base case must be explicitly shown true? The number is determined by how many previous dominoes a general domino depends on. In this case, a_n is defined in terms of the *three* previous terms. Thus, we need to show that the first three specific cases hold. Then, in showing the inductive step, we would be able to conclude that a_3 satisfies the property, having shown a_0, a_1 and a_2 satisfy it. If we were to successfully show that the inductive hypothesis implies the inductive step, yet we only show a_0 and a_1 to have the given property, we could not logically conclude that the property must hold for a_3.

With these in mind, we proceed with the proof.

Proof Suppose $\{a_n\}$ is the sequence with initial terms $a_0 = 1$, $a_1 = 2$, $a_2 = 3$ and defined for $n \geq 3$ by

$$a_n = a_{n-1} + a_{n-2} + a_{n-3}.$$

To prove that $a_n \leq 2^n$ for all non-negative integers n, we proceed via strong induction. The base case must consider the first three terms of the sequence and see that the result holds. Notice that these first three terms, $a_0 = 1$, $a_1 = 2$ and $a_2 = 3$, are less than or equal to, respectively, 2^0, 2^1 and 2^2. The base case consequently holds.

Assume then for the inductive hypothesis that $a_n \leq 2^n$ for all integers n, $0 \leq n \leq k$ with $k \in \mathbb{Z}^+$, $k \geq 3$. We will show that $a_{k+1} \leq 2^{k+1}$.

To that end,

$$\begin{aligned}
a_{k+1} &= a_k + a_{k-1} + a_{k-2} \\
&\leq 2^k + 2^{k-1} + 2^{k-2} \text{ (IH)} \\
&= 2^{k-2}(2^2 + 2^1 + 1) \\
&\leq 2^{k-2}(2^3) \\
&= 2^{k+1},
\end{aligned}$$

as was to be shown. Hence, the result holds: $a_n \leq 2^n$ for all non-negative integers n. □

Before moving on to results involving the Fibonacci sequence, let us discuss a peculiar notational difference between induction and strong induction. Notice the inductive hypothesis of Example 4.1 requires using *two* variables:

n and k. Proofs proceeding by ordinary mathematical induction only require the use of one variable in their inductive hypotheses. Why is this? In a proof by mathematical induction, the result is assumed to be true for just a single general kth case, and then, that single case is used to show the next, the $(k+1)$st case, holds true. Strong mathematical induction, however, assumes not just that the result holds for a single general kth case; it assumes the result holds for the first *through* the kth cases. The additional variable, n in Example 4.1, is playing the role of the "counter" for all these cases, between 0 and k. "When n is any integer from 0 to k, assume the result holds." Because k is the "last" case for which the result is assumed to be true, we must proceed in showing that the result holds for the $(k+1)$st case.

There are many famous results involving the Fibonacci sequence, with Theorem 4.11 being one of them, while others appear in the Exercises of this section. What Theorem 4.11 says is that the *sum* of the first $n+1$ Fibonacci numbers (since the first Fibonacci number is F_0) need not be found by adding all the terms but rather by looking ahead just a couple terms in the Fibonacci sequence and subtracting 1. Where the subtraction of 1 comes from will be apparent in the proof, but that the sum of the terms is a subsequent term in the sequence should be no surprise. The term F_{n+2} is the sum of its two previous terms, each of which is the sum of its two previous terms, etc. You can begin to see the result falling into place:

$$F_{n+2} = F_{n+1} + F_n$$
$$= F_n + F_{n-1} + F_{n-1} + F_{n-2}.$$

This scratch work is in no way a proof of the result, however. Strong induction is the tool we need to prove it explicitly.

Theorem 4.11. *If F_i is the ith Fibonacci number, then for any $n \in \mathbb{Z}^+$,*

$$\sum_{i=0}^{n} F_i = F_{n+2} - 1.$$

Proof Let F_i be the ith Fibonacci number. To prove that the sum of the first n Fibonacci numbers is one less than the $(n+2)$nd Fibonacci number (where $n \in \mathbb{Z}^+$), we proceed by strong mathematical induction. The base case requires us to consider $n = 1$ and $n = 2$:

$$\sum_{i=0}^{1} F_i = 1 + 1$$
$$= 3 - 1$$
$$= F_3 - 1.$$

$$\sum_{i=0}^{2} F_i = 1 + 1 + 2$$

$$= 5 - 1$$

$$= F_4 - 1.$$

Thus, the base case holds. Assume the inductive hypothesis to be true, that for a general $n \in \mathbb{Z}^+$,

$$\sum_{i=0}^{k} F_i = F_{k+2} - 1$$

for $1 \leq k \leq n$, $k \in \mathbb{Z}$. We must show that

$$\sum_{i=0}^{n+1} F_i = F_{n+3} - 1.$$

Then,

$$\sum_{i=0}^{n+1} F_i = \sum_{i=0}^{n} F_i + F_{n+1}$$

$$= F_{n+2} - 1 + F_{n+1} \text{ (IH)}$$

$$= F_{n+3} - 1 \text{ (definition of the Fibonacci sequence)},$$

as was to be shown. Thus, the result holds.   □

Though sequence problems commonly employ the Principle of Strong Mathematical Induction, it is a tool that is used throughout mathematics. The next theorem involves two different proof techniques, with one being strong induction and the other being a proof of uniqueness. If you are familiar with base 2 (or binary) representation, you know that every positive integer has one and only one such representation. Theorem 4.13 proves both the existence and uniqueness of this. Before proceeding with the statement and proof of this, recall the following definition.

Definition 4.12. A positive integer n has a *base b representation* ($b \in \mathbb{Z}$, $b \geq 2$) of

$$a_k a_{k-1} \ldots a_1 a_0,$$

where $a_i \in \mathbb{Z}$, $0 \leq a_i < b$, and $a_k \neq 0$, if

$$n = a_k b^k + a_{k-1} b^{k-1} + \cdots + a_1 b^1 + a_0 b^0.$$

We are used to writing positive integers using base 10 representation, and any reference to positive integers in this text, unless otherwise noted, is in such representation. Various applications, however, require different representations. Example 4.2 gives representations using various bases for a single positive integer.

Example 4.2. The positive integer 72 has the following base representations:

(1) Base 2: 1001000, since

$$72 = 1 \cdot 2^6 + 0 \cdot 2^5 + 0 \cdot 2^4 + 1 \cdot 2^3 + 0 \cdot 2^2 + 0 \cdot 2^2 + 0 \cdot 2^1 + 0 \cdot 2^0.$$

(2) Base 3: 2200, since

$$72 = 2 \cdot 3^3 + 2 \cdot 3^2 + 0 \cdot 3^1 + 0 \cdot 3^0.$$

(3) Base 5: 242, since

$$72 = 2 \cdot 5^2 + 4 \cdot 5^1 + 2 \cdot 5^0.$$

Theorem 4.13. *Every positive integer has a unique base* 2 *representation.*

Our scratch work for this proof leads us to a realization: using the inductive hypothesis requires cases. This is allowed so long as the cases account for all possibilities for the inductive step. Here, the cases are determined by whether $n+1$ is even or odd (impacting the units digit in its base 2 representation).

Note also that in this proof we switch the roles, compared to Example 4.1, of k and n in the inductive hypothesis. This is done for no other reason than to exhibit different methods for writing the inductive hypothesis.

Proof We prove that (1) every positive integer has a binary representation, and (2) such a representation is unique.

(1) Proceed by strong induction on $n \in \mathbb{Z}^+$ to show a binary representation of n exists. The base case, $n = 1$, holds because

$$1 = 1 \cdot 2^0.$$

Then, suppose for our inductive hypothesis that for a general $n \in \mathbb{Z}^+$, every integer k, $1 \leq k \leq n$, has a binary representation. We must show that $n+1$ has a binary representation. Consider separately the cases where (a) $n+1$ is even, and, (b) $n+1$ odd.

(a) If $n+1$ is even, then $n+1 = 2m$, for some $m \in \mathbb{Z}^+$. Because $1 \leq m < n$, the inductive hypothesis applies to m:

$$m = 2^j + a_{j-1}2^{j-1} + \cdots + a_1 2^1 + a_0 2^0,$$

where $j \in \mathbb{Z}^+$ and $a_i = 0$ or $a_i = 1$ for $0 \leq i < j$. Thus,

$$\begin{aligned} n + 1 &= 2m \\ &= 2(2^j + a_{j-1}2^{j-1} + \cdots + a_1 2^1 + a_0 2^0) \\ &= 2^{j+1} + a_{j-1}2^j + \cdots + a_1 2^2 + a_0 2^1, \end{aligned}$$

yielding the base 2 representation for $n + 1$ of

$$1a_{j-1} \ldots a_1 a_0 0.$$

(b) If $n + 1$ is odd, then $n + 1 = 2m + 1$, for some $m \in \mathbb{Z}^+$. As in the previous case, since $1 \leq m < n$, the inductive hypothesis yields

$$m = 2^j + a_{j-1}2^{j-1} + \cdots + a_1 2^1 + a_0 2^0,$$

where $j \in \mathbb{Z}^+$ and $a_i = 0$ or $a_i = 1$ for $0 \leq i < j$. As before,

$$\begin{aligned} n + 1 &= 2m + 1 \\ &= 2(2^j + a_{j-1}2^{j-1} + \cdots + a_1 2^1 + a_0 2^0) + 1 \\ &= 2^{j+1} + a_{j-1}2^j + \cdots + a_1 2^2 + a_0 2^1 + 1, \end{aligned}$$

a binary representation for $n + 1$ of

$$1a_{j-1} \ldots a_1 a_0 1.$$

In either case, the inductive step holds and consequently, every positive integer has a binary representation.

(2) To prove that binary representation of positive integers is unique, suppose $n \in \mathbb{Z}^+$ has two such representations:

$$n = 2^j + a_{j-1}2^{j-1} + \cdots + a_1 2^1 + a_0 2^0,$$

and

$$n = 2^m + b_{m-1}2^{m-1} + \cdots + b_1 2^1 + b_0 2^0,$$

with all coefficients being 0 or 1 and $j, m \in \mathbb{Z}^+$. WLOG, assume $j < m$. Then,

$$\begin{aligned} n &= 2^j + a_{j-1}2^{j-1} + \cdots + a_1 2^1 + a_0 2^0 \\ &\leq 2^j + 2^{j-1} + \cdots + 2^1 + 2^0 \\ &= 2^{j+1} - 1 \quad \text{(Exercise 3)} \\ &< 2^m - 1 \quad \text{since } j + 1 \leq m) \\ &< 2^m \\ &\leq 2^m + b_{m-1}2^{m-1} + \cdots + b_1 2^1 + b_0 2^0 \\ &= n, \end{aligned}$$

a contradiction. Therefore, the binary representation of n is unique. □

To conclude this section we shift gears and look at problems of the *combinatorial* variety. Combinatorics, often referred to as the mathematics of counting,[5] is the study of discrete structures or sets. Questions often investigate patterns or arrangements of the elements of the sets. These questions typically fall into two categories:

1. *Existence*: does an arrangement of the objects exist that fulfills a predetermined set of rules or conditions?

2. *Enumerate*: if such arrangements do exist, how many of them are there?

We focus here on *chessboard* questions. We define the necessary terminology below, with intuitive notions of the definitions following.

> **Definition 4.14.** An $m \times n$ *chessboard* is a rectangular grid of mn squares, consisting of m rows and n columns. A *domino* is a 1×2 or 2×1 chessboard. Figure 4.1 illustrates this concept. A *covering* of an $m \times n$ chessboard by $\frac{mn}{2}$ dominoes is an arrangement of the dominoes on the chessboard so that each domino covers two squares of the grid, no dominoes overlap, and every square of the chessboard is covered.

Intuitively, a chessboard is a a variant of a real-life chessboard (or checkerboard), where the squares are not colored. A domino is two squares connected, and it can be placed vertically or horizontally on the chessboard to cover two adjacent squares. For example, a regular standard chessboard is 8×8. It can be covered by 32 dominoes in many ways; two particularly simple coverings would be 8 rows or columns of 4 dominoes laid horizontally or vertically, respectively.[6]

Questions about chessboard coverings abound, with a few of the results quite surprising. The next example highlights why strong mathematical induction is necessary to prove some of these results.

> **Example 4.3.** Determine how many different ways n dominoes can cover a $2 \times n$ chessboard.
>
> Questions like this require scratch work. We must figure out what

---

[5] A whole area of mathematics devoted to *counting*? Problems in combinatorics may be very simple to state but extremely difficult to answer. Suppose for example you walk into a bakery and wish to purchase a dozen donuts. There are ten varieties to choose from (and enough of each variety so that if you wanted to get all dozen of one type you could). How many different selections could you make?

[6] There are 12,988,816 different coverings of an 8×8 chessboard by 32 dominoes!

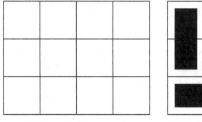 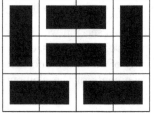

(a) 3 × 4 chessboard   (b) A covering by dominoes

FIGURE 4.1
Chessboards and coverings

what we need to prove before we attempt to prove it! When you are asked to determine a formula involving an arbitrary integer value, it is good practice to look at specific examples. Not only can they often be determined fairly easily and quickly, but in seeing the first few specific cases, you may notice a pattern. In this case, let us consider the first few possible values for n.

$n = 1$: There is a single way to cover a 2 × 1 chessboard with a single domino, since a 2 × 1 chessboard is itself a domino.

$n = 2$: There are 2 possible coverings of a 2 × 2 chessboard; see Figure 4.2.

$n = 3$: How can we go about systematically listing all coverings? If we consider just the first column of the chessboard, there are 2 possible ways that those squares may be covered, as in Figure 4.3. If we can count the coverings that result from each of these two cases, summing the results will yield the total number of coverings for the 2 × 3 board (a sort of "proof by cases" approach).

Notice in Figure 4.3(a) that we are left with a 2 × 2 chessboard to cover, while in Figure 4.3(b) there is a 2 × 1 chessboard to cover. Thus, the number of coverings of the *third case* is the sum of the number of coverings of the *first* and the *second* case. Moreover, we realize that we can generalize this to a 2 × n chessboard. A term defined as the sum of the two previous terms? The resulting number of coverings for a general 2 × n chessboard is precisely determined by the Fibonacci sequence

Theorem 4.15. *The number of coverings of a* 2×n *chessboard by* n *dominoes, where* $n \in \mathbb{Z}^+$, *is* F_n, *the* nth *Fibonacci number.*

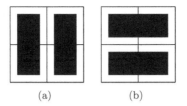

FIGURE 4.2
Coverings of a 2×2 chessboard

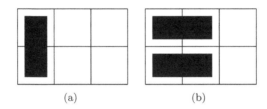

FIGURE 4.3
Covering the first column of a 2×3 chessboard

Proof Let c_n be the number of coverings of a $2 \times n$ chessboard by n dominoes. We proceed by strong induction to prove that $c_n = F_n$.

Observe that $c_1 = 1$, as there is a single way to cover a 2×1 chessboard by 1 domino. Similarly, $c_2 = 2$ as there are two ways to cover a 2×2 chessboard by 2 dominoes (see Figure 4.2). Hence, $c_1 = F_1$ and $c_2 = F_2$.

Assume our inductive hypothesis holds for an arbitrary $n \in \mathbb{Z}^+$:

$$c_k = F_k \text{ for any } k \in \mathbb{Z}, \text{ with } 1 \le k \le n.$$

Next, consider a $2 \times (n+1)$ chessboard. There are two ways to cover the first column of the chessboard, similar to Figure 4.3. In the scenario illustrated in Figure 4.3(a) there are c_n coverings of the remaining chessboard, and the scenario of Figure 4.3(b) yields c_{n-1} coverings. Thus,

$$
\begin{aligned}
c_{n+1} &= c_n + c_{n-1} \\
&= F_n + F_{n-1} \quad \text{(by the inductive hypothesis)} \\
&= F_{n+1},
\end{aligned}
$$

proving the result. □

We conclude this section by looking at the relationship between the two principles of induction. Though they seem very different, particularly in their respective inductive hypotheses, it turns out that the two forms of induction are the *same.* Any result provable using the Principle of Mathematical Induction is provable using the Principle of Strong Mathematical induction, and vice versa! The proofs of equivalency of the two principles are a test in logical thinking and an exercise in understanding what is actually *assumed* at the start of each proof.

Theorem 4.16. *Suppose $P(n)$ is a predicate defined for all positive integers $n \geq a$, for some $a \in \mathbb{Z}^+$. If $P(n)$ is provable via mathematical induction, then $P(n)$ is provable via strong mathematical induction.*

Proof Assume that $P(n)$ is a predicate provable using the Principle of Mathematical Induction (for integers $n \geq a$, where $a \in \mathbb{Z}^+$). Let $b \in \mathbb{Z}^+$ with $b \geq a$.

Because $P(n)$ can be proven via mathematical induction for $n \geq a$, it follows that $P(a), P(a+1), \ldots, P(b)$ are all true. Additionally, for any integer $k \geq b$, assuming $P(a), P(a+1), \ldots, P(k)$ true implies $P(k+1)$ is true (via induction, knowing $P(k)$ true implies $P(k+1)$ is true).

Thus, $P(n)$ is true by the Principle of Strong Mathematical Induction. $\square$

Theorem 4.16 should come as no surprise. The choice of calling the new version "strong induction" implies that it accomplishes all that induction does and more. What should come as a bit of a surprise is the next result. Anything provable via strong induction can be proven using standard induction.

Theorem 4.17. *Let $P(n)$ be a statement that can be proven true using the Principle of Strong Mathematical Induction, where $a, n \in \mathbb{Z}^+$ and $n \geq a$. Then, $P(n)$ can be proven true for $n \geq a$ using the Principle of Mathematical Induction.*

Proof Suppose $P(n)$ is provable via strong mathematical induction, for all integers $n \geq a$, where $a \in \mathbb{Z}$. Then, for $b \in \mathbb{Z}$, $a \leq b$,

1. $P(a), P(a+1), \ldots, P(b)$ are true, and

2. for any integer $k \geq b$, $P(n)$ true for all $n \in \mathbb{Z}$, $a \leq n \leq k$ implies $P(k+1)$ is true.

To proceed in showing that $P(n)$ is true for all $n \geq a$, define $Q(k)$ to be the statement defined for integers $k \geq a$, "$P(n)$ is true for all $n \leq k$." We will prove $Q(k)$ is true for all $k \geq b$ using the Principle of Mathematical Induction.

Base case: Note that $Q(b)$ is the statement, "$P(n)$ is true for all $n \leq b$." This is true by (1).

Inductive hypothesis: Assume for a general integer k with $k \geq b$ that $Q(k)$ is true.

Inductive step: We have that $Q(k+1)$ is the statement, "$P(n)$ is true for all

$n \le k+1$." By (2), along with the inductive hypothesis, we know that $P(k+1)$ is true. Along with the inductive hypothesis, we have shown $Q(k+1)$ is true.

Therefore, $Q(k)$ is true for all $k \ge b$, proving that $P(n)$ is true for all integers $n \ge a$. $\qquad\qquad\qquad\qquad\qquad\qquad\qquad\qquad\qquad\qquad\qquad\qquad\square$

Though these two previous theorems show that a result is provable via induction if and only if it is provable via strong induction, you are cautioned to now not simply brush aside strong induction and think, "I can just always use 'regular' induction!" Though this is true, the proofs will be quite messy. In particular, the approach used in the proof of Theorem 4.17 will have to be used *within* the proof you are constructing. It is much more straightforward (as you write the proof and later for those reading your proofs) to use the Principle of Strong Mathematical Induction.

Exercises

1. Find the first 5 terms of the recursively defined sequence.

    (a) $a_1 = 2$, $a_2 = 3$, and for $n \ge 3$,
    $$a_n = 2a_{n-1} - a_{n-2}$$

    (b) $b_0 = -1$, $b_1 = 2$, $b_2 = -2$, and for $i \ge 3$,
    $$b_i = (-1)^{i+1}b_{i-1} + (-1)^i b_{i-3}$$

    (c) $s_1 = 1$ and when $i \ge 2$,
    $$s_n = \sum_{i=1}^{n-1} s_i$$

    (d) The terms C_i of the sequence of *Catalan numbers*,[7] defined by $C_0 = 1$ and for $n \ge 0$,
    $$C_n = \sum_{i=0}^{n} C_i C_{n-i}$$

2. Explain why each of the following is *not* a properly defined sequence.

    (a) $b_n = b_{n-1}^2$ for $n \ge 1$

    (b) $f_0 = 2$, $f_1 = 2$ and for $n \ge 1$,
    $$f_n = f_{n-1}f_{n-2}$$

    (c) $a_1 = 2$, $a_2 = 0$ and when $n \ge 3$,
    $$a_n = \frac{a_{n-1}}{a_{n-2}}$$

    (d) $s_1 = 5$, $s_3 = 3$ and if $i \ge 4$,

---

[7]The Catalan numbers, named after Eugène Catalan (1814-1894), appear often in counting-related problems. For example, they describe the number of ways a polygon with $n + 2$ sides can be cut into n triangles.

$$s_i = s_{i-1} - s_{i-3}$$

3. Write the inductive hypothesis if each of the following claims is proven via strong mathematical induction.

   (a) Let $b_0 = 1$, $b_1 = 5$ and
   $$b_k = b_{k-2} - 2b_{k-1}$$
   for $k \geq 2$. Then, b_k is odd for all $k \geq 0$.

   (b) The only positive integer that divides consecutive terms of the Fibonacci sequence is 1.

   (c) A sequence is defined by $a_1 = 1$, $a_2 = 4$, $a_3 = 9$ and for $n \geq 4$,
   $$a_n = a_{n-1} - a_{n-2} + a_{n-3} + 4n - 6.$$
   Prove that $a_n = n^2$ for $n \in \mathbb{Z}^+$.

4. Suppose the stated claim is to be proven using a proof by strong induction. State what is *wrong* with the listed inductive hypothesis of such a proof.

   (a) *Claim*: Every term in the sequence given recursively by $s_1 = 3$, $s_2 = 5$, and
   $$s_n = s_{n-2} + 2s_{n-1},$$
   when $n \geq 3$, is odd.
   Inductive hypothesis: Suppose for $n \in \mathbb{Z}$, $n \geq 3$, and $k \in \mathbb{Z}$ with $k < n$ that s_k is odd.

   (b) *Claim*: For all $n \in \mathbb{Z}$ with $n \geq 0$,
   $$10|(n^5 - n).$$
   Inductive hypothesis: Suppose $k, n \in \mathbb{Z}$ with $0 \leq k \leq n$ that
   $$10|(n^5 - n).$$

   (c) *Claim*: For every $n \in \mathbb{Z}^+$, the nth Fibonacci number satisfies
   $$F_n = \frac{(\frac{1+\sqrt{5}}{2})^n - (\frac{1-\sqrt{5}}{2})^n}{\sqrt{5}}.$$
   Inductive hypothesis: Suppose for $k, n \in \mathbb{Z}$ with $0 \leq k \leq n$ that
   $$F_k = \frac{(\frac{1+\sqrt{5}}{2})^k - (\frac{1-\sqrt{5}}{2})^k}{\sqrt{5}}.$$

5. Use strong mathematical induction to prove the following claims about the defined sequences.

   (a) Let $a_1 = 1$, $a_2 = 1$, $a_3 = 3$, and
   $$a_n = a_{n-1} + a_{n-2} + a_{n-3} \text{ for } n \geq 4.$$
   Prove that, for $n \in \mathbb{Z}^+$,
   $$a_n \leq 2^n.$$

(b) If $b_1 = 1$, $b_2 = 8$, and
$$b_n = b_{n-1} + 2b_{n-2}$$
for $n \geq 3$, prove, for every positive integer n, that
$$b_n = 3 \cdot 2^{n-1} + 2 \cdot (-1)^n.$$

(c) If $c_i = 1$ for $i = 1, 2, 3$ and
$$c_{n+3} = c_{n+2} + c_{n+2} + c_n$$
for $n \in \mathbb{Z}^+$, prove that
$$c_n < 2^n$$
for all $n \in \mathbb{Z}^+$.

(d) Define $d_0 = 1$, $d_1 = 2$, $d_2 = 3$ and for $n \geq 3$,
$$d_n = d_{n-1} + 3d_{n-3} + 3.$$
Prove, for $n \geq 5$, that
$$d_n \leq 2^n.$$

(e) Prove for all $n \in \mathbb{Z}^+$ that
$$a_n = 2n - 1,$$
where $a_1 = 1$, $a_2 = 3$, and when $n \geq 3$,
$$a_n = 2a_{n-1} - a_{n-2}.$$

(f) Consider the sequence defined by $f_1 = \frac{8}{9}$, $f_2 = \frac{9}{10}$, $f_3 = \frac{10}{11}$ and
$$f_n = f_{n-1} \cdot f_{n-3},$$
for $n \geq 4$. Prove that
$$0 < f_n \leq 1$$
for all positive integers n.

6. Prove the following results about the Fibonacci sequence. Unless otherwise stated, the claim is for every term in the Fibonacci sequence.

(a) $F_n \in \mathbb{Z}^+$.

(b) For $n \geq 1$,
$$F_n \leq \left(\frac{7}{4}\right)^{n-1}$$

(c) $\displaystyle\sum_{i=0}^{n} F_i^2 = F_n F_{n+1}.$

(d) F_n is even if and only if $3|(n+1)$.

(e) F_n is divisible by 3 if and only if $4|(n+1)$.

(f) $F_n = \dfrac{a^n - b^n}{a - b}$, where

$$a = \frac{1 + \sqrt{5}}{2} \text{ and } b = \frac{1 - \sqrt{5}}{2}.$$

7. Prove each of the claims in Exercise 3.

8. Give the base 2 and the base 3 representation of each of the following.

   (a) 19
   (b) 93
   (c) 129
   (d) 14262

9. What positive integers, in base 10 representation, do the following represent if the representation is in (i) base 2 and (ii) base 3.

   (a) 10
   (b) 1011
   (c) 100110

10. Prove that every positive integer has a unique base 3 representation.

11. Prove that every positive integer has a unique base b representation for a general $b \in \mathbb{Z}^+$, $b \geq 2$.

12. Prove that an $m \times n$ chessboard has a covering by dominoes if and only if at least one of m or n is even.

13. How many coverings of a 4×4 chessboard by dominoes are there?

14. Every square of a $1 \times n$ chessboard is to be colored either black or red. Let C_n be the number of colorings so that no two adjacent squares are the same color. Find a recursive formula for C_n (that is, in terms of colorings of smaller chessboards) and prove the formula indeed holds.

15. Let T_n be the number of colorings of a $1 \times n$ chessboard where every square is colored one of three colors and no two adjacent squares are the same color. Find and prove true a recursive formula for T_n.

16. A *monomino* is a 1×1 chessboard. Let D_n be the number of coverings of a $1 \times n$ chessboard by monominoes and/or dominoes, so that no two dominoes are placed consecutively on the chessboard (monominoes may be adjacent on the board, however). Determine a recursive formula for D_n and prove your formula true.

17. The "Tribonacci" sequence is defined as $T_i = 1$ (for $i = 1, 2, 3$) and

$$T_n = T_{n-1} + T_{n-2} + T_{n-3}$$

for $n \geq 4$. Prove that

$$T_n < 2^n$$

for all $n \in \mathbb{Z}^+$.

18. (a) Use mathematical induction to prove

$$\sum_{i=1}^{n} i!i = (n+1)! - 1,$$

where $n \in \mathbb{Z}^+$.

   (b) Use strong induction to prove that for any $n \in \mathbb{Z}^+$, there exist $b_i \in \{0, 1, \dots, i\}$ such that

$$n = b_1 \cdot 1! + b_2 \cdot 2! + \cdots + b_r \cdot r!,$$

   for some $r \in \mathbb{Z}^+$.

19. Use strong induction to show that any group of 8 or more people can be divided into groups of 4 and 5 people.

20. *Mini-nim* is a two-person game that begins with $n \in \mathbb{Z}^+$ sticks placed on the ground. The players alternate turns, removing 1, 2 or 3 sticks on a turn. The player to remove the last stick loses the game. Prove that the player to go second has a winning strategy when $n = 4k + 1$ for some $k \in \mathbb{Z}^+$ and that the first player has a winning strategy in all other cases.

4.3 Applications of Induction: Number Theory

With a firm grasp on induction, we proceed to prove key results in the realm of number theory. While there are certain tools that appear early and often in elementary number theory that we have already introduced (divisibility, primes and parity), others have been withheld from discussion until this point (with the exception of the Quotient-Remainder Theorem). You undoubtedly are familiar with them, however, as they are critical pieces to understanding the integers. We begin with a tool first encountered when expanding binomial expressions.

Definition 4.18. For $n, r \in \mathbb{Z}$ with $0 \leq r \leq n$, the *binomial coefficient* or *r-combination* is given by

$$\binom{n}{r} = \frac{n!}{r!(n-r)!}.$$

This basic counting tool, pronounced "n choose r," represents the number of ways r objects can be selected from a total of n objects, if the order in which the objects are selected does not matter. For example, if you are allowed to

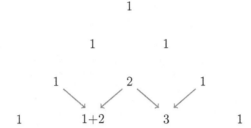

FIGURE 4.4
Pascal's Triangle

choose two pieces of fruit to pack in your lunch and you must choose from one banana, one apple and a single orange, there are

$$\binom{3}{2} = 3$$

ways you can select the fruit: banana and apple, banana and orange, or apple and orange.

Because basic probability and counting go hand-in-hand, it is not surprising that a result involving combinations is named after Pascal, the aforementioned father of probability [11]. Lemma 4.19 is often called Pascal's Formula, and you are surely familiar with it, even if its formulaic presentation appears new. Expansion of binomial expressions is often first taught using Pascal's Triangle. Recall that to construct the triangle, a term in the nth row is gotten by summing the two terms above it in the $(n-1)$st row, as in Figure 4.4. Lemma 4.19 is the formulation of this process, as the rth term in the nth row of Pascal's Triangle (starting our counting at 0 for both the rows and columns) is $\binom{n}{r}$.

Lemma 4.19. (*Pascal's Formula*) *Let* $n, r \in \mathbb{Z}$ *with* $0 \leq r < n$*. Then,*

$$\binom{n}{r} = \binom{n-1}{r-1} + \binom{n-1}{r}.$$

While we leave the proof of Lemma 4.19 as an exercise, we intuitively justify it with the following argument. The left side of the formula, $\binom{n}{r}$, represents how many ways r objects can be chosen from a total of n objects. Imagine such a scenario in choosing a team of r people from a room filled with n people, of which you are one. You, being one of the potential team members, are either on the team or not on the team. In the case that you are on the team, $r-1$ additional teammates must be selected from the other $n-1$ people. This can be done in exactly $\binom{n-1}{r-1}$ ways. Yet, if you are not a member of the team, from the remaining $n-1$ people in the room, all r members of the team must be chosen, which can be done in $\binom{n-1}{r}$ ways. How many total ways can we then select the team? Sum the number of ways with you on the team and the number of ways with you not on the team.

As previously mentioned, Pascal's Triangle is one way to expand a binomial, and Pascal's Formula is intimately related to it. Hence, it should be no surprise that Lemma 4.19 plays a critical role in the proof of the Binomial Theorem, the explicit formula for expanding a binomial expression. You may find yourself overwhelmed when first reading the proof of Binomial Theorem, particularly in proving the inductive step. The majority of the work in the proof is simple algebraic manipulation, using properties of sums and exponent laws. To clarify how the proof works, you are encouraged to recreate the proof with all sums written out using long-hand notation rather than condensed summation form. This may help you see why certain steps took place in the proof.

Presentation of a proof that at first glance seems so involved is a good opportunity to discuss how to *learn* a proof, not just memorize it (you will surely forget it as quickly as you memorized it). Before we do that, why is digesting and understanding the proof of such a result even worthwhile? Should you not be learning how to write proofs so that you can write *your* proofs of *your* results? Of course, that is ultimately the end-goal, but to get to the point where you are proposing results and then proving them, you need to understand the foundations on which the mathematics is built.

Seeing how others have proven results builds your collection of techniques and approaches; your mathematical repertoire expands with every new proof you encounter. The *why* and *how* of an argument are oftentimes as important, or moreso, than the *what* of an argument. Learning proofs builds your appreciation for the discipline: where it has come from and in what direction it is going. Simply put, fully comprehending proofs of others' results makes you a better thinker, a better puzzle-solver and an all-around better mathematician.

So how do you go about *learning* a proof beyond simply reading it and thinking to yourself, "Yes, I believe that step is true, and I understand why that step is there." Let us take the proof of the Binomial Theorem below. While you may be able to read through it and follow every step, after one read-through, you are unlikely to be able to prove the result yourself (i.e., in your own words, not simply regurgitating what is presented below). To *understand* the proof, you need to follow it at multiple levels. First, you should understand the big picture of the proof. What technique is used? What are the main steps of the proof and what are the assumptions of each of those steps? From there, you need to begin to understand the finer details of each major step. What is actually being shown at a particular point in the proof?

To get at these levels of understanding, you are encouraged to not just read and reread the proof, but rather, to work through the proof and completely and explicitly justify every step of the algebraic process. Follow along with the proof by *writing it down*, even if in abbreviated form, and put in all the gory details that do not appear in the pristine final version of the proof: basic algebra, citations and justifications. In doing so, you will much more soundly follow the logic of the proof, and the process of putting pen-to-paper will help the mind make connections with the material [30]. Do this once and you will find yourself understanding leaps and bounds beyond what you would obtain via a quick reading. With every reworking of the proof, you will make it more and more your own, until you will be able to take ownership of the resulting work. While the proof idea is not yours, the proof itself has become something that is.

Theorem 4.20. (The Binomial Theorem) *Let $x, y \in \mathbb{R}$ and $n \in \mathbb{Z}^+$. Then,*

$$(x+y)^n = \sum_{i=0}^{n} \binom{n}{i} x^{n-i} y^i.$$

Proof Let $x, y \in \mathbb{R}$. To prove the result, we proceed by mathematical induction on $n \in \mathbb{Z}^+$. Because

$$\binom{1}{0} = 1 = \binom{1}{1},$$

the base case holds:

$$x + y = \sum_{i=0}^{1} \binom{n}{i} x^{1-i} y^i.$$

Assume then for a general $n \in \mathbb{Z}^+$ that the result holds and consider the $(n+1)$st case:

$$(x+y)^{n+1} = (x+y)(x+y)^n$$
$$= (x+y) \sum_{i=0}^{n} \binom{n}{i} x^{n-i} y^i \text{ (IH)}$$
$$= x \sum_{i=0}^{n} \binom{n}{i} x^{n-i} y^i + y \sum_{i=0}^{n} \binom{n}{i} x^{n-i} y^i$$
$$= \sum_{i=0}^{n} \binom{n}{i} x^{n-i+1} y^i + \sum_{i=0}^{n} \binom{n}{i} x^{n-i} y^{i+1}.$$

Substituting $i = k - 1$ in the second sum and then replacing its index variable k with i yields

$$
\begin{aligned}
(x+y)^{n+1} &= \sum_{i=0}^{n} \binom{n}{i} x^{n-i+1} y^i + \sum_{k=1}^{n+1} \binom{n}{k-1} x^{n-(k-1)} y^{(k-1)+1} \\
&= \sum_{i=0}^{n} \binom{n}{i} x^{n+1-i} y^i + \sum_{k=1}^{n+1} \binom{n}{k-1} x^{n+1-k} y^k \\
&= \sum_{i=0}^{n} \binom{n}{i} x^{n+1-i} y^i + \sum_{i=1}^{n+1} \binom{n}{i-1} x^{n+1-i} y^i.
\end{aligned}
$$

In order to combine the two sums into one, pull out the first and last terms from the first and second sum, respectively:

$$
(x+y)^{n+1} = \binom{n}{0} x^{n+1} + \sum_{i=1}^{n} \binom{n}{i} x^{n+1-i} y^i + \sum_{i=1}^{n} \binom{n}{i-1} x^{n+1-i} y^i + \binom{n}{n} y^{n+1}.
$$

This gives

$$
\begin{aligned}
(x+y)^{n+1} &= x^{n+1} + \left(\sum_{i=1}^{n} \left[\binom{n}{i} + \binom{n}{i-1} \right] x^{n+1-i} y^i \right) + y^{n+1} \\
&= x^{n+1} + \left(\sum_{i=1}^{n} \binom{n+1}{i} x^{n+1-i} y^i \right) + y^{n+1} \quad \text{(Lemma 4.19)} \\
&= \sum_{i=0}^{n+1} \binom{n+1}{i} x^{n+1-i} y^i,
\end{aligned}
$$

as was to be shown, consequently proving the result. $\square$

You may find yourself thinking the next result is obvious and "of course it's true," you may find yourself thinking. Its proof, though, is not as simple. You also may be asking, "Why would such a thing even be necessary?" Surprisingly, the result is a fundamental tool for showing the existence of things in various areas of mathematics.

Theorem 4.21. (Well-Ordering Principle) *Every nonempty subset of positive integers has a least element.*

Proof Let $S \subseteq \mathbb{Z}^+$ be nonempty. We proceed by strong induction on the statement $P(n)$:

For $n \in \mathbb{Z}^+$, if $n \in S$, then S has a least element.

Base case ($n = 1$): If $1 \in S$, then 1 is the least element of S since $1 \leq m$ for any $m \in \mathbb{Z}^+$. Hence, $P(1)$ is true.

Inductive hypothesis: Assume for a general $n \in \mathbb{Z}^+$ that $P(k)$ is true for all $k \in \mathbb{Z}^+$, $k \leq n$.

Inductive step: Assume $n + 1 \in S$. To show S has a least element, we consider two cases: S contains an element smaller than $n + 1$ or it does not. Showing $P(n + 1)$ is true is equivalent to showing that S has a least element in either of these cases; we begin with the latter of the two.

Case 1: Suppose S has no element less than $n+1$, That is, for every $k \in \mathbb{Z}^+$, $k < n + 1$, we have that $k \notin S$. In this case, then, $n + 1$ is the least element of S, showing $P(n + 1)$ is true.

Case 2: Assume for some $k \in \mathbb{Z}^+$ with $k \leq n$, $k \in S$. By the inductive hypothesis, then, $P(k)$ is true, meaning that S has a least element. Hence, $P(n + 1)$ is true.

In either case, $P(n + 1)$ is true, and by strong induction, every nonempty subset of positive integers has a least element. $\square$

At the end of the last section, we saw that the Principle of Mathematical Induction and Principle of Strong Mathematical Induction are equivalent; anything provable with one of them is provable with the other. It turns out that the Well-Ordering Principle is an equivalent statement to both as well! Anything that can be proven with induction can be proven using the Well-Ordering Principle, and vice versa. The proof of this is outlined in the exercises.

Now, we shift our attention to previously stated results, beginning with the Quotient-Remainder Theorem of Section 3.5, whose proof was postponed until now. The theorem is actually two claims: the existence of particular integers and the uniqueness of them. As mentioned before, the Well-Ordering Principle is often used in proving the existence of particular quantities, and it is the tool needed to prove Theorem 3.42.

Theorem 3.42 (Quotient-Remainder Theorem) *Let $a, b \in \mathbb{Z}$ with $b > 0$. Then there exist unique integers q and r, with $0 \leq r < b$ such that $a = bq + r$.*

Proof Suppose $a, b \in \mathbb{Z}$ with $b > 0$. We prove there exist unique integers q and r, with $0 \leq r < b$ such that $a = bq + r$ first by showing the existence of such integers and then showing that they are in fact unique.

Existence: Consider the set

$$S = \{a - mb \mid m \in \mathbb{Z} \text{ and } a - mb \geq 0\}.$$

Note that S is nonempty, because

$$a - (-|a|b) = a + |a|b$$
$$\geq a + |a| \ (\text{ since } b \geq 1)$$
$$\geq 0 \ (\text{Lemma 3.38}).$$

Hence, the Well-Ordering Principle implies that S has a least element r. By the definition of S, we know then that $r = a - qb$ for some $q \in \mathbb{Z}$, and that $0 \leq r$. We claim also that $r < b$. Suppose, for contradiction, that $b \leq r$. Then,

$$a - (q+1)b = a - qb - b$$
$$= r - b$$
$$\geq 0,$$

showing that $a - (q+1)b \in S$. But

$$a - (q+1)b < a - qb$$
$$= r,$$

a contradiction to the minimality of r in S. Hence, $0 \leq r < b$, as required to prove the desired existence.

Uniqueness: To prove the uniqueness of q and r, assume that

$$a = q_1 b + r_1$$

and

$$a = q_2 b + r_2,$$

with $q_i, r_i \in \mathbb{Z}$ and $0 \leq r_i < b \ (i = 1, 2)$. Then,

$$-b < r_1 - r_2 < b,$$

and

$$r_1 - r_2 = (a - q_1 b) - (a - q_2 b)$$
$$= q_2 b - q_1 b$$
$$= (q_2 - q_1)b.$$

Substituting,

$$-b < (q_2 - q_1)b < b,$$

or equivalently,

$$-1 < q_1 - q_2 < 1.$$

But since $q_1 - q_2 \in \mathbb{Z}$ (by closure), we have that $q_1 - q_2 = 0$ and that $q_1 = q_2$. Then,

$$r_1 = a - q_1 b$$
$$= a - q_2 b$$
$$= r_2,$$

proving the uniqueness of both q and r.   □

Another previously stated (and proven) result is Theorem 3.20. It was proven directly, with a key step of the proof being the claim, "the process must stop." While this claim was valid and was the critical observation in the proof, to some, it may not have "felt right." Such a bold assertion surely needs justification, but none was offered. Hence, we put those concerns to rest and provide here an alternate proof utilizing the Well-Ordering Principle.

Theorem 3.20. *Every positive integer greater than 1 has a prime divisor.*

Proof To show that every positive integer greater than 1 has a prime divisor, define the set S as

$$S = \{x \in \mathbb{Z}^+ \mid x > 1 \text{ and } x \text{ does not have a prime divisor}\}.$$

For contradiction, let us assume that S is nonempty. Thus, because S is a nonempty subset of of the positive integers, the Well-Ordering Principle insinuates the existence of a least element s of S. Note that s is not prime, else it would have a prime divisor (itself). Hence, s must be composite. Write, by Lemma 3.19,

$$s = nm,$$

where $n, m \in \mathbb{Z}^+$, $1 < n \leq m < s$. In particular, because s is the least element of S, $n \notin S$. Because $n > 1$, then, the definition of S tells us that n has a prime divisor p. But, $p|n$ and $n|s$, so that $p|s$ (Theorem 3.17), a contradiction to s having no prime divisors. Thus, it must be that S is empty, so that every positive integer greater than 1 does indeed have a prime divisor.

□

Prime divisors are of the utmost importance when considering a single integer, but when considering two integers (at least one of which is nonzero), we often want to know what divisors they have in common, and of those, which is the greatest. Hence, we define the *greatest common divisor*, a concept you are undoubtedly familiar with. The specific definition of it follows from the intuitive notion of the phrase "greatest common divisor" itself. To be *the* greatest common divisor or two integers, a number must be both a common divisor (that is, a divisor of each) and the greatest such number.

Definition 4.22. Let $a, b \in \mathbb{Z}$, at least one of which is nonzero. An integer d is called a *common divisor* of a and b if $d|a$ and $d|b$. A common divisor d of a and b is called the *greatest common divisor* of a and b, denoted by $d = \gcd(a, b)$, if $d \geq c$ for any other common divisor c of a and b. If $\gcd(a, b) = 1$, then a and b are said to be *relatively prime*.

Example 4.4. The integers 20 and 32 have as common divisors

$$\pm 1, \ \pm 2, \ \text{and} \ \pm 4.$$

Of these, 4 is the greatest and consequently, $\gcd(20, 32) = 4$. The integers 20 and 21 have only ± 1 as common divisors, making them a relatively prime pair of integers.

Were 20 and 21 in the previous example just two consecutive integers that *happen* to have a greatest common divisor of 1? Could there possibly be a pair whose greatest common divisor is greater than 1?

As you have familiarized yourself with different approaches to proofs, when questions like the previous one are posed, you are also learning how to attack them. How so? Rather than simply concluding that the statement was either true or false, you probably considered other specific pairs of consecutive integers. Finding that their greatest common divisor was also 1, you chose another pair. Perhaps you tried to conjure up some property that was causing all the chosen pairs to have a greatest common divisor of 1. Regardless, after a few failed attempts at finding a counterexample, you may begin thinking the statement *is* true: any two consecutive integers *do* have a greatest common divisor of 1. What is the next step in the mathematical attack? To brainstorm a possible proof.

With consecutive integers given by n and $n + 1$, a proof technique should jump out at you: contradiction. In assuming that n and $n + 1$ have a greatest common divisor *greater* than 1, we introduce a new tool; call this greatest common divisor d. This integer is both a common divisor of n and $n + 1$, and it is the greatest such common divisor. Which of these facts help us reach a contradiction?

In looking for a contradiction, it is often wise to remember other results related to the information at hand. For example, in our case, we know that $d|n$ and $d|(n+1)$. Theorem 3.31 has a similar notion in its proof. What happened in that proof? We called upon Lemma 3.16, and doing that here tells us that

$$d|[(n + 1) - n],$$

or equivalently, $d|1$. Our contradiction has revealed itself!

Lemma 4.23. *Any two consecutive integers are relatively prime.*

Proof Let $n \in \mathbb{Z}$ and suppose $d \in \mathbb{Z}$ with $d|n$ and $d|(n+1)$. By Lemma 3.16, we know then that d divides $(n+1) - n = 1$, and consequently by Theorem 3.15 that $d = \pm 1$. Thus, $\gcd(n, n+1) = 1$, showing that n and $n+1$ are relatively prime. $\square$

One thing that makes the greatest common divisor of two integers particularly useful is the following result. While it is a nice result, it is much more of a tool. In countless results, it is the crux that makes the proof work. [8]

Theorem 4.24. *Let $a, b \in \mathbb{Z}$, at least one of which is nonzero. If $d = \gcd(a, b)$, then there exist $x, y \in \mathbb{Z}$ so that*

$$d = ax + by.$$

Proof Suppose $a, b \in \mathbb{Z}$, supposing, without loss of generality, a is nonzero, with $d = \gcd(a, b)$. Consider then the set

$$S = \{ax + by \mid x, y \in \mathbb{Z} \text{ and } ax + by > 0\}.$$

S is nonempty, because $a^2 + b^2 \in S$ and at $a \neq 0$. Then, the Well-Ordering Principle guarantees a least element

$$d = ax + by$$

of S. We show that $d = \gcd(a, b)$ by showing first that d is a common divisor of a and b, and then, that it is the greatest of all common divisors.

First, by the Quotient-Remainder Theorem,

$$a = dq + r$$

for some $q, r \in \mathbb{Z}$ and $0 \leq r < d$. Then,

$$
\begin{aligned}
r &= a - dq \\
  &= a - (ax + by)q \\
  &= a(1 - qx) + b(-yq) \\
  &\in S, \text{ by closure and the definition of } S,
\end{aligned}
$$

a contradiction to the minimality of d in S, unless $r = 0$. Since it must be that $r = 0$ we have that $a = dq$, so that d is a divisor of a. Likewise, we can show $d|b$, proving that d is a common divisor of a and b.

To show that d is the greatest common divisor, let c be any positive common divisor of a and b (if $c < 0$, then $c < d$ since $d > 0$). Then,

$$c|(ax + by)$$

---

[8]This result says that the greatest common divisor of two integers can be written as a linear combination of those two integers.

by Exercise 3.2.6. Lemma 3.14 tells us that $c \leq d$, proving $d = \gcd(a, b)$. $\quad \square$

Results such as Theorem 4.24 are often called *existential-* or *existence*-type results. This theorem says that particular integers exist, but it gives us no particular way to find them. For example, $\gcd(56, 266) = 14$. Thus, there are integers x and y so that

$$14 = 56x + 266y.$$

What are those integers? Neither the theorem nor its proof gives us insight into finding them. All we know is that x and y exist.[9]

The Well-Ordering Principle is a key component of each of the previous two proofs. When reading over them, it comes across as a very straightforward tool to use. However, when developing a proof like these from scratch, it can be challenging to define the set on which you want to apply the Well-Ordering Principle. How did we know to let S be defined in *that* way in the proof of Theorem 4.24?

Staying aware of what it is you are trying to prove is crucial. We knew that we wanted to show d was of the form $ax + by$. Knowing this pointed our scratch work in the right direction (that is, defining S correctly). If you are stuck in your proof development, remind yourself of what it is you are trying to prove.

If two integers a and b are relatively prime, Theorem 4.24 tells us that we can find $x, y \in \mathbb{Z}$ so that $ax + by = 1$. It turns out that the converse of this is true as well.

Theorem 4.25. *Two integers a and b, with at least one nonzero, satisfy* $\gcd(a, b) = 1$ *if and only if there exist $x, y \in \mathbb{Z}$ so that*

$$ax + by = 1.$$

Proof Let $a, b \in \mathbb{Z}$ with at least one nonzero. Theorem 4.24 says that if $\gcd(a, b) = 1$, then there exist $x, y \in \mathbb{Z}$ so that $ax + by = 1$. Hence, let us assume that there exist $x, y \in \mathbb{Z}$ so that

$$ax + by = 1$$

and show that a and b are relatively prime.

If $d = \gcd(a, b)$, then by Exercise 3.2.6,

$$d | (ax + by),$$

or equivalently, $d | 1$. Theorem 3.15 tells us that $d = \pm 1$, so that $\gcd(a, b) = 1$. $\square$

---

[9]The Euclidean Algorithm can be used to find the integers x and y of Theorem 4.24. For the example shown, $14 = 5 \cdot 56 + -1 \cdot 266$.

The following powerful result, attributed to Euclid, could be considered a corollary to Theorem 4.25, yet the result is strong enough to stand alone.[10]

Theorem 4.26. (Euclid's Lemma) *If p is a prime and $p|ab$ for $a, b \in \mathbb{Z}$, then $p|a$ or $p|b$.*

Proof Let $p, a, b \in \mathbb{Z}$ with p prime and $p|ab$. If $p|a$, then the result holds. Else, $\gcd(p, a) = 1$, since the only divisors of p are 1 and p. By Theorem 4.25,

$$1 = px + ay$$

for some $x, y \in \mathbb{Z}$. Then,

$$b = b \cdot 1$$
$$= b(px + ay)$$
$$= pbx + aby.$$

Because $p|p$ and $p|ab$, Exercise 3.2.6 tells us that

$$p|(pbx + aby)$$

and consequently $p|b$, proving the result.   □

The following result exhibits a second type of corollary.[11] We have seen corollaries where the result follows immediately from the statement of a previous result, such as the example given in the introduction to Chapter 3 or the to-be-seen Corollary 5.7. Here, it is the *proof* of Corollary 4.27 that follows immediately from the *proof* of Theorem 4.26. The approach of the proof is identical to that of Theorem 4.26.

When this is the case, a corollary may be presented without a proof. The author assumes the reader is astute enough to follow the parallel line of attack. To illustrate the similarities, however, we fully prove Corollary 4.27 as if Theorem 4.26 had not yet been proven.

Corollary 4.27. *Let $a, b, c \in \mathbb{Z}$ with $c|ab$ and $\gcd(a, c) = 1$. Then, $c|b$.*

Proof Assume $a, b, c \in \mathbb{Z}$ with $c|ab$ and $\gcd(a, c) = 1$. We know that there exist $x, y \in \mathbb{Z}$ so that

$$ax + cy = 1.$$

---

[10]If Theorem 4.26 is strong enough to not be listed as a corollary to Theorem 4.25, why is it called Euclid's Lemma and not Euclid's Theorem? Many texts refer to the Euclidean Algorithm as Euclid's Theorem, and so not to confuse the two, Theorem 4.26 is called Euclid's Lemma. You will find, however, places in the literature where it is referred to as Euclid's *First* Theorem.

[11]Some texts will swap the roles of Theorem 4.26 and Corollary 4.27, calling the latter the theorem and the prior the corollary.

Multiplication of this by b yields

$$abx + bcy = b.$$

We know that c divides both ab and bcy. Employing Exercise 3.2.6 (as $c|ab$, and consequently $c|abx$, and $c|bcy$), we get that

$$c|(abx + bcy),$$

or, $c|b$, as was to be shown. □

Theorem 3.20 shows that every positive integer greater than 1 has a prime factor. Repeatedly applying this result allows us to write any such positive integer as a product of primes. Since first learning about prime factorization, you have referred to such a representation as *the* prime factorization rather than *a* prime factorization.

Did you ever question whether such a factorization is unique? You will not have to after proving Theorem 4.29, referred to as the Fundamental Theorem of Arithmetic. It was known by the ancient Greeks, with versions appearing in Euclid's *Elements*. Yet, it is commonly attributed to Carl Friedrich Gauss (1777-1855), often considered amongst the greatest mathematicians of all time [17]. It is in his *Disquisitiones Arithmeticae* that the current notation and proof of the theorem appear [12]. Before stating and proving the Fundamental Theorem of Arithmetic, we present the following necessary definition.

Definition 4.28. The *canonical form prime factorization* of a positive integer $a > 1$ is

$$a = p_1^{m_1} p_2^{m_2} \ldots p_n^{m_n},$$

where each p_i is prime, $m_i \in \mathbb{Z}^+$ $(i = 1, 2, \ldots, n)$, and

$$p_1 < p_2 < \cdots < p_n.$$

The canonical form prime factorization of 169,400 is $2^3 \cdot 5^2 \cdot 7 \cdot 11^2$. We are allowed to say this is *the* canonical form prime factorization because of this next result.

Theorem 4.29. (The Fundamental Theorem of Arithmetic) *Every positive integer greater than 1 has a unique canonical form prime factorization.*

Proof To prove that every positive integer greater than 1 has a unique canonical form prime factorization, we proceed by first proving the existence of such a factorization and then showing the uniqueness of it.

Existence: Using strong induction, we show every positive integers n greater than 1 has a canonical form prime factorization. The base case, $n = 2$, is trivial, because

$$2 = 2^1$$

is a canonical form prime factorization. Assume then for our inductive hypothesis that for any $k \in \mathbb{Z}^+$, $k \leq n$, for some fixed $n \in \mathbb{Z}^+$, $n > 1$, that k has a canonical form prime factorization.

Consider $n + 1$. If $n + 1$ is prime, just as in our base case, $(n + 1)^1$ is itself of the required form. Otherwise, by Theorem 3.20, $n + 1$ has a prime divisor, and consequently, the set

$$S = \{\text{prime divisors of } n + 1\}$$

is nonempty. The Well-Ordering Principle guarantees a least element p_1 of S. We have that $n + 1 = p_1 q$, where $q \in \mathbb{Z}$, $1 < q < n + 1$. By the inductive hypothesis, q has one of two canonical form prime factorizations:

$$q = p_1^{m_1} p_2^{m_2} \cdots p_j^{m_j},$$

or

$$q = p_2^{m_2} p_3^{m_3} \cdots p_j^{m_j},$$

where $p_1 \leq p_2$ (since if $p_2 < p_1$, then $p_2 | q$ and $q | (n + 1)$ implies $p_2 | (n + 1)$, contradicting the minimality of p_1 in S). In either case, the product $p_1 q$ can be written in canonical form prime factorization, proving the existence of such a factorization for all positive integers greater than 1.

Uniqueness: Let $n \in \mathbb{Z}^+$, $n > 1$. To prove canonical form prime factorizations are unique, note that it is enough to show that the prime factorization of n is unique; the canonical form of the factorization follows from properties of algebra. To that end, assume

$$n = p_1 \cdots p_j = q_1 q_2 \cdots q_k,$$

where p_s, q_t are prime for all s and t, and WLOG, assume $j \leq k$. By Euclid's Lemma, $p_1 | q_i$ for some i; let us assume $p_1 | q_1$. Because both p_1 and q_1 are prime, it must be that $p_1 = q_1$. Then,

$$p_2 p_3 \cdots p_j = q_2 q_3 \cdots q_k.$$

The same conclusion will hold for every prime p_j and one of the q_r terms (as before, assume $p_j = q_j$). Because $j \leq k$, this yields

$$1 = q_{j+1} q_{j+2} \cdots q_k,$$

a contradiction unless $j = k$. Consequently, we have shown the two factorizations to be the same, proving the desired uniqueness. $\square$

We conclude this section by pointing out something *missing* from the proof of Theorem 4.29. In various steps, we say things such as, "for all s and t," when technically we should say, "for all $s, t \in \mathbb{Z}$, $1 \le s \le j$ and $1 \le t \le k$, with $j, k \in \mathbb{Z}^+$."

When your audience is assumed to be keen mathematicians and the context of a statement is clear, the absence of over-explanation sometimes makes mathematics read more smoothly. The aforementioned particular line, for example, references the equations

$$n = p_1 p_2 \dots p_j = q_1 q_2 \dots q_k.$$

It is clear that j and k must be integers, as well s and t. As you develop your proof writing skills, you will learn the proper balance between too little and too much explanation.

Quality Tip 21. *Sometimes less is more.*

Exercises

1. Calculate the binomial coefficient $\binom{n}{r}$ for the given values.

   (a) $n = 7$, $r = 2$
   (b) $n = 7$, $r = 5$
   (c) $n \in \mathbb{Z}^+$, $r = 0$

2. Show that Pascal's Formula holds when $n = 10$ and $r = 4$.

3. Prove Theorem 4.19: Pascal's Formula.

4. Expand the binomial using the Binomial Theorem.

   (a) $(x + a)^4$
   (b) $(y - 3)^5$
   (c) $(1 + \sqrt{2})^3$
   (d) $(x^3 - \frac{1}{x})^5$

5. Prove each of the following.

   (a) For integers n and r with $1 \le r \le n$, the binomial coefficient

   $$\binom{n}{r}$$

   is indeed an integer.

   (b) When For $n \in \mathbb{Z}^+$,

   $$\sum_{i=1}^{n} \binom{i+1}{2} = \binom{n+2}{3}.$$

   (c) For $n \in \mathbb{Z}^+$,

   $$\sum_{i=1}^{n} i(i+1) = 2\binom{n+2}{3}.$$

(d) The Generalized Pascal's Formula, where $n \in \mathbb{Z}^+$ and $m \in \mathbb{Z}$, $m > 0$:

$$\sum_{i=1}^{n} \binom{m+i}{i} = \binom{m+n+1}{n}.$$

(e) For integers $i, n \geq 0$,

$$\sum_{k=0}^{n} \binom{k}{i} = \binom{n+1}{i+1}.$$

6. Prove each of the following results, assuming all variables represent integers as defined.

   (a) If $n, r \geq 0$,
   $$\binom{n}{r} = \binom{n}{n-r}.$$

   (b) For $n \geq 1$ and $0 \leq k \leq n$,
   $$\binom{k}{2} + \binom{n-k}{2} + k(n-k) = \binom{n}{2}.$$

   (c) For $n \geq 2$,
   $$\binom{n}{0} - \binom{n}{1} + \binom{n}{2} - \cdots + (-1)^n \binom{n}{n} = 0.$$

   (d) For $n \geq 0$,
   $$\sum_{i=0}^{n} \binom{n}{i} = 2^n.$$

   (e) If $n \geq 0$, then
   $$\sum_{i=0}^{n} 2^i \binom{n}{i} = 3^n.$$

   (f) For $n \geq 1$,
   $$\binom{2n}{n} + \binom{2n}{n+1} = \frac{1}{2}\binom{2n+2}{n+1}.$$

7. (a) Let p be prime and $a \in \mathbb{Z}$. Prove that if $p|a^2$, then $p|a$.
   (b) Let p be prime. Suppose
   $$p|a_1 a_2 \ldots a_n$$
   for integers a_i. Prove that $p|a_i$ for some i.

8. Suppose $a|b$ and $a|c$ for integers a, b, c. Show that if a and b are relatively prime, then $ab|c$.

9. Prove that $\log_2(3)$ is irrational.

10. Explain why the Well-Ordering Principle does not apply to the following sets.

    (a) $\{1, .1, .01, .001, \ldots\}$

(b) $\{x \in \mathbb{Z} \mid x^2 > 10\}$

(c) $\{x \in \mathbb{Z}^+ \mid \sqrt{x} < 0\}$

(d) $\mathbb{Q}^+$

(e) $\{x \in \mathbb{R} \mid x \notin \mathbb{Q} - \mathbb{Z}\}$

11. Use the Well-Ordering Principle to prove the following.

    (a) $\sqrt{2}$ is irrational.

    (b) 8 divides $5^{2n} - 1$ for non-negative integers n.

    (c) $8^n - 3^n$ is divisible by 5 for integers $n \geq 0$.

    (d) If $m \in \mathbb{Z}^+$, then
$$2 \nmid 2m^2 + 5m - 1.$$

12. Find the greatest common divisor of the given pair of integers.

    (a) 158; 364

    (b) 157; 16227

    (c) $p_1(p_2)^3 p_3; (p_4)^2 p_5$, where the p_i are all distinct primes.

13. Prove each of the following claims involving the greatest common divisor.

    (a) For any $n \in \mathbb{Z}$, $\gcd(n, 0) = n$.

    (b) Let $a, b \in \mathbb{Z}$ and $d = \gcd(a, b)$. Prove that if $a = dx$ and $b = dy$, then $\gcd(x, y) = 1$.

    (c) Prove that $d = \gcd(a, b)$ if and only if d is both a common divisor of a and b, and if c is any common divisor of a and b, then $c|d$.

    (d) If $n \in \mathbb{Z}^+$ and $a \in \mathbb{Z}$, then
$$\gcd(a, a + n)|n.$$

    (e) For any integer x, $\gcd(5x + 2, 7x + 13) = 1$.

    (f) If $\gcd(a, b) = 1 = \gcd(a, c)$, then $\gcd(a, bc) = 1$.

    (g) If a and b are relatively prime, then a^2 and b^2 are relatively prime.

14. What is the smallest positive integer value of a such that $2^4 \cdot 3^3 \cdot 5 \cdot 11^3 \cdot a$ is a perfect square? Explain.

15. If $a = p_1^{m_1} p_2^{m_2} \ldots p_n^{m_n}$ is written in canonical form, what is the canonical form of a^4? Explain.

16. Use the Fundamental Theorem of Arithmetic to determine how many 0's the number $7^9 30^{10} 37^{16}$, when written out in standard base 10 representation, ends in. How about 100!?

17. Prove that the Principle of Mathematical Induction, the Principle of Strong Mathematical Induction, and the Well-Ordering Principle are equivalent by proving each of the following (and recalling Theorems 4.16 and 4.17).

    (a) The Principle of Strong Mathematical Induction implies the Well-Ordering Principle.

    (b) The Well-Ordering Principle implies the Principle of Mathematical Induction.

18. Use the Well-Ordering Principle and a proof by contradiction to prove Theorem 4.2.

19. Use the Well-Ordering Principle to prove that the sum of the cubes of three consecutive positive integers is a multiple of 9.

5

Relations

In our everyday conversations, we often discuss pairs of objects. Perhaps we compare them: "University A has more students than University B," "I am older than you," "Tristan wrote a more cohesive paper than anyone else in the class." We may find ourselves describing relationships between objects: "Micah is married to Elsa," "Des Moines is the capital of Iowa," "Meetings only happen on Mondays." It is natural to pair objects based on some property or relationship. Numerous areas of mathematics parallel this same idea with a tool called *relations*.

Though the term *relations* may be new to you, they are surely something you have encountered throughout your mathematical upbringing. In algebra, inequalities provide a way to relate real numbers:

$$\sqrt{2} < 3.5.$$

In fact, given any two distinct real numbers m and n, either $m < n$ or $n < m$. In other words, m and n are *related* in exactly one of these ways. Compare this to the relational symbol $=$. Given any two real number m and n, it is *not* the case that $m = n$ or $n = m$. In the later sections of this chapter we will generalize this concept, where any two objects satisfy some relational property (as in the case of $<$).

Real-valued functions, $y = f(x)$, form a relation. For example, consider $f(x) = x^2$. We could define a rule, that a is *related* to b if and only if $f(a) = b$. Since $f(2) = 4$, we would say 2 is related to 4; we would not say 2 is related to 5. Our rule provides a definitive method for saying when two real numbers *are* or *are not* related.

In this chapter, we will define relations in the broadest possible sense and use that definition to define functions. From there, by requiring relations to fulfill certain criteria, we will define two concepts critically important to a multitude of mathematical areas: equivalence relations and order relations. Finally, we define and briefly investigate a particular relation stemming from the Quotient-Remainder Theorem.

5.1   Mathematical Relations

It is helpful to understand individual elements of a set. In doing so, we get a
strong notion for properties of the set as a whole. However, it is often critical
to understand *relationships* between elements of the set. The idea of relating
two objects, as discussed in the opening paragraphs of this chapter, is crucial
to much of mathematics. How can we go about defining, in the most general
way possible, such a concept?

We do not want to restrict what type of objects are being related. In
particular, the two objects need not be of the same type. Real numbers should
be able to be related to complex numbers, functions to sets, or cats to college
math professors. We have a tool for talking vaguely about objects: sets and
their elements. We can use sets to collect the objects we want to relate.

How then can we talk about two totally different objects? For example,
every person was born during a single month of the year. If we wanted to discuss this relationship using mathematical terms, it would not be very helpful
to consider a single set S consisting both of all people and all months. If we
did, the phrase "$x \in S$" is vague. Is x a person or a month? A better approach
would be to consider two separate sets: the set P of all people, and, the set
M of all months. With this, an expression such as

$$\forall p \in P, \, \exists \, m \in M \text{ such that } p \text{ was born in } m$$

makes complete sense. Our goal of generalizing to any pairing of any types of
objects will use this same approach: consider *two* sets.

Lastly, we must have a method to pair up objects in one of the sets with
objects in the other; in particular, we must insist that the order of the sets
matters. In regards to the previous example, we want to consistently say,
"Person p was born in month m." The set P must somehow be denoted as the
first set with M the *second*. But we have such a tool: the Cartesian product.

Let us take the previous example and consider not the set P of all people
but rather the set S of students in a particular math class and the set M of
months of the year. If we wanted to talk about the relationship given by a
student being related to the month of the year they were born in, we could
not look at the entire set $S \times M$. Why? If there is only one person named
Brad in S, there are twelve different elements of $S \times M$ of the form (Brad, m):

$$(\text{Brad}, \text{January}), (\text{Brad}, \text{February}), \ldots, (\text{Brad}, \text{December}).$$

Only one of these corresponds to the month in which Brad was born. Thus,
we only want to consider a particular *subset* of $S \times M$.

All of this discussion has led to the following definition. It is a specific
mathematical way of defining *relationships*.

Definition 5.1. Let A and B be sets. A *relation* R *from* A *to* B is a subset of $A \times B$. We say $a \in A$ *is related to* $b \in B$ if $(a, b) \in R$, sometimes written as aRb. Should $(a, b) \notin R$ we write $a\cancel{R}b$. If $A = B$, then we say R *is a relation on the set* A and it is a subset of $A \times A$.

Example 5.1. Let $A = \{1, 2, 3, 4, 5\}$ and $B = \{1, 3, 5\}$. Define R as the relation from A to B as aRb if and only if $a > b$.

Because $5 > 3$, we have $5R3$. An equivalent way of denoting this is by saying $(5, 3) \in R$. But, $3\cancel{R}5$ (or, $(3, 5) \notin R$. We can explicitly list the entire relation R:

$$R = \{(2, 1), (3, 1), (4, 1), (4, 3), (5, 1), (5, 3)\}.$$

Note that R is a *set* whose elements are themselves ordered pairs.

It is important to note what Example 5.1 exhibits. Relations are directional; aRb does not imply bRa. In fact, the latter expression may not even make sense. If B is the relation defined from the set of people in a particular math class to the set of months, where pBm if and only if person p was born in month m, then saying

$$(\text{Peter}, \text{August}) \in B$$

makes sense, whereas

$$(\text{August}, \text{Peter}) \in B$$

does not. There are a multitude of ways to define relations, and they need not be defined only on finite sets, as in Example 5.1.

Example 5.2. Consider the relation E on $\mathbb{R}$ defined by aEb if and only if $|a| = |b|$.

The relation here is defined on a single set, meaning

$$E \subseteq \mathbb{R} \times \mathbb{R}.$$

We see xEx for all $x \in \mathbb{R}$ since $|x| = |x|$ for any real number x. Similarly, $xE(-x)$ for every $x \in \mathbb{R}$ by Lemma 3.39.
Does this show that

$$E = \{(x, x) \mid x \in \mathbb{R}\} \cup \{(x, -x) \mid x \in \mathbb{R}\}?$$

We must be careful. This only shows that

$$\{(x,x) \mid x \in \mathbb{R}\} \cup \{(x,-x) \mid x \in \mathbb{R}\} \subseteq E.$$

Remember, when showing that two sets are equal, we must show that they are subsets of one another. To show that E does indeed equal this set, consider $(a,b) \in E$. We must show

$$(a,b) \in \{(x,x) \mid x \in \mathbb{R}\} \cup \{(x,-x) \mid x \in \mathbb{R}\}.$$

Because of how E is defined, we know that $|a| = |b|$. Definition 3.37 yields four possible cases: $a = b$, $a = -b$, $-a = b$, or $-a = -b$. In all cases, $b = \pm a$, showing (a,b) is indeed an element of the desired set and that the two sets are equal.

Relations may be defined between sets whose elements are completely different types of mathematical objects, as in the following example.

Example 5.3. Define R as the relation from the set A to $\mathcal{P}(A)$ by aRS if and only if $a \in S$, where

$$A = \{a,b,c,\ldots,x,y,z\}.$$

Often understanding the notation of a relation is the most difficult part of gaining a feeling for what it looks like. In this case, the relation is from the set A of lowercase letters of the English alphabet to the power set of A. The latter is a a set with 2^{26} elements, the elements of which are subsets of A. We list, using both notations of Definition 5.1, a few elements of R as well as some ordered pairs that are not in R.

(1)
Since $a \in \{a,b,c\}$, we have that

$$aR\{a,b,c\}.$$

(2) The set A is an element of $\mathcal{P}(A)$, and $e \in A$. Thus,

$$(e,A) \in R.$$

(3) For every $\alpha \in A$,

$$xR\{\alpha\}.$$

(4) Because $m \notin \{a,e,i,o,u\}$, we have that

$$m \not\!R \{a,e,i,o,u\}.$$

(5) The empty set contains no elements, so for every $\alpha \in A$,

$$\alpha \not\!R \emptyset.$$

This notation of (3) in Example 5.3 seems awkward and contradictory to some previous mathematical writing tips; why is the Greek letter α used? In this context it is used to simply help the reader realize that it represents a general element of A. We just as easily could have written, "For every $x \in A$," though this then requires the reader to note that x is a variable and not the specific element $x \in A$.

A major application of relations is in the definition of a function, which are relations satisfying particular requirements (see Section 6.1). The following terms are intuitively synonymous with the terms you have come to understand in your study of real-valued functions.

> **Definition 5.2.** Let R be a relation defined from a set A to a set B. The *domain of R* is a subset of A defined as the set
>
> $$\text{Dom}(R) = \{a \in A \mid aRb \text{ for some } b \in B\}.$$
>
> The *range of R* is a subset of B defined as
>
> $$\text{Ran}(R) = \{b \in B \mid aRb \text{ for some } a \in A\}.$$

We can think of the domain and range of a relation R in terms of the specific elements in R. The set R consists of ordered pairs and each ordered pair has a first and second coordinate. The set of all first coordinates of all elements of R is the domain of R; the set of all second coordinates of all the elements in R is the range of R. This is highlighted in the next example.

> **Example 5.4.** List the domain and range for each of the relations defined in Examples 5.1, 5.2 and 5.3.
>
> Example 5.1: We saw that
>
> $$R = \{(2,1), (3,1), (4,1), (4,3), (5,1), (5,3)\}.$$
>
> Thus,
>
> $$\text{Dom}(R) = \{2,3,4,5\} \text{ and } \text{Ran}(R) = \{1,3\}.$$
>
> Example 5.2: The relation E was defined on the set $\mathbb{R}$, and we showed
>
> $$E = \{(x,x) \mid x \in \mathbb{R}\} \cup \{(x,-x) \mid x \in \mathbb{R}\}.$$
>
> In particular, xEx for every $x \in \mathbb{R}$. Thus,
>
> $$\text{Dom}(E) = \mathbb{R} = \text{Ran}(E).$$

Example 5.3: The relation of this example was neither explicitly written out nor summarized in a condensed form, like Example 5.2. To determine its domain and range, we reason through the definition of R.

The domain of R is simple to determine. Because $\alpha R\{\alpha\}$ for every $\alpha \in A$,

$$\mathrm{Dom}(R) = A.$$

But what about $\mathrm{Ran}(R)$? The elements of $\mathrm{Ran}(R)$ are elements of $\mathcal{P}(A)$, namely subsets of A. Thus, $S \in \mathrm{Ran}(R)$ if and only if there exists $\alpha \in A$ such that $\alpha \in S$. But S is a subset of A. Which subsets of A contain an element of A? Precisely the nonempty subsets. Hence,

$$\mathrm{Ran}(R) = \{S \in \mathcal{P}(A) \mid S \neq \varnothing\}$$
$$= \mathcal{P}(A) - \{\varnothing\}.$$

Thus far we have presented relations two ways. If the relation is finite with relatively few elements, we can explicitly list it out, as in Example 5.1. If the relation contains a large number or infinitely many elements, this is not an option. In such a case it is often possible to describe the relation via some predicate:

$$xRy \text{ if and only if } P(x,y),$$

where $P(x,y)$ is written in terms of the variables x and y. Examples 5.2 and 5.3 demonstrate this. An alternative method that is sometimes available is a visual representation of the relation using its *graph*. This should come as no surprise, having mentioned that real-valued functions are a particular type of relation and the graph of a function is highly useful. We will see that the graph of a function, as you are familiar with them, is exactly the graph of the function as a relation, defined below.

Definition 5.3. Let R be a relation from a set A to a set B. The *graph* of R is the collection of points (a,b) in the relation plotted on a coordinate-plane, with the horizontal axis denoting the set A and the vertical axis representative of B.

The graph of the relation

$$R = \{(2,1),(3,1),(4,1),(4,3),(5,1),(5,3)\}$$

of Example 5.1 is pictured in Figure 5.1.

A graph of a relation can provide a geometric representation of a relation with infinitely many elements, as in Example 5.5.

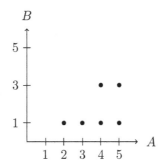

FIGURE 5.1
Graph of R from Example 5.1.

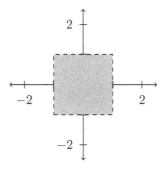

FIGURE 5.2
$S = \{(a, b) \mid |a| < 1 \text{ and } |b| < 1\}$

Example 5.5. Let S be the relation defined on $\mathbb{R}$ where aSb if and only if both $|a| < 1$ and $|b| < 1$. The relation S can be visualized on the coordinate plane as in Figure 5.2.

Before investigating one last visualization tool for relations, we define two specific types of relations. The first, the *identity relation*, is dependent upon the set on which it is defined, while the second, the *inverse relation*, is dependent upon some already defined relation.

Definition 5.4. Let A be a set. The *identity relation on A*, denoted I_A, is defined by aI_Ab if and only if $a = b$, or equivalently,

$$I_A = \{(a, a) \mid a \in A\}.$$

Is it necessary to prove that the identity relation is *actually* a relation on a set A? While we could, it is inherent from the definition:

$$\{(a,a) \mid a \in A\} \subseteq A \times A.$$

Definition 5.5. Let R be a relation defined from a set A to a set B. The *inverse relation to R*, denoted by R^{-1}, is the relation from B to A defined by $bR^{-1}a$ if and only if aRb.

While it was clear that the identity relation was indeed a relation, the same cannot be said about the inverse relation; we must show it is actually a relation. Note the importance of actually doing this. When we claim that something is some mathematical object, it is not enough to say, "it is this object because we defined it to be one." We must argue why that thing actually satisfies the definition of being such an object.

Theorem 5.6. *If R is a relation from A to B, then R^{-1} is a relation from B to A.*

Proof Suppose R is a relation from a set A to a set B. Let

$$(b,a) \in R^{-1}.$$

By Definition 5.5,

$$(a,b) \in A \times B.$$

Thus, $b \in B$ and $a \in A$. Consequently,

$$(b,a) \in B \times A,$$

showing

$$R^{-1} \subseteq B \times A,$$

proving that R^{-1} is indeed a relation from B to A.   □

As a consequence of Theorem 5.6, we have the following corollary.

Corollary 5.7. *Let R be a relation from A to B.*

    1. $\mathrm{Dom}(R) = \mathrm{Ran}(R^{-1})$.

    2. $\mathrm{Dom}(R^{-1}) = \mathrm{Ran}(R)$.

We will prove (1) and leave (2) as an exercise.

Proof Let R be a relation from A to B. To show

$$\mathrm{Dom}(R) = \mathrm{Ran}(R^{-1}),$$

we show both

$$\text{Dom}(R) \subseteq \text{Ran}(R^{-1})$$

and

$$\text{Ran}(R^{-1}) \subseteq \text{Dom}(R).$$

First, let $a \in \text{Dom}(R)$. Then, there exists $b \in B$ so that $(a, b) \in R$. Consequently, $(b, a) \in R^{-1}$, proving $a \in \text{Ran}(R^{-1})$. Hence,

$$\text{Dom}(R) \subseteq \text{Ran}(R^{-1}).$$

Likewise, let $a \in \text{Ran}(R^{-1})$. We have that $(b, a) \in R^{-1}$ for some $b \in B$. Thus, $(a, b) \in R$, giving that $a \in \text{Dom}(R)$, as required to prove

$$\text{Ran}(R^{-1}) \subseteq \text{Dom}(R),$$

and consequently, that

$$\text{Dom}(R) = \text{Ran}(R^{-1}).$$

$\square$

In the proof of Corollary 5.7, the two set-element method arguments are remarkably similar, and it seems like the proof should be able to be shortened in some way without damaging its integrity. Indeed it can be, and this is because the justifications for the steps in either part are not just conditional statements; they are biconditional results. Hence, the entire proof could be condensed to just a few lines:

Proof Let R be a relation from A to B. We know $a \in \text{Dom}(R)$ if and only if there exists $b \in B$ so that $(a, b) \in R$. This is true if and only if $(b, a) \in R^{-1}$. Consequently,

$$\text{Dom}(R) = \text{Ran}(R^{-1}).$$

$\square$

> **Quality Tip 22.** *Using biconditional results, rather than two conditional results, can sometimes make a proof more efficient.*

We conclude this section with a discussion of an area of pure mathematics that extends into countless disciplines beyond mathematics. It traces its roots back to 1736, when Leonhard Euler was asked to address a problem about the city of Königsberg in Prussia. The Pregel River ran through the town, creating two large islands (as pictured in Figure 5.3). Altogether, seven bridges connected the islands to one another and the mainland, and Euler was asked if it was possible for a person to take a walk through the city, cross each bridge exactly once, and return to his or her starting point. In proving that it was not possible, Euler created a new structure in mathematics called a *graph* and consequently an entire branch of mathematics called *graph theory* [39].

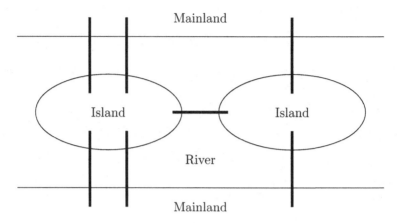

FIGURE 5.3
Bridges of Königsberg

Definition 5.8. A *graph* G is an ordered pair (V, E) of sets, with V being a nonempty set of elements called *vertices* and E a collection of unordered pairs of elements of V called *edges*. The vertices v_1 and v_2 are called the *endpoints* of the edge (v_1, v_2). If the elements of E are considered to be ordered pairs, then G is called a *directed graph* or *digraph*.

Before proceeding with our introduction to mathematical graphs, it is worth noting that these graphs are completely different mathematical structures than the graphs of functions seen throughout algebra. While they are related, take care to not generalize the results developed here to the graphs of functions.

Graphs naturally model relations between objects, with the objects being represented by the vertices of the graph. Two objects are related if and only if there is an edge between them. While this relationship is notational, in terms of sets, it is easily visualized. This visualization is the purpose for including graphs in our introduction to relations.

Example 5.6. Let G be the graph with vertex set $V = \{a, b, c\}$ and edge set $E = \{(a, b), (b, b), (b, c)\}$. The graph in Figure 5.4 represents G.

Had the graph G of Example 5.6 been directed, we simply decorate the edges with arrows to indicate direction, as in Figure 5.5. Digraphs will be helpful in visualizing relations like those of Example 5.1. In that example, $(2, 1) \in R$ but $(1, 2) \notin R$. Some method for visualizing this one way relationship is necessary, and directed edges provide just that.

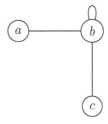

FIGURE 5.4
The unordered graph in Example 5.6

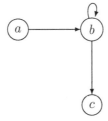

FIGURE 5.5
Example 5.6 as digraph

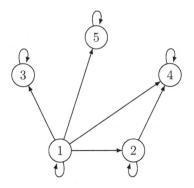

FIGURE 5.6
R defined on $A = \{1, 2, 3, 4, 5\}$

While there are ample introductory concepts from graph theory we could introduce here, that is not our primary goal. Rather, what we have is another tool for visualizing relations, and consequently, for developing proofs. Humans tend to be visual creatures, and graphs allow us to see a relation in action. When this happens, we may discover the crux of a proof or a possible counterexample to a conjecture. Thus, you can think of relations not as a tool to help you write the *final* draft of a proof but rather the *first* draft of a proof.

Example 5.7. Let R be the relation on the set

$$A = \{x \in \mathbb{Z} \mid 1 \le x \le 5\}$$

defined by aRb if and only if $a|b$. List R as a set of ordered pairs and represent it as a digraph.

We have R defined on the set $A = \{1, 2, 3, 4, 5\}$, with an element $a \in A$ related to an element $b \in A$ if and only if a divides b. Thus,

$$R = \{(1,1), (1,2), (1,3), (1,4), (1,5), (2,2), (2,4), (3,3), (4,4), (5,5)\}.$$

The relation R can be represented as in Figure 5.6. Visually, a directed edge comes from a vertex a to a vertex b if and only if $a|b$, just as defined in the relation itself.

We can use the digraph representation of a relation, as in Figure 5.6, to understand the relation without having to know anything about how the relation itself was defined. For example, we can see from the digraph that every element of the set is related to itself, and 1 is also related to every other element in the set. The elements 3, 4, and 5 are only related to themselves.

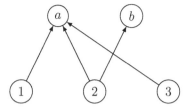

FIGURE 5.7
A bipartite digraph

In Example 5.7, the relation was defined on a single set. For such relations, the directed edges of the graph representing the relation are able to go from any vertex to any other vertex. But what if the relation was defined from a set A to a set B, where A and B have different elements? In that case, there would be no edge from an element of A to another element of A. Graphs of such relations are called *bipartite graphs*.

Definition 5.9. A graph G is called *bipartite* if its vertices can be partitioned into two disjoint sets V_1 and V_2 where the first and second coordinates of every edge of G are elements of V_1 and V_2, respectively.

Figure 5.7 is a bipartite digraph. Its vertices can be partitioned into the sets $\{1, 2, 3\}$ and $\{a, b\}$. If the edges were unordered, it would still be bipartite, but simply a bipartite graph.

Relations between certain sets naturally give rise to bipartite graphs. The domain of the relation and the range of the relation form the two partitioning sets of the graph's vertices. We must be careful, however. Quickly sketching an example such as Figure 5.7 may lead us to think that every relation can be represented with a bipartite graph. Not all can, however. The construction of an example is left as an exercise.

Can we generalize, however, the idea of bipartite graphs representing relations to general graphs? Every relation gives rise to a graph, but it is natural to ask: does every graph induce a relation? This is the notion behind our final theorem of this section.

Theorem 5.10. *Every nonempty relation R can be represented by a digraph G and vice versa.*

Proof Let R be a nonempty relation from a set A to a set B. To find a digraph representation of R, it is necessary to define the vertex set V and edge set E of such a graph. Let

$$V = \text{Dom}(R) \cup \text{Ran}(R).$$

Since R is nonempty, we know V is nonempty. Let $E = R$. Then, $e \in E$ if and only if $e \in R$, meaning the graph with vertex set V and edge set E represents R.

Now we show it is possible to define a relation from a given digraph G. If G has vertex set V and edge set E, then let R be defined on the set V with aRb if and only if $(a, b) \in E$. As before, by construction, $E = R$, so that

$$R \subseteq V \times V,$$

so that R is indeed a relation.   □

Note the use of the words "by construction" in the previous proof. When a proof builds a mathematical object to fulfill certain properties, and it is clear that this object does fulfill said properties, then such a claim is often justified as being by construction.

In the next section we consider relations satisfying particularly natural properties. These relations turn out to be extremely powerful and make appearances in almost every area of mathematics. Moreover, we will prove that they are intimately related to a concept defined in our investigation of sets.

Exercises

1. List all possible relations from A to B.

    (a) $A = \{1\}$, $B = \{m\}$
    (b) $A = \{1, 2\}$, $B = \{m\}$
    (c) $A = \{1\}$, $B = \{m, n\}$
    (d) $A = \varnothing$, $B = \mathbb{Z}$

2. Explicitly list, as ordered pairs, all elements of the given relation, then find the domain and range of the relation.

    (a) The relation R from months of the year to the set of primes, where mRx if and only if month m's name is x letters long.
    (b) If A is the set of odd integers less than 10 and B the set of even integers less than 6, the relation S from A to B given by aSb if and only if $a^2 + 1 = b$.
    (c) The relation T from $\mathbb{Z}$ to the set

    $$\{x \in \mathbb{Z} \mid 2 \leq x \leq 6\}$$

    defined by xTy if and only if x is a prime factor of y.
    (d) The relation E from $\mathbb{Z} \times Y$ to A where $(x, y)Ea$ if and only if $x - y = a$, where

    $$Y = \{y \in \mathbb{Z} \mid 2\sqrt{2} \leq y \leq 2\pi\}, \text{ and}$$
    $$A = \{0, \pm 3\}.$$

    (e) If $S = \{a, b\}$, the relation R from S to $S \times \mathcal{P}(S)$ given by $s_1 R(s_2, T)$ if and only if $s_1 \in T - \{s_2\}$.

3. List the domain and range of the given relation.

   (a) R on $\{1, 2, 3\}$ given by aRb if and only if
   $$(a, b) \in \{(1, 2), (1, 3), (3, 3)\}.$$

   (b) P from $\{1, 2, 3, 4, 5\}$ to $\{0, 2, 4, 6, 8\}$ given by aPb if and only if a is a prime divisor of b.

   (c) f from A to $\mathbb{R}$ given by $x f y$ if and only if $y = x^2$, where
   $$A = \{x \in \mathbb{R} \mid -2 < x < 2\}.$$

   (d) If B is the set of base 2 representations of positive integers, then the relation E from B to $\mathbb{Z}$ by bEx if and only if b is the base 2 representation of x and the rightmost digit of b is a 1.

4. Sketch each relation on the Cartesian plane.

   (a) $L = \{(x, y) \in \mathbb{Z} \times \mathbb{Z} \mid x < y\}$
   (b) $E = \{(x, y) \in \mathbb{R} \times \mathbb{R} \mid x \leq y + 2\}$
   (c) $P = \{(x, y) \in \mathbb{R} \times \mathbb{R} \mid xy > 0\}$
   (d) $A = \{(x, y) \in \mathbb{R} \times \mathbb{R} \mid xy = 0\}$

5. Draw the digraph for each relation in Exercise 2.

6. Define a relation that can be represented via each of the following digraphs.

   (a)

   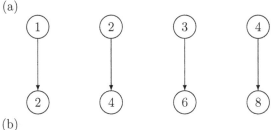

   (b)

   (c)

   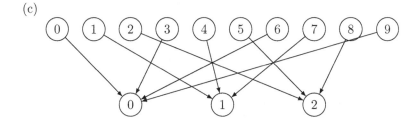

(d)

7. Give an explicit proof that the identity relation is indeed a relation. Describe its digraph representation.

8. Prove that the *complete relation* C on any nonempty set A, defined as aCb for all $a, b \in A$, is indeed a relation. Describe its digraph representation.

9. Prove the second principle of Corollary 5.7.

10. If R is any relation on a set A, prove that $(R^{-1})^{-1} = R$.

11. Let D be defined from A to B by aDb if and only if $a|b$, where

$$A = \{x \in \mathbb{Z} \mid -5 \le x \le 5\}, \text{ and}$$
$$B = \{x \in \mathbb{Z} \mid 1 \le x \le 10\}.$$

   (a) List all elements of D.
   (b) List all elements of D^{-1}.
   (c) Prove or disprove: If xDy and $yD^{-1}x$, then $x = y$.

12. Give an example of a relation whose graphical representation is not bipartite. Explain your reasoning. Can you generalize your specific example to a general property, so that if a relation satisfies this property, then its graphical representation is not bipartite?

13. If A and B are any sets, is there always a relation from A to B? Prove or disprove.

14. If A is a set with n elements and B is a set with m elements, where $n, m \in \mathbb{Z}^+$, how many relations are there from A to B?

15. Define relations R and S on $\mathbb{Z}$ by xRy if and only if $y = x^2$ and xSy if and only if $y^2 = x$. Prove or disprove:

   (a) For every $x \in D$, there exists a unique $y \in D$ such that xRy.
   (b) For every $x \in D$, there exists a unique $y \in D$ such that xSy.

16. Suppose R_1 is a relation from a set A to a set B, and R_2 is a relation from B to a set C. Define the *composition* $R_1 \circ R_2$ as a relation from A to C as the set

$$\{(a, c) \in A \times C \mid (a, b) \in R_1 \text{ and } (b, c) \in R_2 \text{ for some } b \in B\}.$$

(a) Let R be the relation defined from $\{1, 2, 3, 4, 5\}$ to $\mathbb{Z}^+$ by aRb if and only if $2a = b$, and S the relation from $\mathbb{Z}^+$ to $\mathbb{Z}$ by aSb if and only if $a^2 = b$. Find all elements of the relation $R \circ S$.

(b) Prove that the composition of two relations is indeed a relation.

(c) Describe how to form the digraph representation of $R_1 \circ R_2$ given the digraph representations of R_1 and R_2.

(d) Prove or disprove: the relation $<$ defined on $\mathbb{Z}$ is can be composed with itself.

17. A relation R on a set A is *reflexive* if for all $a \in A$, $(a, a) \in \mathbb{R}$. It is *irreflexive* if $(a, a) \notin R$ for every $a \in A$. The relation R is *symmetric* if whenever $(a, b) \in R$, then $(b, a) \in R$. It is *asymmetric* if $(a, b) \in R$ implies $(b, a) \notin R$. Let $A = \{1, 2, 3, 4\}$.

(a) Are the following relations R on A reflexive, irreflexive, symmetric, or asymmetric?

   i. $\{(1, 1), (1, 2), (2, 1), (2, 2), (3, 3), (4, 4), (1, 4), (4, 1)\}$
   ii. $\{(1, 3), (3, 1), (2, 1), (2, 1)\}$
   iii. $\{(1, 2), (2, 2), (2, 3)\}$
   iv. $\{\}$
   v. $A \times A$

(b) Define a relation on A that is neither reflexive nor irreflexive, or explain why such a relation is not possible.

(c) Define a relation on A that is neither symmetric nor asymmetric, or explain why such a relation is not possible.

18. If R is a relation on $\mathbb{Z}$, is R also a relation on $\mathbb{Q}$? $\mathbb{R}$? In general, if $A \subseteq B$ and R is a relation on A, prove or disprove that R can also be considered as a relation on B.

19. If $A \subseteq B$ and R is a relation on B, define the *restriction* of R to A as the relation $R|_A$ on A given by $aR|_A b$ (for $a, b \in A$) if and only if aRb.

(a) Let R be the relation on $\mathbb{Z}$ given by aRb if and only if $a|b$. If
$$A = \{x \in \mathbb{Z} \mid 1 \le x \le 10\},$$
find list all elements, as ordered pairs, of $R|_A$.

(b) Prove that $R|_A$ is indeed a relation.

5.2   Equivalence Relations

Let us consider a particular relation R on the set of integers, defined by aRb if and only if a and b have the same parity. What properties can we observe about

R and the corresponding digraph representing it? There are three particular ones comprising the key definition of this section: an *equivalence relation*.

First, can we determine what other integers an arbitrary integer x is related to? If we knew whether x is even or odd, then we could conclude *all* the integers x is related to. But, x is arbitrary. There is only one integer we can definitively say x is related to: itself. That is, xRx for every $x \in \mathbb{Z}$.

Not every relation on a set has the property that each element is related to itself. The less than relation $<$ defined on $\mathbb{Z}$ has the property that *no* integer x satisfies $x < x$. For those relations, such as R, that do have this property, what does this mean in terms of the digraph representing it? Since an element is related to itself, there must be an edge from every vertex to itself.

Now that we have considered a single arbitrary integer, what if we consider two arbitrary integers, say x and y, such that one is related to the other? That is, if xRy, what else do we know? To start, we know that x and y are both even or both odd but there is no way to determine from the given information which case it is. It may seem that there is nothing else that we know, but keep in mind that relations are in a sense *directional*. For example, the relation $<$ on $\mathbb{Z}$ only works one way. We know that $2 < 3$ but $3 \not< 2$. Some relations, however, work both ways. We know that x and y have the same parity (i.e., xRy). Thus, we know that y and x have the same parity (i.e., yRx). This is true for every pair of integers x and y; if xRy, then yRx.

What does this property mean in terms of the digraph of such a relation? If x is related to y, then there must be an edge from vertex x to vertex y. But since y is also related to x, there also must be an edge from vertex y to vertex x. That is, between two distinct vertices of the digraph, there cannot be a single edge; there is either no edge or exactly two edges between the vertices.

Lastly, let us consider three particular integers x, y and z. In this scenario, let us suppose xRy and yRz. Thus, x and y have the same parity, as do y and z. Logically, then, x and z must have the same parity. This relation R satisfies what we might call *transitivity*, based on previous uses of the word. What does this mean for the digraph representing R? If there is an edge from x to y and an edge from y to z, then there must be an edge from x to z. That is, the digraph satisfies a triangular property.

Relations that exhibit such properties are called *equivalence relations* and are critically important to numerous areas of mathematics. Later in this section we will discuss the usefulness of their corresponding digraphs.

Definition 5.11. A relation R defined on a set A is called an *equivalence relation* if and only if it satisfies all three of the following properties:

    1. R is *reflexive*: for every $a \in A$, aRa.

    2. R is *symmetric*: for every $a, b \in A$, if aRb, then bRa.

    3. R is *transitive*: for every $a, b, c \in A$, if both aRb and bRc, then aRc.

It is important to note that last two properties listed in Definition 5.11 are *conditional* statements. Classifying a relation as symmetric does not mean that every pair of distinct elements of A are related. Such a definition would read, "For every $a, b \in A$, aRb and bRa." Rather, it says, "*if* two are related, that is, aRb, *then* it must be true that bRa." A similar sort of reasoning follows for the transitive property of equivalence relations.

The following examples exhibit how to prove relations are indeed equivalence relations. In short, if you are asked to prove a relation R is an equivalence relation, then you are being told that R is itself a relation. You need not actually begin the proof by showing R is a relation (unless the problem explicitly asks for it). Rather, you have to show R satisfies three things: it is reflexive, symmetric and transitive. This makes your proof naturally broken up into three parts. Be clear with your labeling, so that your reader can easily follow your train of thought. We begin with a fairly straightforward lemma, proving that the I_A on any set A is an equivalence relation.

Lemma 5.12. *The identity relation I_A on any set A is an equivalence relation.*

Let us dissect the scratch work for proving this result. To prove I_A is reflexive, we must show $aI_A a$ for any $a \in A$. To begin, take an arbitrary but general $a \in A$. How can we conclude that a is related to itself? To do this, we need to understand how the relation is defined. In this lemma, elements are only related to themselves; no two distinct elements are related. Thus, reflexivity follows immediately.

To show I_A is symmetric, we begin by assuming two general elements a and b of A are related; $aI_A b$. We must derive then that $bI_A a$. Again, being intuitively familiar with the relation is key. Here, saying $aI_A b$, or equivalently $(a, b) \in I_A$, means

$$(a, b) \in \{(a, a) \mid a \in A\}.$$

Thus, it must be that $a = b$. This yields that I_A is symmetric.

Lastly, to prove that I_A is transitive, we assume $aI_A b$ and $bI_A c$ and show that $aI_A c$. As in the symmetric case, $aI_A b$ means that $a = b$, and likewise, $bI_A c$ means $b = c$. Hence, $a = c$ and transitivity follows.

Putting this scratch work together gives us the following proof.

Proof Let A be a set and I_A the identity relation on it. To prove that I_A is an equivalence relation, we must show it is reflexive, symmetric and transitive.

Reflexive: Let $a \in A$. Because $a = a$, we have $aI_A a$, showing that I_A is reflexive.

Symmetric: Let $a, b \in A$ and assume $aI_A b$. Then, $a = b$, or equivalently $b = a$, and consequently, that $bI_A a$. Thus, I_A is symmetric.

Transitive: Suppose $a, b, c \in A$ with aI_Ab and bI_Ac. Then, $a = b$ and $b = c$, and by transitivity of equality, $a = c$, showing aI_Ac. Hence, I_A is transitive.

Because I_A satisfies these three properties, it is an equivalence relation. $\square$

Notice the organizational structure of the proof of Lemma 5.12. Showing a relation is an equivalence relation requires showing three properties hold, each of which is itself a conditional statement. Because of this it is critical to keep your reader abreast of where you are in your proof. Line breaks and section markers (in this case the words "Reflexive," "Symmetric" and "Transitive," in italicized font) do just that. How you notate your proof, as mentioned numerous times before, is a personal choice. Develop your style, and be consistent. It would be a poor approach to begin proving reflexivity with the marker, *Reflexive* and then begin proving symmetry with *Proof of Symmetry* or *Symmetric case*.

Such techniques may seem trivial here, being that this proof is quite succinct, but showing some relations are equivalence relations takes a good amount of work. Spending a page or more, possibly with cases, to prove that a relation is symmetric will really grab your reader's attention. If you suddenly switch gears, "Assume aRb and $bRc...$," the transition is abrupt. Your reader will pause and try to figure out where the proof is going. Do not let this happen. As the author, you are playing the role of the tour guide. Make sure your readers are never lost!

Also note the use of variable scope (see Quality Tip 7). The variable a appears in multiple places in the previous proof, but each appearance falls outside of the scope of any previous appearance of a. As with proofs using cases, reusing a is actually beneficial to the overall quality of the proof. It is used in a consistent fashion in each of the three parts. Thus, when your readers encounter a in the proof of transitivity, they are comfortable with what role it is playing.

The scratch work in the following example is a bit more involved than that of Lemma 5.12, even though the final draft of the proof is nearly identical in length. The scratch work here is omitted, but you are encouraged to work it out on your own prior to reading through the proof.

Example 5.8. Prove that the relation R defined on the set of integers by aRb if and only if

$$4|(a - b)$$

is an equivalence relation.

Proof Let R be defined on the set of integers by aRb if and only if $4|(a - b)$. We show that R is an equivalence relation.

Reflexivity: Let $a \in \mathbb{Z}$. Since $4|(a - a)$, we have that aRa.

Symmetry: Let $a, b \in \mathbb{Z}$ with aRb. Then, $4|(a - b)$, or equivalently,

$$a - b = 4m$$

for some $m \in \mathbb{Z}$. So,

$$b - a = 4(-m),$$

showing that $4|(b - a)$, proving bRa.

Transitivity: Let $a, b, c \in \mathbb{Z}$ with aRb and bRc. Because $4|(a - b)$ and $4|(b - c)$, there exist $m, n \in \mathbb{Z}$ so that

$$a - b = 4m, \text{ and}$$
$$b - c = 4n,$$

respectively. So,

$$\begin{aligned} a - c &= (a - b) - (c - b) \\ &= 4m - 4(-n) \\ &= 4(m + n). \end{aligned}$$

proving aRc. Consequently, R is an equivalence relation. □

In fact, the relation defined in Example 5.8 is an equivalence relation when 4 is replaced by any positive integer. We will investigate these relations, which are of extreme importance in numerous areas of higher mathematics, thoroughly in Section 5.4.

How do we go about showing a relation is *not* an equivalence relation? It is only necessary to show it fails just one of the three criteria of Definition 5.11. Doing this requires just one specific counterexample, since each of the criteria are universal statements.

Example 5.9. Show that the relation $\leq$ defined on $\mathbb{R}$, defined in the usual fashion, is not an equivalence relation.

We must show $\leq$ is either not reflexive, not symmetric or not transitive. It is clearly reflexive: $a \leq a$ for all $a \in \mathbb{R}$. But is it symmetric? If $a \leq b$, do we have that $b \leq a$? Only if $a = b$.

Now, this is not a proper disproof. This is just our thought process and scratch work. To show $\leq$ is not an equivalence relation, we find a *specific* counterexample:

Disproof: We know $5 \leq 6$ but $6 \nleq 5$. Thus, $\leq$ is not symmetric and consequently not an equivalence relation on $\mathbb{R}$.

You may be wondering why we were not allowed to state generally, in terms of a and b, that $\leq$ is not symmetric, as follows:

Disproof: Let $a, b \in \mathbb{R}$. Then, if $a \leq b$, it is not true that $b \leq a$. Hence, $\leq$ is not symmetric and consequently not an equivalence relation.

Why is this disproof incorrect? It is nearly identical to what was presented in Example 5.9, using a and b in place of 5 and 6, respectively. Where it goes wrong is in its claim that, "it is not true that $b \leq a$." The variables a and b represent arbitrary but specific real numbers. Making a claim about them means you are making a claim about *all possible* choices for a and b. That is, "if $a \leq b$, then it is *never* true that $b \leq a$." Clearly this is not true if $a = b$.

Quality Tip 23. *When possible, make counterexamples specific.*

To further relate relations to our results from set theory, we have the following definition.

Definition 5.13. Let R be an equivalence relation defined on a set A. Then, for $a \in A$, the *equivalence class of a determined by R* is the set

$$[a] = \{x \in A \mid aRx\}.$$

Intuitively, the equivalence class of an element is the collection of all things that the element is related to.

Example 5.10. Determine the equivalence classes of the equivalence relations of (1) the relation R defined at the start of this section, but in this case defined on the set

$$A = \{-3, -2, -1, 0, 1, 2, 3\},$$

(2) Lemma 5.12, and (3) Example 5.8.

(1) If R is the relation on the above set A defined by aRb if and only if a

and b have the same parity, then,

$$[-3] = \{x \in A \mid x \text{ has the same parity as } -3\}$$
$$= \{-3, -1, 1, 3\}.$$

Likewise,

$$\{-3, -1, 1, 3\} = [-1] = [1] = [3],$$

and

$$\{-2, 0, 2\} = [-2] = [0] = [2].$$

(2) If R is the identity relation defined on a set A, then for any $a \in A$,

$$[a] = \{a\}.$$

(3) Let R be defined on $\mathbb{Z}$ as in Example 5.8. While it is impossible to list all elements of any equivalence class, since they are infinite, we can see that, for example,

$$[0] = \{\ldots, -8, -4, 0, 4, 8, \ldots\},$$

and

$$[1] = \{\ldots, -7, -3, 1, 5, 9, \ldots\}.$$

Example 5.10(1) illustrates a key result about equivalence classes that we prove in Lemma 5.14: equivalence classes cannot just partially overlap. They are either equal or they are disjoint. While the proof seems short, it incorporates multiple parts of the definitions of equivalence classes and equivalence relations. Moreover, it utilizes a set element method of proof in showing that two *sets* are equal to one another, since equivalence classes are themselves sets. In short, though this proof seems fairly short and to-the-point, a great deal of scratch work went into its development.

Lemma 5.14. *Let R be an equivalence relation defined on some set A. Then, if*
$[a] \cap [b] \neq \varnothing$, *for some $a, b \in A$, then $[a] = [b]$.*

Proof Let R be an equivalence relation defined on a set A and let $a, b \in A$. Suppose

$$[a] \cap [b] \neq \varnothing$$

and consider $c \in [a] \cap [b]$. By the definition of an equivalence class, aRc and bRc. We will use this fact to show

$$[a] = [b]$$

by showing that the two sets are subsets of one another.

Let $x \in [a]$. Then, aRx. The symmetry of R gives cRa, and by the transitivity of R, we have that cRx. Again, since R is transitive, this together with bRc yields bRx, showing $x \in [b]$ and consequently that

$$[a] \subseteq [b].$$

A similar argument proves that

$$[b] \subseteq [a],$$

proving the desired result that the two equivalence classes are equal. □

Lemma 5.14 inspires the following definition, a special "set of sets" determined by an equivalence relation.

Definition 5.15. The set of all equivalence classes of a set A under an equivalence relation R is called A *modulo* R and is denoted A/R.

Example 5.11. Determine the set A/R for the sets A under the given equivalence relations R in Example 5.10.

(1) A and R defined at the start of this section:

$$A/R = \{\{-3, -1, 1, 3\}, \{-2, 0, 2\}\}.$$

(2) Lemma 5.12: If $A = \{a_i \mid i \in I$, for some index set $I\}$, then

$$A/R = \{\{a_i\} \mid i \in I\}.$$

(3) Example 5.8:

$$\begin{aligned} A/R = \{&\{\ldots, -8, -4, 0, 4, 8, \ldots\}, \\ &\{\ldots, -7, -3, 1, 5, 9, \ldots\}, \\ &\{\ldots, -6, -2, 2, 6, 10, \ldots\}, \\ &\{\ldots, -5, -1, 3, 7, 11, \ldots\}\}. \end{aligned}$$

Notice that in the previous example each set A/R is a partition of the corresponding set A. An equivalence relation on a set A *always* induces a partition of A, and likewise, a partition on A induces an equivalence relation. This is the concept behind the following two theorems.

Theorem 5.16. *The set of equivalence classes of an equivalence relation R on a nonempty set A partitions A.*

Proof Let R be an equivalence relation on a nonempty set A, and suppose

$$A/R = \{[a_i] \mid i \in I \text{ for some index set } I\}.$$

Each element of A/R is nonempty: $a_i \in [a_i]$ for all $i \in I$, since R is reflexive. Every element of A is in some equivalence class: $a \in [a]$ (so that the union of all equivalence classes constitutes A), and Lemma 5.14 proves that the collection of equivalence classes are pairwise disjoint. Hence, by Definition 2.13, A/R is a partition of A. □

Having seen that an equivalence relation induces a partition, we prove Theorem 5.17 and show that every partition induces an equivalence relation.

Theorem 5.17. *Suppose a nonempty set A is partitioned by*

$$\{A_i \mid i \in I \text{ for some index set } I\}.$$

The relation R on A defined by aRb if and only if $a, b \in A_i$ for some $i \in I$ (that is, a and b are in the same partitioning set) is an equivalence relation.

Proof Let A be a nonempty set partitioned by

$$\{A_i \mid i \in I \text{ for some index set } I\}.$$

We claim that the relation R defined on A by aRb if and only if a and b are elements of the same partitioning set A_i is an equivalence relation.

To prove that R is reflexive, let $a \in A$. Because a is in the same partitioning set as itself, it must be that aRa.

We show R is symmetric by considering $a, b \in A$ with aRb. By the definition of R, we have that a and b are in the same partitioning set, or equivalently, that b and a are in the same partitioning set. Hence, bRa.

Lastly, to prove that R is transitive, suppose for some $a, b, c \in A$ that aRb and bRc. Then, $a, b \in A_i$ for some $i \in I$, and, $b, c \in A_j$ for some $j \in I$. If $i \neq j$, since the collection $\{A_i\}$ partitions A, it must be that

$$A_i \cap A_j = \varnothing.$$

But, since $b \in A_i \cap A_j$, it follows that

$$A_i = A_j,$$

so that aRc,

Consequently, the defined relation R is indeed an equivalence relation, as was to be shown. □

What the previous two results say is that there is a one-to-one correspondence (a *bijection*, as to be defined in Section 6.2) between the set of partitions

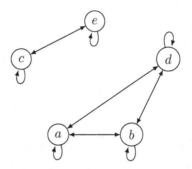

FIGURE 5.8
A graph consisting of two totally connected components

of a set and the set of equivalence relations of that same set. Every partition of a set corresponds to a unique equivalence relation defined on the set, and vice versa. Essentially, they are one and the same!

The digraph representations of equivalence relations, as mentioned at the beginning of this section, satisfy certain properties. In particular, if a digraph G represents an equivalence relation, it has the following three properties:

1. There is a loop at every vertex v in G.

2. If the edge (v_1, v_2) is in G, then the edge (v_2, v_1) is in G.

3. If G contains the edges (v_1, v_2) and (v_2, v_3), then the edge (v_1, v_3) is contained in G.

Specifically, the digraph representation of an equivalence relation has *totally connected components*.

> **Definition 5.18.** Let $G = (V, E)$ be a digraph. A *totally connected component* of G is a pair (V_1, E_1), where $V_1 \subseteq V$ and $E_1 \subseteq E$, such that for all $v_1, v_2 \in V_1$, $(v_1, v_2) \in E_1$. The graph G is said to *consist of totally connected components* if every vertex $v \in V$ lies in some totally connected component of G.

It is not just that equivalence relations induce such graphs; the converse of this statement holds as well. Graphs consisting of totally connected components induce equivalence relations on the set of their vertices. What should not come as a surprise is that the equivalence classes of this induced equivalence relation are precisely determined by the totally connected components of the graph!

For example, the digraph in Figure 5.9 induces an equivalence relation R on the set

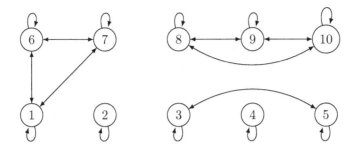

FIGURE 5.9
A digraph inducing an equivalence relation

$$A = \{x \in \mathbb{Z} \mid 1 \leq x \leq 10\}.$$

We have

$$A/R = \{\{1, 6, 7\}, \{2\}, \{3, 5\}, \{4\}, \{8, 9, 10\}\}.$$

The above ideas are summarized in the following theorem, the proof of which is left as an exercise.

Theorem 5.19. *A digraph represents an equivalence relation if and only if it has totally connected components.*

Equivalence relations are intimately related to multiple areas of mathematics, and parallel ideas exist between set theory and graph theory. These ideas are more than just theoretical results, however. They become tools with which we can approach conjectures about equivalence relations, partitions or graphs with totally connected components. In trying to prove a result about partitioning a set, Theorems 5.16 and 5.17 provide another method of attack.

Exercises

1. Determine if the following relations on $\{1, 2, 3, 4\}$ arc reflexive, symmetric, or transitive.

   (a) $\{(1, 1), (1, 2)\}$
   (b) $\{(1, 2), (2, 3), (3, 4), (4, 1)\}$
   (c) $\{(1, 1), (3, 3)\}$
   (d) $\{(1, 2), (2, 2), (2, 3), (1, 3), (2, 1), (3, 1), (3, 2)\}$

2. Let $A = \{a, b\}$. List all relations on A that are

   (a) Not reflexive, not symmetric and not transitive
   (b) Reflexive, not symmetric and not transitive

(c) Not reflexive, symmetric and not transitive

(d) Not reflexive, not symmetric and transitive

(e) Reflexive, symmetric and not transitive

(f) Reflexive, not symmetric and transitive

(g) Not reflexive, symmetric and transitive

(h) Reflexive, symmetric and transitive

3. Suppose a relation R is defined on a general set S. Describe the digraph representation of R if R satisfies the lists of properties in the previous problem.

4. Determine if each of the following relations are reflexive, symmetric and/or transitive, with full justification.

   (a) The divides relation $|$ on $\mathbb{Z}$: $a|b$ if and only if a divides b.

   (b) The relation R on $\mathbb{R}$ given by xRy if and only if
$$x + y = 10.$$

   (c) The relation D on $\mathbb{Z}^+$ given by aDb if and only if a and b have the same number of positive divisors.

   (d) The relation P on $\mathbb{Z}$ given by aPb if and only if a and b have a common prime divisor.

   (e) The relation T on the set of all logical statements given by $(P, Q) \in T$ if and only if $P \Leftrightarrow Q$ is a tautology.

   (f) The relation N on $\mathcal{P}(S)$, where
$$S = \{1, 2, 3, 4\},$$
   given by $(A, B) \in N$ if and only if $A \cap B \neq \varnothing$.

5. Are the following relations equivalence relations? Prove or disprove.

   (a) The relation R on $\mathbb{Q} \times \mathbb{Q}$ given by $(q_1, q_2)R(q_3, q_4)$ if and only if $q_i = q_j$ for some $i \neq j$, $i, j \in \{1, 2, 3, 4\}$.

   (b) The relation P on the set of all lines in $\mathbb{R}^2$ given by $(L_1, L_2) \in P$ if and only if L_1 and L_2 are perpendicular to one another.

   (c) The relation P on the set of all lines in $\mathbb{R}^3$ given by $(L_1, L_2) \in P$ if and only if L_1 and L_2 are perpendicular to one another.

   (d) The relation S on $\{0, 1, 2, 3\}$ given by aSb if and only if
$$a + b \neq 2.$$

   (e) The relation E on $\mathbb{Z}$ given by xEy if and only if 9 divides $x^2 - y^2$.

6. Prove the following relations are equivalence relations, and find 3 elements of the given equivalence classes (or explain why the equivalence class does not contain 3 elements).

   (a) The relation R on $\mathbb{R}$ given by xRy if and only if $x^2 = y^2$; equivalence classes $[0]$, $[1]$, and $[2]$.

(b) If W is the set of all words in the English language, then the relation L on W given by $w_1 L w_2$ if and only if w_1 and w_2 have the same number of letters; equivalence classes [cat] and [equation].

(c) The relation on $\mathbb{R}$ given by $\{(a,b) \mid a - b \in \mathbb{Z}\}$; equivalence classes [3.1] and [π].

(d) If V is the set of all vectors in $\mathbb{R}^2$, then the relation R on V given by $\mathbf{v}_1 R \mathbf{v}_2$ if and only if $|\mathbf{v}_1| = |\mathbf{v}_2|$ ($|\mathbf{v}|$ is the *length* or *magnitude* of $\mathbf{v}$); equivalence classes [$\mathbf{i}$] and [$4\mathbf{i} - 5\mathbf{j}$].

(e) The relation S on $\mathbb{R} \times \mathbb{R}$ given by $(a,b)S(c,d)$ if and only if
$$a + d = b + c;$$
equivalence classes [$(2,3)$] and [$(-1,4)$].

7. If B is the set of binary representations of positive integers, define the relation S on B by xSy if and only if x and y have the same number of 1s. Prove or disprove: B is an equivalence relation.

8. Suppose H is the relation defined on the set of positive integers by aHb if and only if a and b have the same digit in their 100s place.

(a) List 4 elements of [2136].

(b) List 4 elements of [5].

(c) How many elements are in $\mathbb{Z}^+/H$?

9. Define S on $\mathbb{R} \times \mathbb{R}$ by $(a,b)S(x,y)$ if and only if
$$a^2 + b^2 = x^2 + y^2.$$
Describe the sets [$(1,0)$] and [$(2,1)$].

10. Define R on the below set S by aRb if and only if the sum of the digits of a equals the sum of the digits of b. Sketch the digraph determined by S.
$$S = \{x \in \mathbb{Z} \mid 10 \le x \le 30\}$$

11. Let R be an equivalence relation on a set A. Prove

(a) $a \in [a]$ for all $a \in A$.

(b) If $x \in [a]$ for some $a \in A$ and xRb, then $[b] = [a]$.

12. If R is defined on $\mathcal{P}(\{1,2,3\})$ by $(A,B) \in R$ if and only if A and B have the same number of elements, draw the digraph determined by R.

13. Sketch the digraph representing each of the relations R defined on the set
$$\{1, 2, \ldots, 9, 10\}.$$

    (a) aRb if and only if $a \le b$.

    (b) aRb if and only if $a|b$.

    (c) aRb if and only if a and b have the same parity.

    (d) aRb if and only if the base 2 representations of a and b have the same number of digits.

14. Let R be defined on $\mathbb{Z}$ by aRb if and only if $a|b$. Prove or disprove: R is an equivalence relation.

15. Prove Theorem 5.19.

16. Let F be the set of differentiable real-valued functions. Define the relation D on F by fDg if and only if

$$f'(x) = g'(x).$$

    (a) Prove that F is an equivalence relation.

    (b) Describe $[x^3]$, $[3x + 2]$, and $[e^x]$.

17. Define the relation Z on $\mathbb{R}^3$ by $(x, y, z)Z(a, b, c)$ if and only if $z = c$.

    (a) Prove that Z is an equivalence relation.

    (b) Describe the equivalence relation Z geometrically.

18. What is wrong with the following claim and argument?

    Claim: If a relation R on a set A is symmetric and transitive, then it is reflexive.

    Proof Let R be a symmetric and transitive relation on a set A. If aRb, the symmetry of R yields that bRa. Then, the transitivity of R gives that aRa, proving that R is reflexive.   □

19. Suppose R is a relation on a set A that is reflexive, and, for all $a, b, c \in A$, if aRb and aRc, then bRc. Prove that R is an equivalence relation.

5.3   Order Relations

There are endless examples of relations in our day-to-day context that simply exhibit a symmetric relationship. That is, knowing that a is related to b is synonymous with knowing b is related to a. Examples include

1. Persons p_1 and p_2 are roommates

2. States S_1 and S_2 share a border

3. Books b_1 and b_2 were written by the same author.

Such relationships do not in some way compare the objects being related, however. It just says that the two objects share some trait. Yet, there are plenty of relations that do serve this comparison role, such as

1. Cat C_1 weighs more than cat C_2

2. A serving of fruit f_1 has fewer calories than a serving of fruit f_2

3. Set S_1 is a subset of set S_2.

Clearly these types of comparisons are relations. Is there a way, then, to mathematically define, in the most general terms (i.e., an arbitrary relation on an arbitrary set) a relation that exhibits such a comparison? There is, and it comes from first considering the symmetric property of equivalence relations. A relation R that is symmetric, as seen in the first examples above, cannot compare two objects. However, if we were to think about the negation of being symmetric, defined below as being *antisymmetric*, we have exactly what is necessary for comparison.

Definition 5.20. A relation R on a set A is called a *partial order relation* (or *partial order* or *partial ordering on A*) if it is reflexive, transitive and *antisymmetric*: for every $a, b \in A$, if aRb and bRa, then $a = b$. The set A together with R, often written (A, R), is called a *partially ordered set* or *poset*.

Have we encountered any partial order relations? A natural candidate would be $\leq$ defined on $\mathbb{R}$; it clearly defines what our intuition would call an ordering of the real numbers. We saw that it is not an equivalence relation (Example 5.9). However, it *is* a partial ordering on $\mathbb{R}$.

Lemma 5.21. *The relation $\leq$ defined on $\mathbb{R}$ is a partial order relation.*

Proof To prove that $\leq$ is a partial ordering on $\mathbb{R}$, we must show that it is reflexive, antisymmetric, and transitive.

Reflexive: Because $a \leq a$ for every $a \in \mathbb{R}$, $\leq$ is indeed reflexive.

Antisymmetric: Suppose $a \leq b$ and $b \leq a$ for $a, b \in \mathbb{R}$. Properties of the real numbers give $a = b$, showing that $\leq$ is indeed antisymmetric.

Transitive: Suppose $a, b, c \in \mathbb{R}$ with $a \leq b$ and $b \leq c$. It follows from properties of the real number system that $a \leq c$, showing that $\leq$ is transitive.

Hence, $\leq$ is a partial order relation on $\mathbb{R}$, as was to be shown. ☐

Mathematicians are often clever in their use of words[1] and symbols. Sym-

---

[1] Recall the definition of the *greatest common divisor*: the common divisor that is the greatest.

bolically, the relations $<$ and $\leq$ on $\mathbb{R}$ are very similar to the relations $\subset$ and $\subseteq$ on sets. In thinking about the difference within the pairs, saying $a < b$ means that the real number a is *strictly smaller* than the real number b. Likewise, saying $A \subset B$ means that the set A is *strictly smaller* than the set B (it is a *proper* subset). Is this coincidence? No, it is not. The set of real numbers $\mathbb{R}$ under $\leq$ (Lemma 5.21) and the power set of a set under $\subseteq$ (Lemma 5.22) are both posets, though we will see that they differ in one significant major way.

Lemma 5.22. *For any set A, the set $\mathcal{P}(A)$ with the relation $\subseteq$ is a poset.*

Proof Let A be a set. Because any set is a subset of itself, $\subseteq$ is reflexive on $\mathcal{P}(A)$.

Next, the definition of set equality proves that $\subseteq$ is antisymmetric on $\mathcal{P}(A)$: for $A_1, A_2 \in \mathcal{P}(A)$, if

$$A_1 \subseteq A_2$$

and

$$A_2 \subseteq A_1,$$

then $A_1 = A_2$.

Lastly, The transitivity of $\subseteq$ is proven in Theorem 2.8.

Consequently, because these three properties hold, we have shown that $\mathcal{P}(A)$ under the relation $\subseteq$ is a poset. $\square$

Armed with these few examples of posets, we can address a question that may be burning in the back of your mind: why is this new object called a *partial* ordering and not simply just an *ordering*? Let us investigate the relations in the previous two lemmas. In doing so we will discover a major difference between the relations $\leq$ and $\subseteq$.

Consider $\leq$ on $\mathbb{R}$. Take any $a, b \in \mathbb{R}$. What do we know about these two numbers relative to the relation $\leq$? At least one of $a \leq b$ or $b \leq a$ holds. That is, under this relation any two real numbers are *comparable*. This idea of comparability does not carry over to sets, however. Two randomly chosen sets A and B need not satisfy one of $A \subseteq B$ or $B \subseteq A$. For example, taking $\subseteq$ on $\mathcal{P}(S)$, if

$$S = \{a, b, c\},$$

we cannot compare $\{a\}$ and $\{b\}$:

$$\{a\} \nsubseteq \{b\}$$

and

$$\{b\} \nsubseteq \{a\}.$$

Those sets and relations that have the comparability property exhibited by $\leq$ on $\mathbb{R}$ are called *linearly ordered*.[2]

> **Definition 5.23.** A partial order R on a set A is called a *linear ordering* if for any $a, b \in A$, either aRb or bRa (in this case we say a and b are *comparable*). Such a set together with the linear ordering is called a *linearly ordered set*.

As mentioned, $\mathbb{R}$ under $\leq$ is a linearly ordered set. Are there others? Because we have the linear ordering on $\mathbb{R}$, a first choice to consider might be $\mathbb{R} \times \mathbb{R}$. Is there a way to generalize this $\leq$ ordering on $\mathbb{R}$ to an ordering for $\mathbb{R} \times \mathbb{R}$? How might we compare $(5, 7)$ and $(6, 7)$? We are tempted by this single example to say $(5, 7) \leq (6, 7)$ (using the same $\leq$ symbol) because $5 \leq 6$. But this logic fails if we consider another example: $(5, 7)$ and $(5, 8)$? Clearly we must somehow use the second coordinate. We might say, "Because the first coordinates are the same, let's look at the second coordinates. Since $7 \leq 8$, it makes sense to say $(5, 7) \leq (5, 8)$.

You have used this approach before, though not necessarily in dealing with ordered pairs of numbers. It is exactly how we alphabetize words. If we were to look up the word *proof* in the dictionary, we would look after the word *pronoun* but before the word *property*. Why? We compare the first two letters of the words. Because they are the same, we proceed to compare the second letters. They too are the same. Proceed to the third letter, and so on. So, we not only have come up with a linear ordering on $\mathbb{R} \times \mathbb{R}$, but we can generalize this to $\mathbb{R} \times \mathbb{R} \times \cdots \times \mathbb{R}$ by generalizing the following definition.

> **Definition 5.24.** If A and B are linearly ordered sets under $\leq_A$ and $\leq_B$, respectively, then $A \times B$ is a linearly ordered set under the relation $\leq$ where
>
> $$(a_1, b_1) \leq (a_2, b_2)$$
>
> if and only if $a_1 <_A a_2$ (i.e., $a_1 \leq_A a_2$ with $a_1 \neq a_2$), or, if $a_1 = a_2$, then $b_1 \leq_B b_2$. We call $\leq$ the *dictionary order on $A \times B$*.

This definition requires justification. A claim that the dictionary order is a linear order on $A \times B$ must be proven.

Theorem 5.25. *The dictionary order defined on $A \times B$, where A and B are linearly ordered sets under $\leq_A$ and $\leq_B$, respectively, is a linear order.*

Proof Let A and B be linearly ordered sets under $\leq_A$ and $\leq_B$, respectively. To prove that the dictionary order $\leq$ on $A \times B$ is a linear order, we

---

[2]Why a *linear* ordering? Two objects arranged in a line are naturally comparable: one must be in front of the other.

show it fulfills the three requirements to be a partial order and that any two elements of $A \times B$ are comparable.

Reflexive: Let $(a, b) \in A \times B$. Because both A and B are linearly ordered, we know $a \leq_A a$ and $b \leq_B b$. Thus, by the definition of the dictionary order,

$$(a, b) \leq (a, b).$$

Antisymmetric: Suppose $(a_1, b_1), (a_2, b_2) \in A \times B$ with

$$(a_1, b_1) \leq (a_2, b_2)$$

and

$$(a_2, b_2) \leq (a_1, b_1).$$

The first of these, inequalities implies that $a_1 \leq_A a_2$. The latter implies that $a_2 \leq_A a_1$. By the antisymmetry of $\leq_A$, it must be that $a_1 = a_2$.

Now, since $a_1 = a_2$, it must be that $b_1 \leq_B b_2$ (by the first inequality above) and also that $b_2 \leq_B b_1$ (by the second inequality above). The antisymmetry of $\leq_B$ yields $b_1 = b_2$. Thus,

$$(a_1, b_1) = (a_2, b_2),$$

proving that $\leq$ is antisymmetric.

Transitive: This step of the proof is left as an exercise.

Comparability: Let $(a_1, b_1), (a_2, b_2) \in A \times B$. By the comparability of $\leq_A$, we know that either $a_1 \leq_A a_2$ or $a_2 \leq_A a_1$. Without loss of generality, assume $a_1 \leq_A a_2$, so that either $a_1 <_A a_2$ or $a_1 = a_2$. Consider both possibilities.

If $a_1 <_A a_2$, then

$$(a_1, b_1) \leq (a_2, b_2).$$

If $a_1 = a_2$, consider the comparability of $\leq_B$, so that $b_1 \leq_B b_2$ or $b_2 \leq_B b_1$. Again, without loss of generality, assume $b_1 \leq_B b_2$. By the definition of $\leq$, we have that

$$(a_1, b_1) \leq (a_2, b_2).$$

Having shown the dictionary order satisfies all the necessary properties to be a linear order, we have proven the required result. $\quad\quad\quad\quad\quad\quad\square$

Let us conclude our discussion of order relations by investigating properties of the digraphs that represent them. Because partial orderings possess some of the same properties as equivalence relations, their digraphs have similarities. In particular, the graphical properties corresponding to reflexivity (every vertex has a loop) and transitivity (if there is an edge from v_1 to v_2 and from v_2 to v_3, then there is an edge from v_1 to v_3) exist. However, because partial

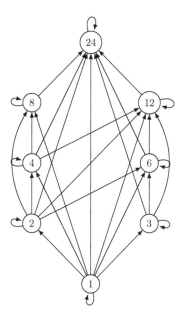

FIGURE 5.10
The relation of Example 5.12

orderings are antisymmetric rather than symmetric, they lack the property that every edge from v_1 to v_2 must have an accompanying edge from v_2 to v_1. In fact, this can never happen. If these edges both existed, say for a relation R, then we have that $v_1 R v_2$ and $v_2 R v_1$. But antisymmetry would tell us that $v_1 = v_2$, a counterexample to the claim of being antisymmetric.

Example 5.12. Let the relation R be defined on

$$A = \{1, 2, 3, 4, 6, 8, 12, 24\}$$

by aRb if and only if $a|b$. The relation R is a partial, but not linear, ordering on A (the proofs of both of which are left as exercises). A digraph representation of R is shown in Figure 5.10.

While the digraph of Figure 5.10 accurately represents the order relation, it is messy and contains redundant information. If we know that the relation the digraph represents is a partial order, then there must be a loop at every vertex. Moreover, the way this particular digraph has been drawn allows us to remove certain edges without losing accompanying information. Notice how every edge points upward on the page. Because of this, if there is an edge pointing up from vertex v_1 to v_2, and then an edge pointing up from v_2 to v_3,

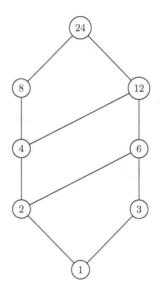

FIGURE 5.11
A Hasse diagram for the poset of Example 5.12

transitivity of the relation means there must be an edge pointing from v_1 to v_3. However, knowing that the digraph represents a poset, this third edge is unnecessary.

A tool for "cleaning up" digraph representations of posets is called a *Hasse diagram*. Though he was not the first to introduce them, Helmut Hasse (1898-1979) incorporated them into a variety of applications [4].

> **Definition 5.26.** A *Hasse diagram* representing a partial ordering R on a finite set A is obtained from a digraph representing R by the following.
>
> 1. Place all vertices on the page so that all edges (excluding loops) point upward (relative to the page) and draw edges as line segments rather than arrows.
>
> 2. Eliminate all loops as well as all edges guaranteed by the transitivity of R.

Figure 5.11 shows a Hasse diagram representing the partially ordered set of Example 5.12 (that is, the Hasse diagram obtained from the digraph of Figure 5.10).

What is the benefit of creating Hasse diagrams, other than making the graph somewhat easier to read? It more clearly exhibits the partial ordering

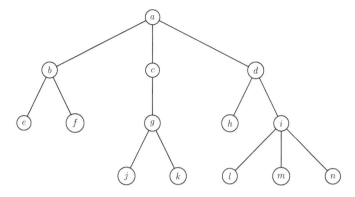

FIGURE 5.12
Hasse diagram on a general set

of the relation. Progressing along branches of the diagram, we can quickly
determine what elements are related to others.

Figure 5.12 demonstrates this readability factor more clearly. Suppose the
Hasse diagram in Figure 5.12 represents a relation $\le$ on the set

$$S = \{a, b, c, \ldots, m, n\}.$$

What elements of S are related to others? A quick glance at the diagram tells
us that $e \le b$ and consequently $e \le a$, and these are the only elements of S
that e is related to. Likewise, we know that while m and d are comparable, m
and c are not. We can also see that a plays a special role; for every $x \in S$, we
have that $x \le a$. We call a the *greatest* element of this poset.

> **Definition 5.27.** If S is any subset of a partially ordered set under a
> partial order relation $\le$, then an element $s \in S$ is called the *greatest
> element* (resp. *least element*) of S if $x \le s$ (resp. $s \le x$) for all $x \in S$.

Consider the relation of Example 5.12 along with the accompanying Hasse
diagram of Figure 5.11. We see that 24 is the greatest element of A, while 1 is
the least element. Not every collection of elements of A must have a greatest
or least element, as demonstrated by $\{2, 3, 4, 6\}$.

Hasse diagrams represent partial orderings, but what would the Hasse
diagram of a *linearly* ordered finite (Hasse diagrams are defined only on finite
sets) set S look like under a relation $<$?

Assume S has two or more elements (otherwise the Hasse diagram is a
single vertex with no edges). Take $a_1, a_2 \in S$. By the comparability property
of linearly ordered sets, either $a_1 < a_2$ or $a_2 < a_1$. In either case, we begin
to construct the Hasse diagram by forming a a vertical segment between a_1

FIGURE 5.13
Constructing a Hasse diagram of a linearly ordered set

and a_2. Supposing there is a third element a_3, we use the comparability property to appropriately place a_3 into the Hasse diagram. How? Proceed from the bottom up. Compare a_3 to the element lowest in the diagram; say this is a_1. If $a_3 < a_1$, then a_3 becomes the lowest element in the Hasse diagram. Otherwise, move up to the next element in the diagram, a_2. If $a_3 < a_2$, then we place a_3 in the Hasse diagram between a_1 and a_2. Continue this process (which is possible because S is finite). Once all elements are placed, we have a vertical linear segment, such as in Figure 5.13, called a vertical *branch*.

Exercises

1. Describe how to recreate a digraph representation of a partial ordering if you are given the Hasse diagram.

2. Use Figure 5.12 to determine the following, assuming the diagram represents a relation $\leq$ on the set $\{a, b, c, \ldots, m, n\}$.

    (a) All elements comparable to a.
    (b) All elements comparable to c.
    (c) All elements that are greater than or equal to exactly two other elements.
    (d) If $\{a, b, c, d\}$ under the given relation is a poset.
    (e) If $\{a, b, e, f\}$ under the given relation is a poset.
    (f) If $\{a, e, f, c, g, h, i\}$ under the given relation is a poset.

3. For which sets A, if any, is the relation $\subseteq$ on $\mathcal{P}(A)$ a linear order?

4. Prove or disprove.

    (a) $\mathbb{R}$ under $<$ is a linearly ordered set.
    (b) $\mathbb{R}$ under the relation $\neq$ is a poset.
    (c) $\mathbb{Z}^+$ under the relation does *not* divide relation $\nmid$ is a poset.
    (d) The relation in Example 5.12 is a partial but not linear order.

5. Consider the divides relation on $\mathbb{Z}^+$.

   (a) Prove that the relation is a partial order.

   (b) In this partially ordered set, are 5 and 15 comparable? 10 and 15? 15 and 15?

   (c) Describe all positive integers comparable to 2.

6. Define the relation R on $\mathbb{Z}^+$ by aRb if and only if both $a + b$ is even and $a \leq b$.

   (a) Prove that R is a partial ordering.

   (b) Are 5 and 15 comparable under R? 10 and 15? 15 and 15?

   (c) Describe all positive integers comparable to 6.

7. Prove the transitivity portion of Theorem 5.25.

8. Let S be a partially ordered set under the relation $\leq$. Say that an element $a \in S$ *covers* an element $b \in S$ ($b \neq a$) if both $b \leq a$ and if there does not exist $c \in S$ such that $b \leq c \leq a$, with $c \neq a$, $c \neq b$.

   (a) Determine all pairs of elements (a, b) from the set $\{2, 4, 6, 8, 10, 12\}$ where a covers b under the divides relation.

   (b) Explain how to determine which elements of a set cover which other elements of a set given the Hasse diagram representing the partial order.

   (c) If $A = \{1, 2, 3\}$, consider the relation $\subseteq$ on $\mathcal{P}(A)$. Determine all sets $B, C \in \mathcal{P}(A)$ such that B covers C.

9. Describe all elements (a, b) of $\mathbb{Z}^+ \times \mathbb{Z}^+$ satisfying the following, where $\leq$ is considered to be the dictionary order, or explain why no such elements exist.

   (a) $(2, 3) \leq (a, b)$ with $a = 2$.

   (b) $(2, 3) \leq (a, b)$ with $a < 2$.

   (c) $(2, 3) \leq (a, b)$ with $2 < a$.

   (d) $(2, 3) \leq (a, b)$ with $b = 3$.

   (e) $(2, 3) \leq (a, b)$ with $b < 3$.

   (f) $(2, 3) \leq (a, b)$ with $b > 3$.

10. Generalize Definition 5.24 by writing a precise definition for the dictionary order defined on the Cartesian product of n linearly ordered sets, $A_1 \times A_2 \times \cdots \times A_n$, where A_i is defined with the linear ordering $\leq_i$.

11. Under the dictionary order on $\mathbb{R}^3$, determine the order of the following pairs of elements.

   (a) $(\pi, \pi, -\pi)$, $(3.14, -3.14, \pi)$

   (b) $(0, 1, 2)$, $(0, 2, 1)$

   (c) $(1, 2, 3)$, $(0, 1, 2)$

12. Let $A = \{0, 1, 2\}$. Draw the digraph on $A \times A$ under the dictionary order and convert it to a Hasse diagram.

13. Determine which of the following digraphs represent a poset.

(a)

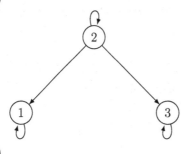

(b)

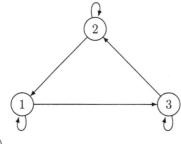

(c)

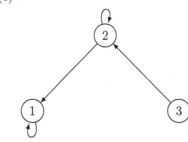

(d)

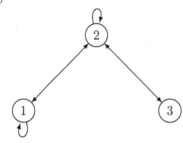

14. Draw the Hasse diagram for the relation $\subseteq$ on $\mathcal{P}(\{a, b, c\})$.

15. Find the greatest element of each of the following posets, if they exist.

    (a) The relation $\leq$ on $\mathbb{Z}^-$.
    (b) The relation $\leq$ on $\{x \in \mathbb{Z} \mid 1 < x < 100\}$.
    (c) The relation $\leq$ on $\{x \in \mathbb{R} \mid 1 < x < 100\}$.
    (d) The relation $\leq$ on $\{\dfrac{x}{x+1} \mid x \in \mathbb{Z}^+\}$.
    (e) The relation $\subseteq$ on $\mathcal{P}(A)$, where A is any nonempty set.

16. Given two distinct points (x_1, y_1) and (x_2, y_2) on the Cartesian plane, describe how to tell graphically if $(x_1, y_1) < (x_2, y_2)$ or $(x_2, y_2) < (x_1, y_1)$.

17. The set $\mathcal{P}(\mathbb{Z}^+)$ under $\subseteq$ is a partially ordered set (see Lemma 5.22).

    (a) If E is the set of all even positive integers, which of the following sets are comparable to E?
        i. $\{2, 4, 6, 8\}$
        ii. $\mathbb{Z}^+ - \{2, 4, 6, 8\}$
        iii. $\{2^n \mid n \in \mathbb{Z}, n \geq 0\}$
    (b) Describe all sets comparable to E.

18. Let R be a partial order on a set A, and suppose $a_1 R a_2$, $a_2 R a_3$, ..., $a_n R a_1$. Prove that

$$a_1 = a_2 = \cdots = a_n.$$

19. Define a *weak partial order* to be a relation R on a set S that is transitive, antisymmetric, and *irreflexive*: it is not the case that aRa for any $a \in S$.

    (a) Show that $\subseteq$ is not a weak partial order on $\mathcal{P}(A)$ for any nonempty set A.
    (b) Define the relation D on $\mathbb{R}^2$ by $(P, Q) \in D$ if and only if P and Q are the same distance from the origin. Prove that D is a weak partial order.

20. The *dual* of a poset (A, R) is the poset (A, R^{-1}).

    (a) Prove that the dual of a poset (A, R) is indeed a poset.
    (b) Find the dual of each of the following posets.
        i. $\mathbb{Z}^+$ under $\leq$,
        ii. $\mathcal{P}(\{a, b, c\})$ under $\subseteq$.
        iii. $\mathbb{Z}^+$ under the divides relation.
    (c) Prove that the dual of the dual to (A, R) is (A, R).

21. Let P under $\leq$ be a poset. Define the *sequence space* as the set of all infinite sequences of elements of P.

(a) Prove that the sequence space of a poset under the dictionary order is a partial ordered set.

(b) Prove or disprove: the relation $<$ on the sequence space of a poset $(P, <)$ given by

$$(a_1, a_2, \dots) < (b_1, b_2, \dots)$$

if and only if $a_i < b_i$ for all $i \in \mathbb{Z}^+$ is a partial order.

22. Suppose a poset S has a greatest element. Prove that it is unique.

23. If a poset has a greatest element, what can you say about the placement of that element in the Hasse diagram of the poset?

24. Let $A = \{1, 2\}$ and $B = \{a, b\}$. Consider the dictionary order defined on $\mathcal{P}(A) \times \mathcal{P}(B)$, where $\subseteq$ is the partial order defined on either power set.

(a) Find all $(S_1, S_2) \in \mathcal{P}(A) \times \mathcal{P}(B)$ such that

$$(\{2\}, \{a, b\}) \leq (S_1, S_2).$$

(b) Does $\mathcal{P}(A) \times \mathcal{P}(B)$ have a least element?

(c) Does $\mathcal{P}(A) \times \mathcal{P}(B)$ have a greatest element?

5.4   Congruence Modulo m Relation

Anyone familiar with computer science has undoubtedly encountered concepts such as sorting, trees and task scheduling. These are just some of the applications of partial and linear orderings. Within mathematics, they play a critical role in abstract algebra, number theory and even topology. But they are just general types of relation; there are a lot of different partial orderings and linear orderings, so it does not come as such a surprise that their applications vary so widely.

What is surprising is that there is a specific relation that is used as much, if not more, throughout both pure and applied mathematics, and it is the focus of this section. As we will see in Theorem 5.29, it is intimately related to a major theorem that has served us well already: the Quotient-Remainder Theorem.

Definition 5.28. For $m \in \mathbb{Z}^+$, we say $a, b \in \mathbb{Z}$ are *congruent modulo m*, denoted $a \equiv_m b$ (or $a \equiv b \bmod m$), if $m \mid (a - b)$. The integer m is called the *modulus* of the congruence.

Congruence modulo m, or the mod m congruence for short, is a relation defined on the integers that we will refer to simply as the *mod relation*. Though it may seem somewhat non-intuitive, the following theorem tells us it is enough to consider only the remainders when divided by m to tell if two integers are congruent modulo m.

Theorem 5.29. *Suppose $a, b, m \in \mathbb{Z}$ with $m > 0$. The following statements are equivalent.*

    1. $a \equiv_m b$.

    2. a and b have the same remainder when divided by m.

    3. $a = b + km$, for some $k \in \mathbb{Z}$.

Note that to prove these three results are equivalent, it is only necessary to show that each statement implies the next, as well as the first following from the last. In each of these parts of the proof, we will have both the definition of congruence modulo m as well as the assumption of each statement to use.[3]

Proof Let $a, b, m \in \mathbb{Z}$ with $m > 0$. We show the three statements below are equivalent by showing that each implies the next, as well as the first following from the last.

1. $a \equiv_m b$.

2. a and b have the same remainder when divided by m.

3. $a = b + km$, for some $k \in \mathbb{Z}$.

(1) implies (2): Suppose $a \equiv_m b$. Consider the remainders when a and b are divided by m. The Quotient-Remainder Theorem implies that $a = mq_1 + r_1$ and $b = mq_2 + r_2$, for some unique $q_i, r_i \in \mathbb{Z}$, with $0 \le r_i < m$. WLOG, assume

$$0 \le r_2 \le r_1 < m,$$

so that

$$0 \le r_1 - r_2 < m.$$

Our assumption of $a \equiv_m b$ gives that $a - b = mk$ for some $k \in \mathbb{Z}$. Substituting,

$$(mq_1 + r_1) - (mq_2 + r_2) = mk,$$

so that

$$r_1 - r_2 = m(k - q_1 + q_2).$$

showing that m divides $r_1 - r_2$. By Lemma 3.14, if $r_1 - r_2 > 0$, we have that $m \le r_1 - r_2$, a contradiction. Hence, $r_1 = r_2$, as was to be shown.

(2) implies (3): Assume $a = mq_1 + r$ and $b = mq_2 + r$, where $q_1, q_2, r \in \mathbb{Z}$ with $0 \le r < m$. Then, this means

---

[3]Claims such as those in Theorem 5.29 sometimes use the acronym "TFAE" for "the following are equivalent."

$$a - mq_1 = b - mq_2,$$

or equivalently,

$$a = b + m(q_1 - q_2),$$

as required.

(3) implies (1): If $a = b + km$ for some $k \in \mathbb{Z}$, then $a - b = km$, proving m divides $a - b$, or by definition, $a \equiv_m b$, as desired. $\square$

Before investigating properties of the mod m relation, we look at the format of the proof of Theorem 5.29. Conjectures that make a claim about a series of statements being equivalent can be proven in a multitude of ways. Theorem 5.29 reworded is actually three biconditional statements: (1) if and only if (2), (2) if and only if (3), and (1) if and only if (3). Proving any one of these biconditionals individually would require proving two conditional statements. When the entire collection of biconditionals is considered, however, we need not prove two conditional statements for each and every biconditional statement. What is necessary is proving that any one of the statements is implied (either directly or indirectly) no matter which of the other statements is assumed true.

For example, if we wanted to prove that five statements were equivalent (say, (1) through (5)), proving either of the following series of implications would prove the result:

$$(1) \Rightarrow (2) \Rightarrow (3) \Rightarrow (4) \Rightarrow (5) \Rightarrow (1),^{4}$$

or,

$$(2) \Rightarrow (4) \Rightarrow (1) \Rightarrow (5) \Rightarrow (3) \Rightarrow (2).$$

Likewise, proving

$$(1) \Leftrightarrow (3) \text{ and } (1) \Rightarrow (2) \Rightarrow (5) \Rightarrow (4) \Rightarrow (1)$$

would prove the result. However, proving

$$(1) \Leftrightarrow (3) \text{ and } (1) \Rightarrow (2) \Rightarrow (4) \Rightarrow (5) \Rightarrow (2)$$

would not prove the result. Why? If we assume (2) to be true, there is no way to deduce that (3) is true (that is, there is no sequence of conditional statements leading from (2) to (3)).

Is one way better than another in proving a long list of equivalences? Not particularly, though some approaches are more elegant than others. Readers

---

[4]This notation means we prove that (1) implies (2), (2) implies (3), ..., and (5) implies (1).

will often anticipate seeing a series of equivalent statements being proven in the presented order. If your proof does not proceed in that order, simply rearrange the statements in the statement of the result in the order you will prove them, or, you can inform your reader of how the proof will proceed. Like many of our proof tips, you do not want your reader scratching her head and wondering why your work is presented in the way it is.

Quality Tip 24. *Strive to keep readers focused on the result when proving that "the following are equivalent:" by organizing the proof in a natural way or informing the readers how the proof will proceed.*

Ideally, you can prove results like this with a minimal number of sub-proofs. Why do more work than is necessary? For example, to prove three equivalent statements, it would be silly to prove the first if and only if the second, the second if and only if the third, and the third if and only if the first. Six sub-proofs would indeed prove the result but so would three: the first implies the second, the second implies the third and the third implies the first.

Quality Tip 25. *Eliminate unnecessary work by attempting to prove a list of statements are equivalent using the fewest number of implications possible.*

Theorem 5.29 provides alternate ways to state Definition 5.28. Indeed, if you flip through various number theory, discrete mathematics or abstract algebra books, you will find that the mod relation is defined differently by different authors. Because of Theorem 5.29, though, all are actually defining the same thing.

While seemingly simple (how useful can knowing two integers have the same remainder when divided by m be?), it turns out that mod m congruence is an extremely useful tool. Before seeing a few of those uses in action, Theorem 5.30 proves that the mod relation satisfies three familiar properties.

Theorem 5.30. *Let $m \in \mathbb{Z}^+$. Then, for any $a, b, c \in \mathbb{Z}$,*

1. *$a \equiv_m a$;*
2. *if $a \equiv_m b$, then $b \equiv_m a$;*
3. *if $a \equiv_m b$ and $b \equiv_m c$, then $a \equiv_m c$.*

While we could prove these results using only the definition of congruence modulo m (this is left as an exercise), utilizing Theorem 5.29 expedites the process.

Proof Let $m \in \mathbb{Z}^+$, with $a, b, c \in \mathbb{Z}$.

1. Since a has the same remainder as a when divided by m, Theorem 5.29 yields $a \equiv_m a$.

2. Assume $a \equiv_m b$. Then, a and b have the same remainder when divided by m, or equivalently, b and a have the same remainder when divided by m. Thus, $b \equiv_m a$.

3. Suppose $a \equiv_m b$ and $b \equiv_m c$. Then, by Theorem 5.29, a and b have the same remainder when divided by m, as do b and c. Consequently, a and c must have the same remainder upon division by m as well, showing $a \equiv_m c$, as required.

□

Corollary 5.31 follows immediately.

Corollary 5.31. *For any $m \in \mathbb{Z}^+$, the relation $\equiv_m$ is an equivalence relation on $\mathbb{Z}$.*

Once a relation on a set S is determined to be an equivalence relation, it is natural to ask how it partitions S into equivalence classes. For a given $m \in \mathbb{Z}^+$, what then are the equivalence classes of the mod m relation? Our interpretation of the relation in terms of remainders makes it somewhat clear. However, Theorem 5.29, which provides multiple statements that are equivalent to saying "$a \equiv_m b$," solidifies this intuition.

The following corollary to Theorem 5.29 emphasizes the importance of the *unique remainder* guaranteed by the Quotient-Remainder Theorem when an integer is divided by another. Moreover, it provides a mechanism for quickly determining the equivalence classes mod m for any integer m.

Corollary 5.32. *Let $a, m \in \mathbb{Z}$ with $m > 1$. Then, a is congruent modulo m to precisely one of 0, 1, 2, ..., or $m - 1$.*

Proof Suppose $a, m \in \mathbb{Z}^+$ with $m > 1$, and for contradiction, that both $a \equiv_m r_1$ and $a \equiv_m r_2$ with

$$0 \leq r_1 < r_2 \leq m - 1.$$

Because $\equiv_m$ is an equivalence relation, we have that $r_1 \equiv_m r_2$. Moreover, Theorem 5.29 yields that r_1 and r_2 have the same remainder when divided by m. Because these remainders, which are themselves r_1 and r_2 (since both are less than $m - 1$), respectively, are unique (Theorem 3.42), we have reached a contradiction. Consequently, it must be that a is congruent modulo m to precisely one of $0, 1, 2, \ldots, m - 1$. □

Thus, the equivalence classes of the mod m relation are easy to find. Consider the mod 5 relation. All integers that have a remainder of 2 when divided by 5 will be in the same equivalence class. Which integers are these? Precisely those that are 2 more than multiples of 5:

$$2, 7, 12, 17, \text{etc.}$$

The following definition provides common notation.

Definition 5.33. For a given $m \in \mathbb{Z}^+$, denote the equivalence class of an integer a under $\equiv_m$ as $[a]_m$.

Then, Theorem 5.29, along with the discussion preceding Definition 5.33, tells us that for any $a, m \in \mathbb{Z}$ with $m > 0$,

$[a]_m = \{x \in \mathbb{Z} \mid x$ and a have the same remainder when divided by $m\}$.

Example 5.13. Determine the equivalence classes of $\equiv_4$ on $\mathbb{Z}$.

By the Quotient-Remainder Theorem, there are four possible remainders (0, 1, 2, 3) when an integer is divided by 4. The equivalence classes then are the sets of all integers having these same remainders when divided by 4:

$$[0]_4 = \{\ldots, -8, -4, 0, 4, 8, \ldots\}$$
$$[1]_4 = \{\ldots, -7, -3, 1, 5, 9, \ldots\}$$
$$[2]_4 = \{\ldots, -6, -2, 2, 6, 10, \ldots\}$$
$$[3]_4 = \{\ldots, -5, -1, 3, 7, 11, \ldots\}$$

Moreover,

$$\mathbb{Z}/\equiv_4 = \{[0]_4, [1]_4, [2]_4, [3]_4\}.$$

Let us expand the theory of the mod relation by introducing what is known as *modular arithmetic*. Because you know how to read a clock, you are familiar with the basics of modular arithmetic, or as some people call it, wrap-around arithmetic. Why is the clock an example of such? Though days are 24 hours long, we count time on a 12-hour cycle. Typically we do not say, "Dinner is served at 18 o'clock." When we've reached 12 o'clock, we start our counting over again (and we serve dinner at 6 o'clock).[5]

Modular arithmetic is a system of operations defined on the integers modulo some positive integer m. It consists of rules and properties that allow us to compute in this new system. Note that the arithmetic rules stated and proven below are completely natural. They tell us that the order in which we perform various operations is not important. For example, given two integers, we can sum them then determine their remainder when divided by m, or, we can determine each of their remainders when divided by m, sum them, and then find the remainder of that result when divided by m.

---

[5]This is a version of modular arithmetic. While modular arithmetic starts counting at 0, we never say 0 o'clock.

Theorem 5.34. *Let $m \in \mathbb{Z}^+$, and assume $a_i \equiv_m b_i$ for $i = 1, 2$. Then,*

    1. $a_1 + a_2 \equiv_m b_1 + b_2$,

    2. $a_1 a_2 \equiv_m b_1 b_2$, and

    3. $a_1^n \equiv_m b_1^n$, for all $n \in \mathbb{Z}$, $n \geq 2$.

Proof Let $m \in \mathbb{Z}^+$ with $a_i \equiv_m b_i$ for $i = 1, 2$.

1. We know m divides both $a_1 - b_1$ and $a_2 - b_2$, by the definition of modular equivalence. Lemma 3.16 implies, then, that m divides

$$(a_1 - b_1) + (a_2 - b_2) = (a_1 + a_2) - (b_1 + b_2),$$

so that $a_1 + a_2 \equiv_m b_1 + b_2$.

2. Note that

$$\begin{aligned} a_1 a_2 - b_1 b_2 &= a_1 a_2 - a_2 b_1 + a_2 b_1 - b_1 b_2 \\ &= a_2(a_1 - b_1) + b_1(a_2 - b_2). \end{aligned}$$

Since $m|(a_1 - b_1)$ and $m|(a_2 - b_2)$, we know that $m|a_2(a_1 - b_1)$ and $m|b_1(a_2 - b_2)$ (Exercise 3.2.8a). By Lemma 3.16,

$$m|\left(a_2(a_1 - b_1) + b_1(a_2 - b_2)\right),$$

or equivalently,

$$m|(a_1 a_2 - b_1 b_2),$$

as desired.

3. We prove this result via mathematical induction. The base case, $n = 2$, is precisely (2). Assume for a general $n \in \mathbb{Z}$, $n \geq 2$, that the inductive hypothesis holds:

$$a_1^n \equiv_m b_1^n.$$

Then, because we have $a_1 \equiv_m b_1$, it follows from (2) that

$$a_1^n a_1 \equiv_m b_1^n b_1,$$

or equivalently,

$$a_1^{n+1} \equiv_m b_1^{n+1}.$$

Hence, the result holds for all $n \in \mathbb{Z}$, $n \geq 2$.

□

Example 5.14. Note that $22 \equiv_6 4$ and $35 \equiv_6 5$. We demonstrate Theorem 5.34 by computing a sum and a product (modulo m) two different ways:

First, to computer the remainder of $22 + 35$ (modulo 6), we can sum first and then compute the remainder upon division by 6:

$$22 + 35 = 57 \equiv_6 3.$$

Otherwise, we can compute the remainders of 22 and 35 upon division by 6 (4 and 5, respectively), then sum and compute its remainder:

$$4 + 5 = 9 \equiv_6 3.$$

We can take the same approach with multiplication.

$$(22)(35) = 770 \equiv_6 2,$$

or

$$(4)(5) = 20 \equiv_6 2.$$

In both of these examples, we see that we can operate (summing or multiplying) before computing the remainder, or, computing the remainder and then operating.

Beyond these basic calculations, modular arithmetic is a tool for handling extremely large integers. After working the following examples, you may ask yourself why anyone would have a use for such things, beyond fun mathematical computations? The applications are countless. They can be used to check for errors within serial number identifiers, such as book identifiers (ISBN[6] and user-entered bank account numbers[7]). Many of the most secure cryptographic systems have modular arithmetic as their backbone. Musicians utilize a modulus of 12 in the system of twelve-tone equal temperament. Politicians employ modular arithmetic when considering how to apportion fairly. But our goals here are strictly mathematical.

Example 5.15. Find the remainder when 2^{500} is divided by 7.

Note that 2^{500} is *big*; its size is nearly incomprehensible (it is over

---

[6] International Standard Book Numbers (ISBNs) are based on modulo 10.
[7] International Bank Account Numbers (IBANs) use modulo 97 arithmetic.

150 digits in length). Can we turn the problem into 1 raised to a power? Indeed we can, if we are able to find some number of 2's whose product is equivalent mod 7 to 1, then Theorem 5.29(3) could be used to significantly reduce the problem.

This is often the key idea in problems like this. Is there a power of 2 that is one more than a multiple of 7? That is a much easier problem to comprehend. Once we find it, we can eliminate most of the factors of 2, since they will equate to simply multiplying by 1 (modulo 7).

Right away, we can find the key to our computation:

$$2^3 \equiv_7 1.$$

Using this, we can simplify the quantity:

$$2^{500} = (2^3)^{166}(2^2)$$
$$\equiv_7 (1^{166})(2^2)$$
$$\equiv_7 4.$$

Thus, the remainder when 2^{500} is divided by 7 is 4.

A similar approach can be used when dealing with large sums.

Example 5.16. What is the remainder when

$$1! + 2! + \cdots + 500!$$

is divided by 20?

Using the thought process of the previous example, we look for a way to reduce the expression using modular arithmetic. In Example 5.15 we were multiplying, so we identified factors of 1 (the multiplicative identify). We are adding in the expression here, so we look for occurrences of 0 (the additive identify). Note that for $n \geq 5$,

$$n! = n \cdot (n-1) \cdots \cdot 5 \cdot 4 \cdot 3 \cdot 2 \cdot 1.$$

Thus, $20 | n!$ when $n \geq 5$. Hence, Theorem 5.34(1) yields

$$1! + 2! + \cdots + 500! \equiv_{20} 1! + 2! + 3! + 4! + 0 + \cdots + 0$$
$$= 1 + 2 + 6 + 24$$
$$= 33$$
$$\equiv_{20} 13.$$

Therefore, the desired remainder is 13.

The approaches of the previous two examples can be generalized.[8]

Theorem 5.35. *If p is prime and $a, b \in \mathbb{Z}$, then $(a + b)^p \equiv_p a^p + b^p$.*

Example 5.16 exhibits the key idea involved in the proof of this result. In the expansion of $(a + b)^p$, can we find terms that are equivalent mod p to 0? We know that the Binomial Theorem yields the terms a^p and b^p in the expansion of $(a + b)^p$, so our hope is that every other term of this expansion satisfies this desired equivalence. That is indeed the case, which we see in the proof below.

Proof Let p be prime and $a, b \in \mathbb{Z}$. Then, the Binomial Theorem gives

$$(a + b)^p = a^p + b^p + \sum_{n=1}^{p-1} \binom{p}{n} a^{p-n} b^n.$$

It suffices to show that

$$\binom{p}{n} \equiv_p 0$$

for $0 < n < p$. To this end, let $0 < n < p$. Then,

$$\binom{p}{n} = \frac{p!}{n!(p-n)!},$$

or equivalently,

$$n!(p-n)! \binom{p}{n} = p!$$

$$= p(p-1)!.$$

Thus, p divides $n!(p-n)!\binom{p}{n}$, and since p is prime, Euclid's Lemma tells us that p must divide one of the factors of this term. The factors of $n!$ are

$$1, 2, \ldots, \text{or } n$$

and the factors of $(p-n)!$ are

$$1, 2, \ldots, \text{or } p - n.$$

All of these are less than p, meaning p cannot divide them (Lemma 3.14). Hence, p must divide $\binom{p}{n}$, meaning

$$\binom{p}{n} \equiv_p 0,$$

---

[8]Theorem 5.35 is often referred to as the *Freshman's Dream*; can you guess why?

proving that $(a + b)^p \equiv_p a^p + b^p$. $\qquad\qquad\qquad\qquad\qquad\qquad\qquad\qquad\qquad$ □

The next result, known as *Fermat's Little Theorem*, is a fundamental result of elementary number theory. Though attributed to Fermat, due to his communication of the result to a fellow mathematician in 1640, the actual first published proof was given by Euler in 1736 [32].[9]

Theorem 5.36. (Fermat's Little Theorem) *For any $a, p \in \mathbb{Z}$ with p prime,*

$$a^p \equiv_p a.$$

Proof Let $p \in \mathbb{Z}$ be prime. To show

$$a^p \equiv_p a$$

for any integer a, consider two cases: $a \geq 0$ or $a < 0$.

Case 1: Assume $a \geq 0$ and proceed by induction on a to show the result holds. The base case, $a = 0$, is trivial, as

$$0 \equiv_p 0.$$

Assume our inductive hypothesis to be true for a general $a \geq 0$:

$$a^p \equiv_p a.$$

Then,

$$(a + 1)^p \equiv_p a^p + 1^p \text{ (Theorem 5.35)}$$
$$\equiv_p a + 1 \text{ (IH)}.$$

Consequently, the result holds for every integer $a \geq 0$.

Case 2: Suppose $a < 0$. Then, $-a > 0$ and by the previous case,

$$(-a)^p \equiv_p (-a).$$

Next, we consider the two possibilities for p: either $p = 2$ or p is odd.

Subcase 1: Assume $p = 2$. Note that two integers are equivalent modulo 2 if and only if they have the same parity. Then, since a^2, a and $-a$ are all of the same parity, we have that

$$a^2 = (-a)^2$$
$$\equiv_2 -a$$
$$\equiv_2 a,$$

---

[9]Leibniz, of calculus fame, is credited with presenting an unpublished manuscript that included a proof of the result in 1683 [18].

proving the result.

Subcase 2: Suppose p is odd. Then,

$$a^p = (-1)^{p+1}a^p$$
$$= -1(-1)^p a^p$$
$$= -1(-a)^p$$
$$\equiv_p -1(-a)$$
$$= a.$$

Thus, when $a < 0$, and consequently for all a, the result holds. $\square$

Another application of modular arithmetic is one that you are probably familiar with. Suppose you were asked if 3,564,102 is divisible by 3. If you did not have a calculator handy, perhaps you used the "divisibility trick" of summing the digits of 3,564,102 and seeing if that sum was divisible by 3. In this case, the digits sum to 21, and consequently, 3,564,102 is indeed divisible by 3. But *why* does this technique work? It is a corollary to Fermat's Little Theorem.

Corollary 5.37. *A positive integer is divisible by 3 if and only if the sum of its digits is divisible by 3.*

How do we go about proving such a result? It should come as no surprise that we use properties of modular arithmetic to simplify certain expressions. In this case, because we are looking at the digits in the decimal representation of a number, we will call upon the definition base 10 representation. The base 10 representation of an integer represents a sum of multiples of powers of 10, and the critical observation needed is that

$$10 \equiv_3 1.$$

Proof Let $n \in \mathbb{Z}$ with the decimal representation of n given by $d_m d_{m-1} \ldots d_1 d_0$; that is,

$$n = d_m 10^m + d_{m-1} 10^{m-1} + \ldots d_1 10 + d_0.$$

Note that $10 \equiv_3 1$, and hence by Theorem 5.34,

$$10^k \equiv_3 1$$

for all $k \in \mathbb{Z}^+$. Then,

$$3|n \Leftrightarrow n \equiv_3 0$$
$$\Leftrightarrow (d_m 10^m + d_{m-1} 10^{m-1} + \ldots d_1 10 + d_0) \equiv_3 0$$
$$\Leftrightarrow (d_m(1) + d_{m-1}(1) + \ldots d_1(1) + d_0) \equiv_3 0 \text{ (Theorem 5.34)}$$
$$\Leftrightarrow 3|(d_m + d_{m-1} + \cdots + d_1 + d_0),$$

as was to be shown. □

Up to this point, a majority of the presented modular arithmetic results seem natural. This next result, however, shows us that we have to be careful when working with new mathematical objects and operations. You may have had an inkling of it when reading the statement of Fermat's Little Theorem; you may ask why p was required to be prime? It follows from investigating the converse statement to one of the modular arithmetic properties of Theorem 5.34.

We know that we can multiply through modular equivalence without changing the relation: for any $c \in \mathbb{Z}$,

$$\text{if } a \equiv_m b, \text{ then } ca \equiv_m cb.$$

Are we allowed, then, to undo this operation with division? A quick counterexample shows us that we cannot:

$$3 \cdot 4 \equiv_9 3 \cdot 1,$$

but

$$4 \not\equiv_9 1.$$

As curious mathematicians, can we determine either for what integers c the result holds (that is, if $ca \equiv_m cb$, then $a \equiv_m b$), or, can we adjust the result to a similar statement? It is the latter we choose to present here, and it is considered a method for division over modular arithmetic. In this case, it is not just that the numbers being divided that change but also the modulus itself.[10]

Theorem 5.38. *If $ca \equiv_m cb$ for $a, b, c, m \in \mathbb{Z}$ with $m \geq 1$, then*

$$a \equiv_{m/d} b$$

where $d = \gcd(c, m)$.

Proof Assume for $a, b, c, m \in \mathbb{Z}$ with $m \geq 1$ that

$$ca \equiv_m cb.$$

Then, by definition, we know that

$$ca - cb = mk$$

for some $k \in \mathbb{Z}$. If $d = \gcd(c, m)$, then there exist $x, y \in \mathbb{Z}$, with x and y relatively prime, so that $c = dx$ and $m = dy$ (Exercise 4.3(13b)). This yields

---

[10]Note that presenting a result like Theorem 5.38 is very different than saying, "Here are the cases for which the converse to the second result of Theorem 5.34 does hold." Such an approach is given in Corollary 5.39.

$$dxa - dxb = dyk,$$

showing that

$$y|(xa - xb),$$

or equivalently,

$$y|x(a - b).$$

Since $\gcd(x, y) = 1$, Euclid's Lemma tells us that $a - b$ is divisible by y, meaning

$$a \equiv_y b,$$

which is equivalent to

$$a \equiv_{m/d} b,$$

as was to be shown. $\square$

A special case of Theorem 5.38 is when the modulus is a prime not dividing a common factor of the two terms.

Corollary 5.39. *If*

$$ca \equiv_p cb$$

for $a, b, c, p \in \mathbb{Z}$ with p prime and p not dividing c, then

$$a \equiv_p b.$$

Proof Suppose we have that

$$ca \equiv_p cb$$

for $a, b, c, p \in \mathbb{Z}$ with p a prime not dividing c. Then, $\gcd(p, c) = 1$, and the result follows precisely from Theorem 5.38. $\square$

We have barely skimmed the surface of modular arithmetic. Number theory, cryptography, combinatorics and numerous other areas of mathematics, as well as countless applications beyond mathematics, develop and apply it far beyond the basic properties here. Armed with just the definition and these few results, however, you are able to gain a thorough appreciation for its theoretic power.

Exercises

1. Show that a and b, with the given value of m, satisfy the equivalent statements of Theorem 5.29.

    (a) $a = 11$, $b = 19$, $m = 4$

    (b) $a = 148$, $b = -113$, $m = 9$

    (c) $a = -15$, $b = 0$, $m = 5$

    (d) $a = 563$, $b = -1392$, $m = 23$

2. Find the least positive integer m satisfying each of the following.

    (a) $16 \equiv_9 m$

    (b) $18 + 11 \equiv_9 m$

    (c) $18 \cdot 11 \equiv_9 m$

    (d) $9 \cdot 2017 + 4 \equiv_9 m$

    (e) $-4 \equiv_9 m$

    (f) $19^{206} \equiv_9 m$

    (g) $200! \equiv_9 m$

    (h) $4444^{4444} \equiv_9 m$

3. Find the remainder when a is divided by m.

    (a) $a = 5^{201}$, $m = 24$

    (b) $a = 71^{379}$, $m = 379$

    (c) $a = 2^{44}$, $m = 89$

4. List two positive and one negative element of each equivalence class of the mod m relation on $\mathbb{Z}$ for the given values of m

    (a) $m = 2$

    (b) $m = 5$

    (c) $m = 6$

5. Find a number x satisfying each of the following.

    (a) $2x \equiv_7 1$

    (b) $11x \equiv_{13} 1$

    (c) $x \equiv_3 2$ and $x \equiv_4 3$

6. If x and 42 are relatively prime, show that

$$x^6 \equiv_{168} 1.$$

7. Prove Theorem 5.30 using only the definition of congruence modulo m.

8. Let $m \in \mathbb{Z}^+$ and $a \in \mathbb{Z}$. Prove that

$$[a]_m = \{a + xm \mid x \in \mathbb{Z}\}.$$

9. Disprove each of the following results. Assume all variables represent integers, with $m, n > 0$.

    (a) If $a \equiv_m b$, then $a = b + m$.

(b) If $ac \equiv_m bc$, then $a \equiv_m b$.

(c) If $a \equiv_m b$ and $a \equiv_n b$, then $a \equiv_{mn} b$.

10. Prove each of the following. All variables represent integers, with each modulus being positive, unless otherwise noted.

   (a) If $ma \equiv_n mb$ with $\gcd(m, n) = 1$, then $a \equiv_n b$.

   (b) If $ma \equiv_p mb$ for a prime p with $p \nmid m$, then $a \equiv_p b$.

   (c) If $p > 2$ is a prime, then
   $$1^{p-1} + 2^{p-1} + \cdots + (p-1)^{p-1} \equiv_p -1.$$

   (d) If $\gcd(a, m) = 1$, then there exists b such that $ab \equiv_m 1$ (b is called an *inverse* of a modulo m).

11. Find an inverse (see Exercise 10(d)) of

   (a) 7 modulo 9.

   (b) 10 modulo 13.

   (c) 17 modulo 21.

12. Prove that if $x \in \mathbb{Z}$, then $x^2 \equiv_4 0$ or $x^2 \equiv_4 1$.

13. Prove:

   (a) An integer is divisible by 4 if and only if the integer formed by its tens and units digit is divisible by 4 (for example, 15,654,780 is divisible by 4 since $4|80$).

   (b) An integer is divisible by 9 if and only if the sum of its digits is divisible by 9.

   (c) An integer is divisible by 11 if and only if the alternating sum of its digits, from left to right, is divisible by 11 (for example, 7,317,607 is divisible by 11 since the alternating sum of its digits is $7 - 3 + 1 - 7 + 6 - 0 + 7 = 11$).

14. Determine, without actually computing the division, which of the following are divisible by 4, 9, and/or 11.

   (a) 1268

   (b) 17,314

   (c) 8,218,410

15. A *palindrome* is an integer that reads the same forward as backward (such as 131 or 2442). Prove that 11 divides every four-digit palindrome.

16. Prove or disprove each of the following. All variables represent integers, with each modulus positive.

   (a) If $a \equiv_m b$ and $d|m$, then $a \equiv_d b$.

   (b) If $a \equiv_m b$ and $c > 0$, then $ca \equiv_{cm} cb$.

   (c) If $a^2 \equiv_m b^2$, then $a \equiv_m b$.

(d) If $a \equiv_m b$, then $\gcd(a, m) = \gcd(b, m)$.

(e) If a is odd and $m \in \mathbb{Z}^+$, then
$$a^{2^m} \equiv 1 \bmod 2^{m+2}.$$

17. (a) Prove Wilson's Theorem: if p is prime, then
$$(p - 1)! \equiv_p -1.$$

(b) Use Wilson's Theorem to find the remainder when $97!$ is divided by 101.

18. Prove that there does not exist an integer n such that n^2 has a units digit of 2, 3, 7 or 8.

19. Let $m \in \mathbb{Z}^+$. Prove that $\equiv_m$ partitions $\mathbb{Z}$ into m distinct equivalence classes.

6

Functions

Quickly consider the expression,

$$f(x) = x^2 + 1.$$

What came to mind? Did you think, "$f(1) = 2$?" Without exerting much mental sweat, you may have started to ponder the *graph* of this expression. Maybe you considered its *domain* or *range*. If you recently took calculus, perhaps the words *continuous* or *differentiable* popped into your brain. Others still may have instantly thought, "This is a translation of another function, in particular, of the function $g(x) = x^2$."

Regardless of what you thought, you thought *something* grounded in your mathematical background. Having worked with functions for years, you have developed a deep, visceral notion for them. This intuition is critical to becoming a strong mathematician, as functions exist in just about every corner of the field (or computer science, economics, finance, physics ...). Much of the mathematical theory you will encounter in later studies will revolve around functions and their properties or applications. But while a strong feeling for functions is critical in building a firm foundation of mathematical knowledge, it is not enough. Consider the question:

How is a mathematical function actually defined?

The chances are high that you are thinking about how to sharpen your intuitive notions. Words such as "input," "output," "takes" and "sends" are vague. The first chapters of this book have made you long for precision. Thus, you are left wondering, "How can these vague notions be replaced with precise mathematical terminology?" In particular, *what* mathematical objects should be part of the definition?

Our discussion of functions finds its roots with Euler. He provided the first thorough investigations of them as stand-alone algebraic objects. He combined them to form complex expressions involving exponents, logarithms and trigonometric functions, in addition to polynomial and rational expressions. With the newly introduced calculus of Newton and Leibniz, he examined differentiating and integrating them. The study of these objects and their properties was not what took the mathematical world by storm, however. It was that Euler investigated them without reliance on their mathematical graphs [6]! Even today, centuries after the introduction of functions and Euler's demonstration that they, without reference to their graphs, are worthy of investigation, much of our time is spent considering the graphs of functions and gathering information from them. We will see here that *both* approaches are beneficial and worthy of our time.

The goal of this chapter is to not just define functions and the terms related to them. Rather, we want to investigate what role those mathematical objects take in certain proofs. Knowing that, then, we will aim to structure proofs involving functions in the most elegant way possible. While the final drafts of the proofs will often not reference the graph of a function, those same graphs turn out to be a powerful tool used in figuring out how a proof works.

6.1   Functions Defined

Most readers of this book are familiar with *functions* and many of the terms associated with them: *domain, co-domain, range, inverse*, etc. By this point in your proof development, however, you know that such familiarity requires precision and that precision comes in the form of definitions.

Before jumping right into the requisite definitions for this section, let us consider the functions $f(x) = x^3$ and $g(x, y) = x^2 + 3y$. What questions can we ask of them that may lead us to construct a precise mathematical definition of a function?

First off, what is it that these functions actually do? The first, f, takes as input any real number x and assigns it another real number as an output, x^3. Similarly, g takes an ordered pair (x, y) of real numbers and partners it with a real number output $x^2 + 3y$. The words "assigns" and "partners" are synonymous with "relates." Because of this, the precise definition of a function should come as no surprise. A function is a relation that satisfies certain particular properties.

Definition 6.1. A *function* (or *map* or *mapping*) f from a set X to a set Y, denoted

$$f : X \to Y,$$

is a relation from X to Y so that for every $x \in X$, there exists a unique $y \in Y$ such that $(x, y) \in f$ (or equivalently, $x f y$). If $(x, y) \in f$, we write $f(x) = y$ and say y is the *image* of x under f (or that x is *mapped* to y by f). The set X is called the *domain* of f, denoted $\mathrm{Dom}(f)$. The set Y is called the *co-domain* of f, while the *range* of f is the set

$$\mathrm{Ran}(f) = \{f(x) \mid x \in X\}.$$

As with many mathematical objects, it is often helpful to have an imprecise way of thinking about definitions. How, then, can we reword the definition of a function into something more intuitive? A relation between two sets is a function if every input is assigned exactly one output. In algebra, you may have come to learn a graphical interpretation of this: the Vertical Line Test. A graph on the Euclidean plane represents a function (of x) if there is no vertical line that intersects the graph more than once. Here, however, we are not restricting ourselves to real-valued functions (nor have we defined the graph of a function). Thus, we need a different method, and the aforementioned description does just that.

Are there approaches to talking about functions using language other than that of relations? Your mathematical background may be suggesting one of a few different possible options. Each has its own advantages and disadvantages. For functions whose domain and/or co-domain are not numeric, verbal descriptions are often appropriate.

Example 6.1. Let S be the set of all full-time students at a particular college and C the set of all courses being offered at that college this current semester or term. Consider two relations, F_1 and F_2 from S to C:

F_1: sF_1c if and only if student s is taking course c.

F_2: sF_2c if and only if course c is the first class of the first week for student s.

To determine if either is a function, revert to our intuitive notion. *Does each input have exactly one output?* This question is enough to determine if the relation is or is not a function. If it is not a function, because the definition of a function is a universal statement, its negation is an existential statement. Hence, a single counterexample would suffice in showing either F_1 or F_2 is *not* a function (in other words, it is often much easier to show a relation is not a function versus showing it is a function). In the first case, if there is a student s taking more than one course, say c_1 and c_2, then we would have both sF_1c_1 and sF_1c_2. If this is indeed true, then F_1 is *not* a function.

The relation F_2, though, is a function. Each full-time student s has exactly one course that is the first class attended the first week of school. The domain of F_2 is the set of all full-time students at the college, the co-domain is the set of all classes at the college, and $\mathrm{Ran}(F_2)$ is the set of all courses at the college so that for some student in the course, it is the first course of the week. While it is possible that the co-domain of F_2 equals the range, it is rather unlikely. Of the hundreds of courses offered at the college, if there is just one in which it is *not* the first class of the

week for every student in the class, then that class is in the co-domain of
F_2 but not the range.

While we have seen the importance of precisely knowing definitions, the
previous example highlights the usefulness of having an instinctive understanding of them, especially when showing a relation is or is not a function.[1]
Though proofs rely on precision, our intuition is often extremely helpful in
determining if something is or is not a certain mathematical object.

Verbal descriptions of functions like F_2 in Example 6.1 are natural. If we
attempt to use our standard function notation, it is awkward:

$$F_2(s) = \text{the first class of the first week for student } s.$$

Though they appear frequently in the real-world, we will not often work with
such functions. When we want to describe a particular function, we will use
the word "define," meaning we make it explicitly clear how each element of
the domain is treated by the function. Verbal explanations, though they can
explicitly define functions, are cumbersome and inefficient. The following three
methods address this.

Example 6.2. Define a function

$$f : \{1, 2, 3, 4, 5\} \to \mathbb{Z}$$

via the following table.

x	$f(x)$
1	-1
2	2
3	101
4	-71
5	0

This table explicitly defines how every element of the domain is
mapped under f. Moreover, we clearly see the range of f is a proper
subset of its co-domain $\mathbb{Z}$:

$$\mathrm{Ran}(f) = \{-71, -1, 0, 2, 101\}.$$

When the domain of a function is finite with relatively few elements, tables

---

[1] It is worth noting that at most colleges and universities, the relation F_1 of Example
6.1 would not be a function. However, for schools that are on the block calendar system
(students take a single course at a time), F_1 would indeed be a function.

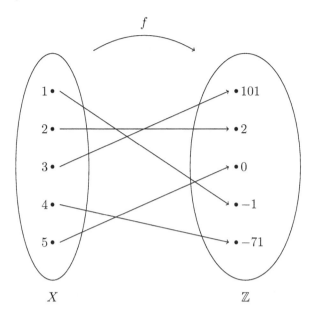

FIGURE 6.1
Example 6.3, $f : X \to \mathbb{Z}$

are satisfactory ways to exhibit how the function is defined. But imagine trying to create such a table for a function whose domain has hundreds of elements. The following method for defining a function has the same limitations. What sets it apart, however, is the visual nature of it.

Example 6.3. If $X = \{1, 2, 3, 4, 5\}$, then the function

$$f : X \to \mathbb{Z}$$

of Example 6.2 is definable via Figure 6.1.

When drawing diagrams to represent functions, it is not necessary to illustrate every element of the co-domain, as in Figure 6.1. Indeed, in this case, the co-domain is infinite and thus every element of it could not be explicitly labeled in the diagram. However, the co-domain is listed; note the "$\mathbb{Z}$" below the rightmost ellipse in Figure 6.1. All the parts of the function do appear in the diagram: the name of the function (f), the domain (X) and the co-domain $(\mathbb{Z})$, and how the functions maps every element of the domain. But what if the domain contained many more than 5 elements, or, if it were infinite? While drawing similar illustrations of such functions can be challenging, there are other ways to express them.

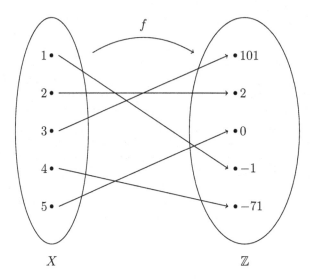

FIGURE 6.2
Example 6.3, $g : X \to \mathbb{R}$

Before getting to our last method of defining functions (which happens to be the method you are likely most familiar with), let us consider what it means for two functions to be *equal*. Consider

$$g : \{1, 2, 3, 4, 5\} \to \mathbb{R}$$

defined via the table below as well as in Figure 6.2.

x	$g(x)$
1	-1
2	2
3	101
4	-71
5	0

The function g seems to behave identically to the function f of Examples 6.2 and 6.3; the visual representations of each are nearly identical as well. For any $a \in \{1, 2, 3, 4, 5\}$, we have $g(a) = f(a)$. Does that mean the two functions are equal? While we are inclined to say so, we have not explicitly defined what it means for functions (mathematical objects that we have just defined) to be equal. Though f and g behave similarly, it turns out that the functions are not equal, according to the following definition of function equality. It reiterates what we mean when we say, in general, that two things are *equal*: everything about the two must be identical

> **Definition 6.2.** Two functions f and g are said to be *equal*, written $f = g$, if all of the following hold.
>
> 1. $\mathrm{Dom}(f) = \mathrm{Dom}(g)$.
>
> 2. The co-domain of f equals the co-domain of g.
>
> 3. $f(x) = g(x)$ for every $x \in \mathrm{Dom}(f)$.

Thus, because they do not have the same co-domain, the aforementioned functions f and g are not equal. In the visual representations of f and g, the only difference, $\mathbb{Z}$ versus $\mathbb{R}$ below the rightmost ellipses, proved to be the determining factor that f and g are indeed different functions.

Now, let us move to the common algebraic representation of functions, the likes of which you undoubtedly are comfortable with.

> **Example 6.4.** Define $f : \mathbb{R} \to \mathbb{R}$ by $f(x) = 3x + 2$.
>
> The function f, even with an infinite domain and co-domain, is completely defined in this succinct statement. How f maps any element of the domain is given in terms of a variable x. This provides a quick method for mapping specific elements.

Determining the range of a function f can sometimes be challenging. However, the tools you have developed throughout your mathematical career can be quite useful. The function f of Example 6.4 has as its range the set of all real numbers. How do we know this? Consider its graph in the Euclidean plane. It is a linear function of positive slope. However, when we consider g defined as $g : \mathbb{Z} \to \mathbb{Z}$ by $g(x) = 3x + 2$, determining $\mathrm{Ran}(g)$ is much more involved. The range of g must be a subset of its co-domain $\mathbb{Z}$. Because of this, we can no longer consider the graph of g.

Perhaps then our thought process is, "What sorts of outputs does g have?" Or more concretely, we must determine all integers y so that $y = 3x + 2$ for some integer x. This gives us something to work with:

$$y = 3x + 2$$

means

$$\frac{y - 2}{3} = x$$

In other words, $\dfrac{y - 2}{3}$ must be an integer. That is,

$$y - 2 \equiv_3 0,$$

or equivalently,

$$y \equiv_3 2.$$

Thus,

$$\begin{aligned}
\mathrm{Ran}(g) &= \{y \in \mathbb{Z} \mid y \equiv_3 2\} \\
&= \{\ldots, -7, -4, -1, 2, 5, 8, \ldots\} \\
&= [2]_3.
\end{aligned}$$

In mathematical literature, a function is sometimes described as *well-defined*. To say that a function is well-defined simply means that it is indeed a function. The statement, "The function is not well-defined" is actually an oxymoron; if it is not well defined, then it is not a function. Why? A function $f : X \to Y$ is not well-defined if there is some $x \in X$ that maps to two different elements of the co-domain:

$$\begin{aligned}
f(x) &= y_1 \\
&\neq y_2 \\
&= f(x.)
\end{aligned}$$

Example 6.5. Let $f : \mathbb{Q} \to \mathbb{Z}$ by $f(\frac{a}{b}) = a$.

While f may appear to be a function, it is not. Elements of the domain have different representations (i.e., $\frac{1}{2} = \frac{2}{4}$):

$$f(\frac{1}{2}) = 1 \neq 2 = f(\frac{2}{4}).$$

Thus, we may say f is not well-defined. Simply put, f is not a function. It is still a relation, however.

We continue with terminology specific to functions. The first definition is a natural way to restrict the domain of a function to allow only certain inputs, while the latter is an extension of a previous definition addressing a particular question.

A method exists for talking about how a certain element of the domain is mapped under a function, but what if we want to consider mapping multiple elements of the domain? Definition 6.3 addresses both of these issues.

Definition 6.3. Let $f : X \to Y$ be a function and $A \subseteq X$. The *restriction* *of f to A* is the function $f_A : A \to Y$ defined by $f_A(x) = f(x)$. The *image of A under f*, denoted $f(A)$, is the set

$$f(A) = \{b \in Y \mid \exists\, a \in A \text{ such that } f(a) = b\}$$
$$= \{f(a) \mid a \in A\}.$$

The *image of f* is defined to be $f(X)$.

In other words, the image of a subset A of X under $f : X \to Y$ is the subset of the co-domain consisting of all the images (or outputs) coming from elements of A. Similarly, the image of f is the subset of the co-domain consisting of the images of *all* the elements of the domain. We have previously defined this set: $\mathrm{Ran}(f)$.

In Definition 6.3, we start with a subset of the domain and map all of its elements. Is it possible to start with a subset of the co-domain and unmap all of its elements? Definition 6.4 defines such a notion.

Definition 6.4. Let $f : X \to Y$ be a function and let $B \subseteq Y$. The *inverse image of B under f*, denoted $f^{-1}(B)$, is the set

$$f^{-1}(B) = \{a \in X \mid f(a) \in B\},$$

a subset of X. In other words, $a \in f^{-1}(B)$ if and only if $f(a) \in B$. The *inverse image of y*, where $y \in Y$ is the inverse image of the set $\{y\}$.

What is the intuitive notion of $f^{-1}(B)$? It is the set of all elements of the domain that are mapped into B (i.e., whose image is in B). That is, an element is in $f^{-1}(B)$ only if that element is sent by f into B. A previously mentioned suggestion is worth repeating here: understand this definition viscerally. Working with functions, images of sets, and inverse images of sets can be a challenge in notation, and that notational struggle often leads to frustrations when trying to figure out how proofs with these objects work. Discern how the proof proceeds at this intuitive level, and then sketch it out using proper notation.

Example 6.6. Consider the function f of Example 6.2. Then,

(1) If $A = \{1, 2, 4\}$, then $f(A) = \{-71, -1, 2\}$.
(2) If $A = \varnothing$, then $f(A) = \varnothing$.
(3) If $B = \{0, 2\}$, then the inverse image of B is $\{2, 5\}$.
(4) The inverse image of 101 is 3.

(5) If $B = \{4, 5\}$, then $f^{-1}(B) = \varnothing$.

Images and inverse images of sets are straightforward to compute when a function is defined via a table or visually. When defined algebraically, determining these things takes a bit of work.

Example 6.7. Let $f : \mathbb{R} \to \mathbb{R}$ be the function defined by

$$f(x) = 3x^2 + 1.$$

If $A = \{-1, 0, 1, 2\}$, find both $f(A)$ and $f^{-1}(A)$.

To find the image of A under f, we simply evaluate $f(a)$ for all $a \in A$. Since $f(-1) = 4 = f(1)$, $f(0) = 1$ and $f(2) = 13$, we have that

$$f(A) = \{1, 4, 13\}.$$

To find $f^{-1}(A)$, use your intuition. We ask, "What elements of the domain map into A?" In particular, what $x \in \mathbb{R}$ satisfy $f(x) = -1$, $f(x) = 0$, $f(x) = 1$ or $f(x) = 2$? The problem now has become simple algebra.

If $f(x) = -1$, then

$$3x^2 + 1 = -1,$$

or equivalently,

$$3x^2 = -2.$$

This has no solution in the real numbers. Thus, there is no element $x \in \mathbb{R}$ mapping to -1. The same is true for 0. However,

$$3x^2 + 1 = 1$$

and

$$3x^2 + 1 = 2$$

yield solutions of $x = 0$ and $x = \pm\dfrac{1}{\sqrt{3}}$, respectively. Therefore,

$$f^{-1}(A) = \left\{ -\frac{1}{\sqrt{3}}, 0, \frac{1}{\sqrt{3}} \right\}.$$

Just as the image of a set is itself a *set*, the inverse image of a set is also a *set*. This is a critical idea to keep in mind, especially when trying to figure out the steps of a proof. If you find yourself stuck when trying to prove results dealing with multiple types of mathematical objects, step back from the "busy work" of the problem and look holistically at the objects. Is the object you are working with a set? An element of a set? Perhaps it is something else? If you know what type of object is at hand, you can think about the tools you have developed that are related to them.

> **Quality Tip 26.** *Stay keen to what types of mathematical objects you are working with.*

Because the image of a set is a set, when attempting to prove results involving them, call upon the tools learned in Chapter 2 and Section 3.3.

Example 6.8. Let $f : \mathbb{R} \to \mathbb{R}$ be the function defined by

$$f(x) = -3x + 4.$$

Compute $f(A)$ where A is the interval

$$[-2, 5] = \{x \in \mathbb{R} \mid -2 \le x \le 5\},$$

and prove that your answer is correct using only the tools we have developed in this text.

Determining $f(A)$ is straightforward, as we can call upon tools learned in previous mathematics courses. Knowing linear functions are continuous, we know that intervals map to intervals. Then, because $f(-2) = 10$ and $f(5) = -11$, we know that

$$f(A) = [-11, 10].$$

But how do we *prove* that this is indeed the result? The use of continuity only helped us figure out what it was we have to prove. Continuity is not a term we have defined; it cannot become part of our proof. However, because $f(A)$ and $[-11, 10]$ are sets, to show that they are equal, we must show that they are subsets of one another. In this case, the two arguments can be combined into one:

$$a \in A \Leftrightarrow -2 \le a \le 5$$
$$\Leftrightarrow -3(-2) + 4 \ge -3a + 4 \ge -3(5) + 4$$
$$\Leftrightarrow 10 \ge f(a) \ge -11.$$

This shows the two sets are indeed equal.

A minor detail is hiding in Example 6.8. In showing

$$f(A) = [-11, 10],$$

we did not start by saying, "Let $y \in f(A)$." In general, if we are showing one set is a subset of another, we begin by considering an arbitrary but general element of the first set (here we would be showing $f(A)$ is a subset of $[-11, 10]$). Had we started by taking a general $y \in f(A)$, we would then have, "$y = f(a)$ from some $a \in A$, by the definition of $f(A)$." While beginning with "$y \in f(A)$" is not incorrect, it introduces the unnecessary variable y into the proof. The structure of images and inverse images of sets allows us to be more efficient with our proofs.

Quality Tip 27. *Unnecessary variables congest a proof.*

Mathematicians define objects and investigate the properties of them. We started that process here with functions. But mathematicians also consider the relationships between newly and previously defined objects. We know that images of sets *are* sets. It is thus natural to ask how certain set-theoretic objects and functions interact. The following theorem relates images of unions and intersections.

Theorem 6.5. *Let $f : X \to Y$ be a function and let $A, B \subseteq X$.*

1. $f(A \cup B) = f(A) \cup f(B)$
2. $f(A \cap B) \subseteq f(A) \cap f(B)$

How would we go about proving such results? If you are stuck, revert back to Quality Tip 26. What sorts of objects are we working with? Sets. And what is it we are trying to do with them? Prove that two of them are equal: $f(A \cup B)$ and $f(A) \cup f(B)$. The definition of set equality tells us to use subset arguments. As the proofs are fairly straightforward, we leave out the scratch work and jump straight into the final draft of the proof.

Proof Let $f : X \to Y$ be a function between sets X and Y, and take $A, B \subseteq X$. We prove each result separately.

(1) To prove $f(A \cup B) = f(A) \cup f(B)$, we must show both

$$f(A \cup B) \subseteq f(A) \cup f(B)$$

and

$$f(A) \cup f(B) \subseteq f(A \cup B).$$

To show the first, consider $x \in A \cup B$, so that $f(x) \in f(A \cup B)$. Because $x \in A \cup B$, it follows that $f(x) \in A$ or $f(x) \in B$. Hence, $f(x) \in f(A) \cup f(B)$, showing that

$$f(A \cup B) \subseteq f(A) \cup f(B).$$

Now, take $y \in f(A) \cup f(B)$, so that $y \in f(A)$ or $y \in f(B)$. Since $A \subseteq A \cup B$ (Theorem 2.8), we have that $y \in f(A \cup B)$, proving

$$f(A) \cup f(B) \subseteq f(A \cup B)$$

and consequently the desired result.

(2) In order to prove

$$f(A \cap B) \subseteq f(A) \cap f(B),$$

take $x \in A \cap B$. We must show $f(x) \in f(A) \cap f(B)$. Since $x \in A$ and $x \in B$, it follows that $f(x) \in f(A)$ and $f(x) \in f(B)$, respectively, proving

$$f(x) \in f(A) \cap f(B).$$

$\square$

Note that the two results of Theorem 6.5 are significantly different. The first states that two sets are equal, while the second simply shows that one set is a subset of the other. In Exercise 22, you are asked to find a specific example of a function showing that this result can only be stated this way. That is, you will find a counterexample to show that

$$f(A \cap B) \neq f(A) \cap f(B).$$

In the coming sections we will determine a property so that, if a function f has this property, then

$$f(A \cap B) = f(A) \cap f(B).$$

For now, we consider inverse images of sets and their relationships with unions and intersections. A similar result holds to Theorem 6.5, though in this case, the inverse image preserves equality over both unions and intersections.

Theorem 6.6. *Let $f : X \to Y$ be a function and let $A, B \subseteq Y$.*

  1. $f^{-1}(A \cup B) = f^{-1}(A) \cup f^{-1}(B)$
  2. $f^{-1}(A \cap B) = f^{-1}(A) \cap f^{-1}(B)$

Let us think through the scratch work of (1) (the proof of (2) is left as an exercise). We must show either set is a subset of the other. To show

$$f^{-1}(A \cup B) \subseteq f^{-1}(A) \cup f^{-1}(B),$$

take an element x in $f^{-1}(A \cup B)$. What do we know about x? To be in the inverse image of a set, that element must get mapped by the function into the set. That is, $f(x) \in A \cup B$.

Where do we go from here? Keep our eyes on the prize. We want to show that x is an element of $f^{-1}(A)$ or $f^{-1}(B)$. How do we show this? By showing that x maps into either A or B. More concretely, we want to show $f(x) \in A$ or $f(x) \in B$. Can we conclude this? Of course we can, since we have $f(x) \in A \cup B$.

A similar argumentative approach builds the machinery for the other subset argument. With this intuitive scratch work, we are prepared to construct the final draft of our proof. While numerous linebreaks and whitespace could be used to make this proof extra readable, we choose to leave many symbolic statements in-line, as they are not overly lengthy and do not impact the readability of paragraphs. Visually, this approach may seem dense. However, remember that different audiences have different needs. If preparing a manuscript for publication in a journal (or a solution to a problem section of a journal, as previously mentioned), sometimes brevity is a necessity.

Proof To show (1), we show $f^{-1}(A \cup B)$ and $f^{-1}(A) \cup f^{-1}(B)$ are subsets of one another.

First, let $x \in f^{-1}(A \cup B)$. By Definition 6.4, $f(x) \in A \cup B$. WLOG, we can assume $f(x) \in A$. Thus, $x \in f^{-1}(A)$ and by Theorem 2.8, $x \in f^{-1}(A) \cup f^{-1}(B)$. This shows the first desired inclusion.

Now, take $x \in f^{-1}(A) \cup f^{-1}(B)$. WLOG, if we assume $x \in f^{-1}(A)$, we have that $f(x) \in A$; as above, it follows that $f(x) \in A \cup B$, proving that $x \in f^{-1}(A \cup B)$.   □

If we begin with a subset of a function's domain, let us consider its image. This is a subset of the co-domain. Would the inverse image of that resulting set yield the initial set we began with? That is, if we map a set and then unmap the resulting set, are we right back where we started? Not precisely. The following theorem proves that we can only conclude the resulting set is a subset of the original set.

Theorem 6.7. *Let $f : X \to Y$ be a function.*

> *1. For every subset B of Y, $f(f^{-1}(B)) \subseteq B$.*
>
> *2. For every subset A of X, $A \subseteq f^{-1}(f(A))$.*

While the proof of (1) is left as an exercise, it is worthwhile to examine how the construction of the proof of (2). It clearly should utilize the set element method of proof, as we are showing one set is a subset of another. Take $a \in A$. We want to show $a \in f^{-1}(f(A))$. Simply knowing $a \in A$ does not give us much to work with, but examining the components (i.e., $f^{-1}(f(A))$) leads us in the right direction.

The set $f^{-1}(f(A))$ requires the use of two definitions. First, we want to show a is an element of the inverse image of *some set*. How does a get into such a set? It must be mapped into *that set* via f. Playing the role of *that set* here is $f(A)$. Thus, we need to show that $f(a) \in f(A)$. Since $a \in A$, this is immediate (from the definition of the image of A under f) and we have what we need to construct our rather short proof.

Proof Let $f : X \to Y$ be a function between sets X and Y, and take $A \subseteq X$, with $a \in A$. Because $f(a) \in f(A)$, it follows that $a \in f^{-1}(f(A))$, proving (2). □

There are situations where the sets of Theorem 6.7 turn out to be equal, rather than one a subset of the other. That is, functions $f : X \to Y$ that have a certain property will yield

$$f(f^{-1}(B)) = B$$

for every $B \subseteq Y$, and similarly, for $A \subseteq X$, a different property will guarantee

$$A = f^{-1}(f(A)).$$

As with the second result of Theorem 6.5, these properties are determined and investigated in the coming sections.

Throughout this section we have considered the basics of functions and proofs involving them. As our theory of functions grows, we rely upon these same proof techniques and approaches, regardless of what new definitions we come across. Some of those definitions, which are likely familiar, appear in the next two sections.

Exercises

1. Find the domain and range for each of the functions described. List at least two possible co-domains for each function.

   (a) The function f mapping every integer x to $x + 2$.
   (b) The function g mapping every integer x to its square x^2.
   (c) The function h mapping every integer to its remainder upon division by 17.
   (d) The function F mapping every ordered pair of real numbers (x, y) to $\sqrt{x^2 + y^2}$.
   (e) If L is the set of all letters in the English language, then the function C giving the number of elements in an element of $\mathcal{P}(L)$.

2. Explain why each of the following is *not* a function.

   (a) $f : \mathbb{Z} \to \mathbb{R}$, $f(x) = \pm\sqrt{x}$.
   (b) $g : \mathbb{R} \to \mathbb{R}$, $g(x) = \sqrt{x + 3}$.
   (c) $h : \mathbb{Z} \times \mathbb{Z}^* \to \mathbb{Z}$, $h((a, b)) = \dfrac{a}{b}$.
   (d) $w : W \to \mathbb{Z}^+$, where W is the set of words in the English language, $w(x)$ equals the number of occurrences of the letter e in x.

3. The table below exhibits three relations, f, g, and h, from $\{a, b, c, d, e\}$ to $\{1, 2, 3, 4, 5\}$. Determine if f, g and h are functions.

	a	b	c	d	e
f	3	1, 2	3	5	1
g	2	1	3	5	4
h	2	2	2		2

4. Define a function $g : \mathbb{Z} \to \{1, 2, 3\}$ that satisfies the following.

   (a) $\text{Ran}(g) = \{1\}$.

   (b) $\text{Ran}(g) = \{1, 2\}$.

   (c) The range of g equals the co-domain of g.

5. Give examples of three different functions with domain and co-domain of $\{1, 2, 3, 4\}$.

6. Consider the function $f : X \to Y$ illustrated below.

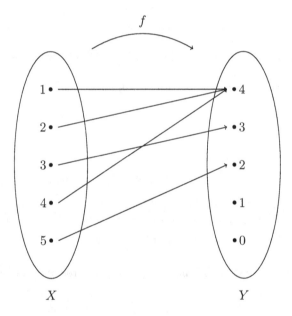

   (a) Find the domain of f.

   (b) Find the co-domain of f.

   (c) Find the range of f.

   (d) Find $f(1)$ and $f(\{1\})$.

   (e) Find $f^{-1}(\{0, 3, 4\})$.

7. Consider the function $g : X \to \mathbb{Q}$ illustrated below.

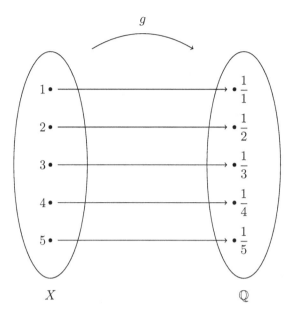

(a) Find Dom(g).

(b) Find the co-domain of g.

(c) Find Ran(g).

(d) Find $g(\{2,4\})$.

(e) Find $g^{-1}(\{\frac{1}{4}\})$ and $g^{-1}(\{\frac{1}{8}\})$

8. Determine if the defined relation f is a function.

(a) From $\mathbb{R}$ to $\mathbb{R}$, xfy if and only if $x^2 + 3 = y + 2$.

(b) From $\mathbb{Z}$ to $\mathbb{Z}$, xfy if and only if $x^2 = y^2$.

(c) From $\mathbb{Z}^+$ to $\mathbb{Z}$, xfy if and only if $x^2 = y^2$.

(d) From $\mathbb{Z}$ to $\mathbb{Z}^+$, xfy if and only if $x^2 = y^2$.

(e) From a nonempty set A to $\mathcal{P}(A)$, afS if and only if $a \in S$.

(f) From $\mathbb{R}^+ \times \mathbb{R}^+$ to $\mathbb{R}$, $(a,b)fx$ if and only if $x = \log_a(b)$.

9. For the given function f, find $f(1)$, $f(2)$, $f(\{1,2\})$ and $f^{-1}(\{1,2\})$.

(a) $f : \mathbb{R} \to \mathbb{R}$, $f(x) = 2x$.

(b) $f : \mathbb{R} \to \mathbb{R}$, $f(x) = |2x|$.

(c) $f : \mathbb{R} \to \mathbb{R}$, $f(x) = \sqrt{2}x$.

10. (a) Let $f : \mathbb{R} \to \mathbb{R}$ be given by $f(x) = x^2$. Find $f((-2,4])$ and prove that your result is indeed true.

(b) If $f : \mathbb{Z} \to \{-1,0,1,2,3\}$ by $f(x) = x^2 \bmod 3$, find the range of f.

(c) If $f : \{1,2,3,4,5\} \to \{1,2,3,4,5,6\}$ by $f(x) = 2x \bmod 7$, what is the range of f?

11. Suppose $h : \mathbb{R} \to \mathbb{R}$ by $f(x) = 1 + x^2$. Find each of the following, proving your result is the desired quantity.

   (a) $h(\{-2, 2\})$
   (b) $h([-2, 2])$
   (c) $h^{-1}(\{1, 5, 10, 17\})$
   (d) $h^{-1}([2, 5])$

12. Suppose $g : \mathbb{R} \to \mathbb{R}$ by $g(x) = (x + \pi)^2$. Find

   (a) the image of $-\pi$, 0, and π.
   (b) the image of $\{-\pi, 0, \pi\}$.
   (c) the inverse image of $-\pi$, 0, and π.
   (d) the inverse image of $\{-\pi, 0, \pi\}$.
   (e) the inverse image of $[0, 9]$.

13. Let $f : \mathbb{Z}^+ \times \mathbb{Z}^- \times \mathbb{R} \to \mathbb{R}$ be defined by $f(x, y, z) = z$. Find

   (a) $f(3, -2, e)$.
   (b) $f(\mathbb{Z}^+ \times \mathbb{Z}^- \times \mathbb{R})$.
   (c) $f(\{2, 5, 7\} \times \mathbb{Z}^- \times \{\pi\})$.
   (d) $f(\mathbb{Z}^+ \times \{-1\} \times \mathbb{R})$.
   (e) $f^{-1}(\mathbb{R})$.
   (f) $f^{-1}(\{\pi\})$.
   (g) $f^{-1}(\mathbb{R} - \mathbb{Q})$.

14. If $f : X \to Y$ is a function, show that $f^{-1} : Y \to X$ given by $f^{-1}(y) = x$ if and only if $f(x) = y$ need not be a function.

15. Let $p : \mathbb{R} \to \mathbb{R}$ be defined as

$$p(x) = \begin{cases} x + 1 & \text{if } x < 0 \\ \pi & \text{if } x = 0 \\ 2x & \text{if } x > 0. \end{cases}$$

   (a) Find $p(\mathbb{Z}^+)$.
   (b) Find $p^{-1}([-5, 5])$.

16. A function f, whose domain and co-domain are both $\mathbb{R}$, is defined below.

$$f(x) = \begin{cases} 2x & \text{if } x < 0 \\ 5 & \text{if } x = 0 \\ 3x & \text{if } 1 < x < 2 \\ x^2 & \text{if } 2 < x. \end{cases}$$

   (a) Find Dom(f).

    (b) Find Ran(f).

    (c) Find $f(\{0\})$.

    (d) Find $f(\{-6, -3, 0, 3, 6\})$.

    (e) Find $f^{-1}(\{-6, -3, 0, 3, 6\})$.

    (f) Show that $2\pi \in f^{-1}(\mathbb{R})$.

17. If $f : \mathbb{Z} \to \mathbb{R}$ is given by $f(x) = \sin(x)$, find $f^{-1}([0, 2\pi])$.

18. If E is the set of even integers and O the set of odd integers, prove that $g(E) = O$ and $g(O) = E$ if

$$g(x) = \begin{cases} x - 3 & \text{if } x \text{ is even} \\ 2x & \text{if } x \text{ is odd.} \end{cases}$$

19. Let $f : X \to Y$ be a function and $A \subseteq X$. Prove that f_A is indeed a function.

20. If $f : X \to Y$ is a function and A a nonempty subset of X, prove that $f(A)$ is nonempty. If B is a nonempty subset of Y, must $f^{-1}(B)$ be nonempty? Prove or disprove.

21. Generalize Theorem 6.5(1) to an arbitrary union of sets and prove that the generalization holds.

22. Provide a specific example of a function $f : X \to Y$ and subsets $A, B \subseteq X$ such that $f(A \cap B) \neq f(A) \cap f(B)$.

23. Let $f : X \to Y$ be a function and $a \in X$. Prove or disprove: $f(a) \neq f(\{a\})$.

24. Prove Theorem 6.6 (2).

25. Prove Theorem 6.7 (1).

26. Give an example of a function $f : X \to Y$ and a subset B of Y for which $f(f^{-1}(B)) \neq B$.

27. Give an example of a function $f : X \to Y$ and a subset A of X for which $f^{-1}(f(A)) \neq A$.

28. Let $f : X \to Y$ with $C, D \subseteq Y$. Prove each of the following claims.

    (a) If $C \subseteq D$, then $f^{-1}(C) \subseteq f^{-1}(D)$.

    (b) $f(f^{-1}(C)) = f(X) \cap C$.

29. Suppose $f : A \to B$ is a function between A and B, each of which has a finite number of elements.

    (a) If A has fewer elements of B, what can we say about Ran(f) and the co-domain of f?

    (b) If A has more elements of B, can we conclude anything about Ran(f) and the co-domain of f?

    (c) If Ran(f) equals the co-domain of f, can we conclude anything about the size of A and B?

30. Let A be a set in a universal set U. Define the function[2] $\chi_A : U \to \{0, 1\}$ given by

$$\chi_A(u) = \begin{cases} 0 & \text{if } x \in U \\ 1 & \text{if } x \notin U. \end{cases}$$

    (a) Let $A = \{1, 3, 5\}$ in the universe $U = \{1, 2, 3, 4, 5\}$. Find $\chi_A(1)$, $\chi_A(2)$, and $\chi_A(\{1, 2\})$.

    (b) When does $\chi_A(U) = \{0\}$?

    (c) When does $\chi_A(U) = \{1\}$?

    (d) When does Ran(χ_A) $= \{0, 1\}$?

6.2 Properties of Functions

Consider for a moment how our investigation of relations progressed. Chapter 5 began by defining a relation as generally as possible. We considered the structure of these objects and their various properties. From there, we moved on to more specific types of relations: equivalence relations and order relations. These are relations that satisfy particularly useful properties.

When a mathematician approaches theory from the generic to the specific, as we did going from general relations to equivalence and order relations, there are often multiple routes such an investigation may take. We just as easily could have moved from considering equivalence relations to the mod relation. From there we could have gone deeper into specific types of these relations, such as those where the modulus is prime (rather than any positive integer).

We choose to take the same approach with functions. In the previous section we considered properties of *any* function. For example, Theorem 6.5 is stated in terms of any function $f : X \to Y$, for any sets X and Y, and any subsets A and B of X. Now, as we did with relations, there are particularly useful kinds of functions that we choose to investigate: those called "surjective" and "injective". The concepts will be familiar, appearing first in algebra and pre-calculus. As we shall see, such functions allow for some of the results in the previous section to be sharpened. Then, continuing along the path of general to specific, in Section 6.3 we investigate special properties of functions that are both injective and surjective.

Let us begin with functions that are called *surjective*. What question or idea spurs defining such a term? It comes from an observation noted in Section 6.1. It is not necessary that every element of the co-domain of a function is an

---

[2]The function χ_A is often called the *characteristic function* of a set.

output of that function. For example, take $f : \mathbb{R} \to \mathbb{R}$ defined by $f(x) = x^2$. Every negative real number is an element of the co-domain of f yet none are in the range of f. In other words, the range of the function is a proper subset of the co-domain. Functions for which this does not happen are called *surjective*.

> **Definition 6.8.** A function $f : X \to Y$ is *surjective* (or *onto*) if for every $b \in Y$, there exists an $a \in X$ such that $f(a) = b$.

Notice that the definition of a function being surjective involves a statement with multiple quantifiers of the form

$$\forall x, \, \exists \, y \text{ so that } P(x, y).$$

Recall that to prove such statements, one begins with an arbitrary but particular element x and then show that some y exists, usually dependent upon that choice of x, so that $P(x, y)$ is true. The following example exhibits this.

Example 6.9. The function $g : \mathbb{R} \to \mathbb{R}$ defined by $g(x) = x^3$ is surjective.

To verify this, consider a general element of the co-domain, $b \in \mathbb{R}$. We must show there exists $a \in \mathbb{R}$ so that $g(a) = b$.

Notice what we must do at this step of the exercise. We must show that there is *some* element of the domain that gets mapped to b by the function g. This is where the "usually dependent upon that choice of x" comment above comes into play. Intuitively and using loose language, we ask ourselves, "What must we put into the function so that b comes out?" Determining the answer amounts to solving a simple algebraic equation: what value of a satisfies $g(a) = b$? We need only solve

$$a^3 = b$$

for a. Because $\sqrt[3]{b}$ is defined for all b, and $(\sqrt[3]{b})^3 = b$, it follows that g is indeed surjective.

If the function g of Example 6.9 were defined as $g(x) = x^2$, proceeding in this manner would yield $a = \sqrt{b}$. While we are tempted then to say, "This is the domain element that maps to a general b," we must be careful, as $\sqrt{b}$ is not defined for all $b \in \mathbb{R}$.

The graphs of real-valued functions allow us to determine if real-valued functions are onto (though it does not actually prove the result). The function of Example 6.9, $g(x) = x^3$, is surjective because every y-value on the Cartesian plane is an output of the function (that is, draw a horizontal line through any point $(0, y)$ and that line will intersect the graph of $y = x^3$). Use caution when employing this technique, however. Example 6.10 highlights this.

Example 6.10. Show that $f : \mathbb{R} \to \mathbb{R}$ defined by $f(x) = 2x + 3$ is surjective but $g : \mathbb{Z} \to \mathbb{Z}$ defined by $g(x) = 2x + 3$ is not.

The only difference between the functions f and g are their domains and co-domains. Let us first show f is surjective. To do so, choose an arbitrary but particular $y \in \mathbb{R}$. What element of the domain gets mapped to y under f? As in the previous example, we solve a simple equation, in this case $2x + 3 = y$ for x:

$$x = \frac{y - 3}{2}.$$

Because $\frac{y-3}{2}$ is a real number for every choice of y, and

$$f\left(\frac{y - 3}{2}\right) = y,$$

it follows that f is onto.

How can we show g is *not* surjective? A single counterexample will suffice. In this case, this means finding an element of the co-domain that is not in the range of the function. A little scratch work reveals that for every value x of the domain, $g(x)$ is odd. Let us show then that 2 is not in the range of the function. Suppose it were. Then,

$$2x + 3 = 2$$

for some $x \in \mathbb{Z}$. Algebra yields

$$x = -\frac{1}{2}.$$

In other words, the only possible element x of the domain that g could map to 2 is $x = -\frac{1}{2}$. But the domain of g is the set of all integers, and $-\frac{1}{2} \notin \mathbb{Z}$. Thus, 2 is not in the range of g and consequently g is not surjective.

Because of the discussion prior to Definition 6.8, Theorem 6.9 should come as no surprise.

Theorem 6.9. *A function $f : X \to Y$ is surjective if and only if $f(X) = Y$.*

The overarching structure of this proof is straightforward. Because we want to prove a biconditional statement, our proof will likely require proving two conditional statements. One of them, supposing $f(X) = Y$ and showing f is surjective, requires only showing f satisfies Definition 6.8. The other, however, will require showing that two sets, $f(X)$ and Y, are equal.

Proof Let $f : X \to Y$ be a surjective function. Because the range of a function is, by definition, a subset of the co-domain, we know that

$$f(X) \subseteq Y.$$

Thus to prove $f(X) = Y$, it suffices to show

$$Y \subseteq f(X).$$

To that end, let $y \in Y$. By the surjectivity of f, there exists $x \in X$ so that $f(x) = y$. This shows that $y \in f(X)$, and consequently

$$Y \subseteq f(X).$$

Now, let us assume $f(X) = Y$ and prove that f is surjective. Take $y \in Y$. Because $f(X) = Y$, it must be that $y = f(x)$ for some $x \in X$, fulfilling Definition 6.8.

Consequently, the result holds: $f : X \to Y$ is surjective if and only if $f(X) = Y$. $\qquad \square$

Because Theorem 6.9 is a biconditional statement, it provides an alternative method for defining a surjective function. Pick up various textbooks that define surjective functions and you will find those that define it as we did in Definition 6.8 and those that define it equivalently using Theorem 6.9. No matter the definition of surjectivity we accept, we ultimately now have two ways to show a function is indeed onto.

We previously mentioned that Theorem 6.7 could be strengthened if the function in it was assumed to satisfy additional properties. For one of the cases, namely 6.7(1), surjectivity is the necessary property.

Theorem 6.10. *Let $f : X \to Y$ be a surjective function with $B \subseteq Y$. Then,*

$$f(f^{-1}(B)) = B.$$

We are asked to prove that two sets are equal. Thus, at first glance it may seem that two results must be shown:

$$f(f^{-1}(B)) \subseteq B$$

and

$$B \subseteq f(f^{-1}(B)).$$

If we did both, then yes, we have proven the result, but we also will have done too much work. Remain conscious to what we have previously proven. Theorem 6.7 says that one of these inclusions is true for *any* function. What we are assuming here is that we have a specific type of function, meaning we can simply call upon the prior result. Had we forgotten about our previous work, then the proof of this case would simply be identical to that portion of the proof in Theorem 6.7.

Proof Assume $f : X \to Y$ is a surjective function and $B \subseteq Y$. We know by Theorem 6.7 that

$$f(f^{-1}(B)) \subseteq B.$$

Hence, we need only show that

$$B \subseteq f(f^{-1}(B))$$

to prove the desired result.

To that end, assume $b \in B$. Because f is surjective, $b = f(x)$ for some $x \in X$. By Definition 6.4, $x \in f^{-1}(B)$. Since $f(x) = b$, then, we have that $b \in f(f^{-1}(B))$. Thus,

$$B \subseteq f(f^{-1}(B)),$$

proving the desired result.   □

Certain specific functions tend to appear throughout various areas of higher mathematics. *Projections*, though very simple to define, are one type. We introduce them here because they are, as Lemma 6.12 shows, surjective.

Definition 6.11. For sets $X_1, X_2, \ldots, X_n$ $(n \in \mathbb{Z}^+)$, define the *ith projection function* $(1 \leq i \leq n)$ to be

$$\pi_i : X_1 \times X_2 \times \cdots \times X_n \to X_i$$

given by

$$\pi_i(x_1, x_2, \ldots, x_n) = x_i.$$

Though the notation may be new to you, you likely have encountered projection functions in calculus, linear algebra or various other courses. In multivariable calculus, for example, three-dimensional objects are often projected to a two-dimensional plane when determining the bounds on a triple integral.[3] The point $(4, 6, 5)$ is projected to the the point $(4, 6)$ in the xy-plane. We just as easily could project such an object to only the x-axis. In this case, $(4, 6, 5)$ projects to 4 on the x-axis. Written in our new terminology, we have that $\pi_1(4, 6, 5) = 4$.

Example 6.11. If $A = \{1, 2, 3\}$ and $B = \{a, b\}$, the projections

$$\pi_1 : A \times B \to A$$

and

---

[3]Note that such a projection is not of the form in Definition 6.11 but are a simple generalization.

$$\pi_2 : A \times B \to B$$

are demonstrated below.

$\pi_1(1, a) = 1$	$\pi_2(1, a) = a$
$\pi_1(2, a) = 2$	$\pi_2(2, a) = a$
$\pi_1(3, a) = 3$	$\pi_2(3, a) = a$
$\pi_1(1, b) = 1$	$\pi_2(1, b) = b$
$\pi_1(2, b) = 2$	$\pi_2(2, b) = b$
$\pi_1(3, b) = 3$	$\pi_2(3, b) = b$

The proof of the following lemma is left as an exercise.

Lemma 6.12. *For sets* $X_1, X_2, \ldots, X_n$, *the* ith *projection function*

$$\pi_i : X_1 \times X_2 \times \cdots \times X_n \to X_i$$

is surjective.

We change our attention to another important property of functions called *injectivity*. By definition, a function maps each element of its domain to a single element of the co-domain. However, it is possible that two different elements of the domain map to the same element in the co-domain. For example, if $f : \mathbb{R} \to \mathbb{R}$ is defined by $f(x) = x^2$, then $f(2) = 4$ and $f(-2) = 4$. Functions for which this does not happen are called *injective*.

Definition 6.13. A function $f : X \to Y$ is *injective* (or *one-to-one*) if and only if for every $a, b \in X$, if $a \neq b$, then $f(a) \neq f(b)$.

Showing a function is not injective is straightforward; the discussion preceding Definition 6.13 did exactly that! We need only find two different elements of the domain that map to the same element of the co-domain. To show that a function is injective, however, requires a bit more work. Computationally, it is often easier to use the contrapositive of Definition 6.13:

A function $f : X \to Y$ is injective if and only if for every $a, b \in X$, if $f(a) = f(b)$, then $a = b$.

Why is this a simpler alternative? The contrapositive statement allows us to assume two things are equal and requires us to show that two things are equal. To show objects are equal is often straightforward but it can be a challenge to argue why things are unequal (as we would have to do if we used Definition 6.13). Thus, to show a general function $f : X \to Y$ is injective, follow these steps.

1. Assume $f(a) = f(b)$ for arbitrary but particular $a, b \in X$.

2. Show $a = b$.

Let us exhibit this process via an example.

Example 6.12. Show that $f : \mathbb{R} \to \mathbb{R}$ defined by $f(x) = 2x^3 - 7$ is injective.

Using the steps above, assume $f(a) = f(b)$ for arbitrary $a, b \in \mathbb{R}$:

$$2a^3 - 7 = 2b^3 - 7$$
$$2a^3 = 2b^3$$
$$a^3 = b^3$$
$$a = b$$

Because we were able to conclude $a = b$, it must be that f is injective.

Note the presentation of the result in Example. 6.12. We began by assuming $f(a) = f(b)$:

$$2a^3 - 7 = 2b^3 - 7$$

and proceeded to show that $a = b$. We do *not* assume $a = b$ and simply say, "By algebra, then, it must be that $2a^3 - 7 = 2b^3 - 7$." This is because of the conditional statement we are trying to prove:

$$\text{If } f(a) = f(b), \text{ then } a = b.$$

Just as there is a graphical test for the surjectivity of real-valued functions, there is a similar test for testing if a real-valued function f is one-to-one. It, too, involves considering horizontal lines and is often referred to as the Horizontal Line Test. If some horizontal line intersects the graph of f more than once, then f is *not* injective. Why? At such a y-value, there would be two distinct domain values mapping to that particular y. See Figure 6.3.

Surjectivity was the tool needed to strengthen one of the results of Theorem 6.7; injectivity provides an avenue for addressing the other. When the function is assumed to be injective, the conclusion of this second case of the theorem is stronger. As in Theorem 6.10, call upon Theorem 6.7 so that you are not reproving something that has already been shown true.

Theorem 6.14. *Let $f : X \to Y$ be an injective function with $A \subseteq X$. Then,*

$$A = f^{-1}(f(A)).$$

Proof Let $f : X \to Y$ be injective with $A \subseteq X$. Theorem 6.7(2) tells us that

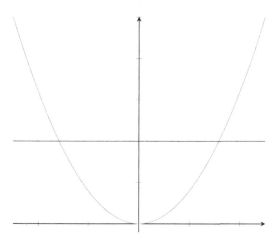

FIGURE 6.3
$f(x) = x^2$ failing the Horizontal Line Test

$$A \subseteq f^{-1}(f(A));$$

to prove

$$A = f^{-1}(f(A)),$$

we need only show

$$f^{-1}(f(A)) \subseteq A.$$

Thus, take $x \in f^{-1}(f(A))$. By Definition 6.4, we know that $f(x) \in f(A)$. Thus, $f(x) = f(a)$ for some $a \in A$. Since f is injective, it must be that $x = a$, showing $x \in A$, as desired. $\square$

Arbitrary functions preserve the union of sets (Theorem 6.5), but they need not preserve the intersection of sets (see Exercise 6.1.22). The preservation does hold when the function is assumed to be injective, however.

Theorem 6.15. *Let $f : X \to Y$ be an injective function and let $A, B \subseteq X$. Then,*

$$f(A \cap B) = f(A) \cap f(B).$$

Proof Let $f : X \to Y$ be an injective function and let $A, B \subseteq X$. It is only necessary to show that

$$f(A \cap B) \subseteq f(A) \cap f(B),$$

as Theorem 6.5 proves that

$$f(A) \cap f(B) \subseteq f(A \cap B).$$

Consider $y \in f(A) \cap f(B)$. Because $y \in f(A)$, we have that $y = f(a)$ for some $a \in A$. Likewise, $y = f(b)$ for some $b \in B$. Therefore

$$f(a) = f(b),$$

and invoking the injectivity of f yields $a = b$. Hence, $a \in A \cap B$, so that $y \in f(A \cap B)$, proving the desired equality. □

In general, knowing that a function is either surjective or injective does not affect its ability to have (or not have) the other property.[4] It is possible for a function to be neither injective nor surjective, exactly one of the two or both injective *and* surjective. Those that are both are called *bijective*.

Definition 6.16. A function that is both injective and surjective is called a *bijection*.

What is the intuitive notion of a function being bijective? Each element of the co-domain is mapped to by exactly one element of the domain. A fairly simple example of a bijective function is called the *identity function*. Regardless of the set X on which it is defined, it is always a bijection.

Definition 6.17. Let X be any set. The *identity function* is the function $\mathrm{id}_X : X \to X$ defined by

$$\mathrm{id}_X(x) = x.$$

Perhaps you have seen the term "identity" in other mathematical contexts. In the real numbers, 1 is what we call a *multiplicative identity*:

$$1 \cdot x = x$$

for all $x \in \mathbb{R}$. Similarly, 0 is the *additive* identity in the real numbers:

$$0 + x = x$$

for any x. Is this at all related to the choice to call the function in Definition 6.17 the *identity* function? Indeed it is, but it requires a certain operation be defined on functions. That is forthcoming in the next section. In the meantime, we proceed in showing that the identity function is a bijection.

Lemma 6.18. *For every set X, id_X is a bijection.*

---

[4]The same cannot always be said when the domain and/or co-domain is finite. See Exercise 2.

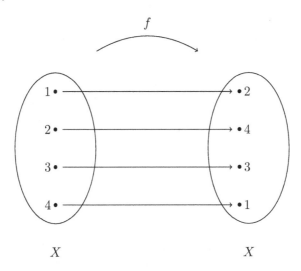

FIGURE 6.4
A permutation of $\{1, 2, 3, 4\}$

Proof Let X be a set. First, if

$$\mathrm{id}_X(a) = \mathrm{id}_X(b)$$

for $a, b \in X$, then $a = b$ by the definition of id_X. Thus, id_X is injective.

Next, let $x \in X$. Because $\mathrm{id}_X(x) = x$, we have that id_X is surjective, proving that id_X is a bijection. $\qquad\qquad\square$

Bijections where the domain and the co-domain are the same set are of particular interest in various areas of mathematics. Consider, for example, if $X = \{1, 2, 3, 4\}$ and $f : X \to X$ is a bijection. Is there an interpretation for what exactly f is doing to the set X? Figure 6.4 exhibits this for a particular function f. Note that the figure is drawn so that all function arrows are horizontal (and consequently the elements of the co-domain are not necessarily written in the same order as in the domain; compare to Figure 6.1).

Bijections such as these are interpreted as reorderings, called *permutations*, of the elements of the domain.

Definition 6.19. A bijection $f : X \to X$ is called a *permutation* of the set X.

The term *permutation* is synonymous with the concept from elementary counting of the same name. In that context, permutations dealt with ordering objects. In particular, how many ways can n objects be ordered? If we start to arrange them, there are n potential first objects . Once chosen, there are $n-1$ elements available to be placed in the second position. Choose that object, and $n-2$ remain for the remaining positions. Continuing, then, yields $n!$ possible orderings; there are $n!$ different *permutations* of the n objects.

Thinking then of reordering the elements of an n-element set via some bijection f, there are n possible elements of f for which the first element can be mapped to. If we are thinking of lining up the elements of the set, f is directing the first element where to stand in line. Where can the second element of the set stand in line, then? There are only $n-1$ options where it could be placed, or equivalently, where it could get mapped to, since f is injective (and thus cannot map this second element to the same element as the first). Continuing, then, there are $n!$ distinct bijections of a set with n elements.

Definition 6.19 does not require the domain of a bijection to be finite. The bijection $f : \mathbb{Z} \to \mathbb{Z}$ defined by $f(x) = x + 1$ is a permutation of the integers. Such bijections between infinite sets yield quite interesting results, some of which we will investigate more fully in Chapter 7.

Example 6.13. Find a bijection

$$f : \mathcal{P}(\{1,2,3\}) \to \{0,1\} \times \{0,1\} \times \{0,1\}$$

and show it is indeed a bijection.

While we could simply list the 8 elements of both the domain and co-domain and randomly create a bijection between the sets, can we be a bit more methodical about our process? Is there a function that in some way feels like a *natural* way to define f?

Recall that $\mathcal{P}(\{1,2,3\})$ is the set of all subsets of $\{1,2,3\}$, a set itself with 3 elements. The co-domain, $\{0,1\} \times \{0,1\} \times \{0,1\}$, consists of ordered 3-tuples. We can exploit this commonality. Consider how we go about forming a subset of $\{1,2,3\}$. We must decide if each of 1, 2 or 3 is in the subset. In other words, each subset of $\{1,2,3\}$ corresponds to an ordered three-word sequence of "yes" or "no;" "yes" if the element is in the subset and "no" if it is not. *Yes-yes-no* (i.e., $\{1,2\}$) would correspond to the subset $\{1,2\}$. If we think about "yes" as a 1 and "no" as a 0, we have the *natural* bijection we were looking for; *yes-yes-no* corresponds to the element $(1,1,0)$ in the co-domain.

How then can we mathematically define the bijection? Defining it symbolically is a good exercise in terminology (see Exercise 22). Because our domain has only 8 elements, it is just as simple to exhibit how the function acts on every element of $\mathcal{P}(\{1,2,3\})$, which also verifies that it is a bijection.

$$
\begin{array}{ll}
f(\{\}) = (0,0,0) & f(\{1\}) = (1,0,0) \\
f(\{2\}) = (0,1,0) & f(\{3\}) = (0,0,1) \\
f(\{1,2\}) = (1,1,0) & f(\{1,3\}) = (1,0,1) \\
f(\{2,3\}) = (0,1,1) & f(\{1,2,3\}) = (1,1,1)
\end{array}
$$

In the previous example, we constructed a bijection from the power set of a set with 3 elements to ordered triples. What if we would have asked if it were possible to find a bijection from the set A to $\mathcal{P}(A)$, for any set A? In quickly scratching out a few examples we may guess that it is not possible when A has only a finite number of elements. But what about if $A = \mathbb{Z}$? It is a much more daunting task. Just because *we* cannot find a bijection does not mean that *somebody else* may not find one. There are infinitely many functions to consider; maybe we are just not looking at the right one. Either we must get clever, or, we need to show that such a bijection could not possibly exist. We will determine the answer to that question in Chapter 7.

Exercises

1. Explain intuitively the notion of

   (a) surjective function
   (b) injective function

2. Let $A = \{1,2,3,4\}$, $B = \{5,6,7,8,9\}$. If possible, define a function satisfying the following. If it is not possible, explain why.

   (a) $f : A \to B$ that is injective.
   (b) $f : A \to B$ that is surjective.
   (c) $f : B \to A$ that is injective.
   (d) $f : B \to A$ that is surjective.
   (e) $f : A \to A$ that is injective but not surjective.
   (f) $f : A \to A$ that is surjective but not injective.

3. Repeat the previous problem with

   (a) $A = \mathbb{Z}$ and $B = \mathbb{Z}^+$.
   (b) $A = \mathbb{Z}^+$ and $B = \mathbb{Z}$.

4. Determine, with justification, if each of the following functions is injective and/or surjective.

   (a) $f : \mathbb{R} \to \mathbb{R}$ given by $f(x) = x^2 + 4$.
   (b) $g : \mathbb{Z}^+ \to \mathbb{Z}$ given by $g(x) = x^2 + 4$.
   (c) Any function $h : S_1 \to S_2$, where S_1 is a set with 10 elements and $S_2 \subset S_1$.
   (d) $d : \mathbb{R} \to \mathbb{R}$ given by $d(x) = \dfrac{x+8}{9}$.

(e) $l : \mathbb{Z}^+ \to \mathbb{Z}^+$ given by $l(x) = \left\lfloor \dfrac{x}{3} \right\rfloor$.

(f) $\chi_A : \mathbb{R} \to \{0,1\}$, where A is a nonempty subset of $\mathbb{R}$ and χ_A is the characteristic function (see Exercise 30).

(g) $g : \mathcal{P}(\mathbb{Z}^+) - \varnothing \to \mathbb{Z}^+$ by $g(S) =$ the least element in S.

(h) $h : [-\frac{\pi}{2}, \frac{\pi}{2}] \to [-1,1]$ given by $h(x) = \sin(x)$.

5. Find the given projection function values.

   (a) $\pi_1(a,b)$.
   (b) $\pi_3(3,2,1)$.
   (c) $\pi_6(2^2, 2^1, 2^0, \ldots, 2^{-8})$.
   (d) $\pi_{10}(a,b,c,\ldots,x,y,z)$.

6. Find $\mathrm{Ran}(\pi_i)$ for the given domain and value of i.

   (a) Domain: $\mathbb{R}^- \times \mathbb{R}^+$; $i = 2$.
   (b) Domain: $\{a,b,c,d,e\} \times \{4,5\}$; $i = 1$.
   (c) Domain: $\mathbb{Z}^* \times \mathbb{R} \times \mathbb{Q} \times \mathbb{Z}^+$; $i = 4$.
   (d) Domain: $\mathbb{R}^+ \times \varnothing \times \mathbb{Z}$; $i = 1$.
   (e) Domain: $\mathcal{P}(\mathbb{Z} \times \mathbb{Z}^+) \times \mathcal{P}(\mathbb{Q}^- \times \mathbb{R})$; $i = 2$.

7. A function $f : \mathbb{R} \to \mathbb{R}$ is defined. Is f a bijection? Justify.

   (a) $f(x) = 2x + 3$.
   (b) $f(x) = 5 - 6x^2$.
   (c) $f(x) = x^3 + 15$.
   (d) $f(x) = e^x$.

8. A function $g : \mathbb{Z} \to \mathbb{Z}$ is defined. Is g a bijection? Justify.

   (a) $g(x) = x + 3$.
   (b) $g(x) = (x - 1)^3$.
   (c) $g(x) = \left\lfloor \dfrac{2x}{3} \right\rfloor$.

9. Determine if the below functions, whose domains involve Cartesian products, are injective or surjective. Assume X and Y are sets with at least 2 elements.

   (a) $f : X \times Y \to X \times Y$ given by $f(x,y) = (x,y)$.
   (b) $f : X \times Y \times Z \to Y \times X$ given by $f(x,y,n) = (y,x)$.
   (c) $f : X \times Y \times Z \to X \times \times Y \times Z$ given by $f(x,y,n) = (x,y,2)$.

10. Let $f : \mathbb{Z}^+ \to \mathbb{Z}^+$ be given by $f(n) = F_n$, where F_n is the nth Fibonacci number.

    (a) Is f injective or surjective? Justify.
    (b) If the domain of f is changed to the set of non-negative integers, is f injective or surjective? Explain.

11. Prove that the mod function, $\mathrm{mod}_n : \mathbb{Z} \to \mathbb{Z}$ given by

$$\mathrm{mod}_n(x) = x \bmod n$$

is neither injective nor surjective for any integer $n \geq 2$.

12. Let S be the set of all strings of 0s and 1s. Define $c : S \to S$ by $c(s) = s1$. For example,

$$c(1100) = 11001.$$

Is c injective or surjective? Prove or disprove.

13. Prove that $\mathrm{Ran}(\mathrm{Id}_X) = X$ for any set X.

14. Determine how many elements $f(\{a, b, c, d\})$ has, or explain why it cannot be determined, if

   (a) f is injective.
   (b) f is surjective.
   (c) f is bijective.

15. Given two real-valued functions $f, g : \mathbb{R} \to \mathbb{R}$, define a new function $f + g : \mathbb{R} \to \mathbb{R}$ by

$$(f + g)(x) = f(x) + g(x).$$

   (a) If f and g are both injective, is $f+g$ necessarily injective? Prove or disprove.
   (b) If f and g are both surjective, must $f + g$ be surjective? Prove or disprove.

16. Let $P : \mathbb{Z}^+ \to S$, where S is all the subsets of $\mathbb{Z}^+$, given by $P(n)$ is the set of all prime factors of n. For example,

$$P(24) = \{2, 3\}.$$

Determine with justification if P is either injective or surjective.

17. Let Q be the set of all strings of primes, ordered from low to high and allowing repetition (for example, $2237 \in Q$ but $273 \notin Q$). Define $F : \mathbb{Z}^+ \to Q$ where $F(n)$ is defined to be the string of all prime factors of n, ordered from low to high and allowing repetition. For example,

$$F(20) = 225.$$

Determine, with justification, if F is either injective or surjective.

18. Define a function $G : \mathbb{R} \times \mathbb{R} \to \mathbb{R}$ that is

(a) injective but not surjective

(b) surjective but not injective

(c) neither injective nor surjective

and justify that your function satisfies the required properties.

19. Prove Lemma 6.12.

20. Let $f : X \to Y$ be a function and $A \subseteq X$.

   (a) If f is injective, prove that f_A is injective.

   (b) Show that if f is surjective, then f_A need not be surjective.

   (c) Prove that $f_A : A \to f(A)$ is surjective.

21. Let A and B be sets. Under what hypotheses are the first and second projections π_1 and π_2 on $A \times B$ injective?

22. Let $n \in \mathbb{Z}^+$. Generalize the bijection of Example 6.13 to a bijection

$$f : \mathcal{P}(\{1, 2, \ldots, n\}) \to \{0, 1\} \times \{0, 1\} \times \cdots \times \{0, 1\},$$

where the co-domain is a Cartesian product of n copies of $\{0, 1\}$, and prove that it is indeed a bijection.

23. Recall that a function $f : \mathbb{R} \to \mathbb{R}$ is called *increasing* if $f(x) < f(y)$ whenever $x < y$. Prove that an increasing real-valued function is injective. Must it be surjective? Justify.

24. Let $f : X \to Y$ with $A, B \subseteq X$ and $C, D \subseteq Y$. Prove the following.

   (a) If f is injective, then $f(A - B) = f(A) - f(B)$.

   (b) If f is surjective and $f^{-1}(C) = f^{-1}(D)$, then $C = D$.

25. Let A be a nonempty set with n elements $(n \in \mathbb{Z}^+)$. Find a function f satisfying each of the following, or explain why no such function exists.

   (a) $f : A \to \mathcal{P}(A)$ that is injective.

   (b) $f : A \to \mathcal{P}(A)$ that is surjective.

   (c) $f : \mathcal{P}(A) \to \mathcal{P}(A)$ that is injective.

   (d) $f : \mathcal{P}(A) \to A$ that is surjective.

   (e) $f : \mathcal{P}(A) \to \mathcal{P}(A)$ that is bijective.

26. Let C be the set of continuous real-valued functions.

   (a) Show $F : C \to \mathbb{R}$ given by

$$F(f) = \int_a^b f(x)\, dx$$

   is surjective but not injective (where $a, b \in \mathbb{R}$, $a < b$).

   (b) Show $D : C \to C$ given by $D(f(x)) = f'(x)$ is not injective.

6.3 Composition and Invertibility

The previous section began with an overview of how mathematical objects are typically investigated. Statements, sets, relations and functions were introduced for specific reasons, but how we have approached them is fairly consistent. After definition, the first natural question is when two of those objects are equal. Their basic properties are considered after these definitions are established.

These two concepts, equality and properties, lay the foundation for the next step: combine the objects in some way to create new objects. Logical connectives were the tools needed to create new statements from old. Unions, intersections and differences of previously defined sets created new sets (as did the power set and Cartesian product). The composition of two relations appeared in Exercise 16 following 5.1, and considering that functions are themselves relations, it should come as no surprise that this gives rise to a concept you are likely familiar with: the *composition* of two functions.

In our algebraic youth, we learned arithmetic ways to combine functions. For example, given functions f and g, we can form other functions: $f+g$, $f-g$, fg and f/g. There is a key flaw with thinking that this is how we can combine functions, however. If f and g are numeric-type functions, whose co-domains consist of sets for which addition is defined (such as the integers or complex numbers), then $(f+g)(x) = f(x) + g(x)$ makes mathematical sense. But what if the co-domain consists of sets, phone numbers or fluffy kittens? Because of this, the general idea of algebraic combinations of functions is fruitless.

A fifth way of combining real-valued functions, function *composition*, is definable for general functions whose domains consist of non-numeric objects, however.

> **Definition 6.20.** Let $f : X \rightarrow Y$ and $g : Y \rightarrow Z$ be functions. The *composition of g with f* is the function $g \circ f : X \rightarrow Z$, pronounced "$g$ compose f," defined for every $x \in X$ by
>
> $$(g \circ f)(x) = g(f(x)).$$

Notice that for the composition of f with g to be defined, the co-domain of f and the domain of g must be the same set. The resulting function, $g \circ f$, has the same domain as f and the same co-domain as g. There are no requirements that f or g be either injective or surjective, as exhibited in Figure 6.5. Neither f nor g in Figure 6.5 is injective or surjective. The function $g \circ f$ turns out to be neither as well.

In light of this, you may be wondering if there are conditions that force the composition of two functions to be either injective or surjective. Theorem

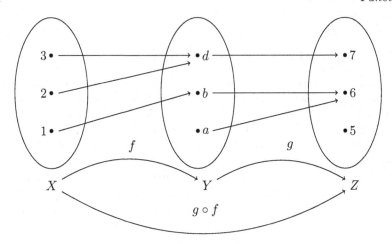

FIGURE 6.5
Function composition

6.23 states relationships between the injectivity or surjectivity of f and/or g together with that of $g \circ f$.

Example 6.14. Define functions $f : \mathbb{R} \to \mathbb{R}$ and $g : \mathbb{R} \to \mathbb{R}$ by

$$f(x) = 2x + 5$$

and

$$g(x) = 3x^2 - 4x - 1.$$

(1) Compute, compare, and interpret $(g \circ f)(2)$ and $(f \circ g)(2)$.

We compute:

$$\begin{aligned}
(g \circ f)(2) &= g(f(2)) \\
&= g(9) \\
&= 206,
\end{aligned}$$

and

$$\begin{aligned}
(f \circ g)(2) &= f(g(2)) \\
&= f(3) \\
&= 11.
\end{aligned}$$

We see that

$$(g \circ f)(2) \neq (f \circ g)(2).$$

This means that, in general, function composition is *not* commutative. This should not be surprising. The function $f \circ g$ need not exist just because the function $g \circ f$ is defined. For this to be the case, both f and g must have the same domains and co-domains.

(2) Find a general formula for both $g \circ f$ and $f \circ g$.

To find a general formula, proceed by algebraically evaluating $(g \circ f)(x)$ and $(f \circ g)(x)$:

$$\begin{aligned}(g \circ f)(x) &= g(f(x)) \\ &= g(2x + 5) \\ &= 12x^2 + 52x + 54,\end{aligned}$$

and

$$\begin{aligned}(f \circ g)(x) &= f(g(x)) \\ &= f(3x^2 - 4x - 1) \\ &= 6x^2 - 8x + 3.\end{aligned}$$

Again, notice that these functions are generally not equal, though they may be for some specific values of their domains.

One of the goals of mathematicians is to describe the *structures* of sets of newly defined objects under particular operations. More generally, if the sets of objects satisfy certain properties under these operations, then these collections form *groups*, *rings*, *fields* or a variety of other options. Tracing its roots to Joseph-Louis Lagrange, Augustin-Louis Cauchy, and Évariste Galois, the area of mathematics concerned with this *structure*, often called *abstract algebra* (or simply *algebra*), is amongst the purest of pure mathematics. Why? Because abstract algebra is often applied to other areas of pure mathematics (rather than being applied to biology, chemistry, economics, etc.). The algebraist aims to develop structures and prove laws about them. Developing theory such as this allows for deeper investigations of the mathematical objects in more general terms. Then, if a specific set of mathematical objects is one of these general structures, all the results that were proven in the general case will hold for that object in the specific case.

Our attempt to mirror this approach to mathematics begins with the next result, Theorem 6.21. It shows that identity functions play the role of an *identity* under the operation of function composition, much like 1 is a multiplicative identity for the real numbers. We prove only the first of the two claims, leaving the second as an exercise.

Recall that to show two functions are equal, one must show they have the same domain and co-domain in addition to showing that they operate the same on every element of the domain.

Theorem 6.21. *If* $f : X \to Y$, *then*

       1. $f \circ \mathrm{Id}_X = f$, *and*

       2. $\mathrm{Id}_Y \circ f = f$.

Proof Suppose $f : X \to Y$. Then, since $\mathrm{Id}_X : X \to X$, it follows that f and $f \circ \mathrm{Id}_X$ have the same domain and co-domain. Then, for $x \in X$, we have

$$(f \circ \mathrm{Id}_X)(x) = f(\mathrm{Id}_X(x)) = f(x),$$

proving that $f \circ \mathrm{Id}_X = f$.   □

While proofs such as these seem straightforward, developing them clearly and precisely, as in Theorem 6.22, is an exercise in good notation.

Theorem 6.22. *If* $f : W \to X$, $g : X \to Y$, *and* $h : Y \to Z$ *are functions, then*

$$h \circ (g \circ f) = (h \circ g) \circ f.$$

Proof Let $f : W \to X$, $g : X \to Y$, and $h : Y \to Z$ be functions. Note that $h \circ (g \circ f)$ and $(h \circ g) \circ f$ have domain and co-domain of W and Z, respectively. Then, for $w \in W$, we have

$$\begin{aligned}
(h \circ (g \circ f))(w) &= h((g \circ f)(w)) \\
&= h(g(f(w))) \\
&= (h \circ g)(f(w)) \\
&= ((h \circ g) \circ f)(w),
\end{aligned}$$

proving that function composition is associative, as required.   □

We continue to relate this newly defined operation of function composition to other properties, in this case injectivity and surjectivity, of functions.

Theorem 6.23. *Let* $f : X \to Y$ *and* $g : Y \to Z$ *be functions. Then,*

       1. *If* f *and* g *are injective, then so is* $g \circ f$.

       2. *If* f *and* g *are surjective, then so is* $g \circ f$.

       3. *If* $g \circ f$ *is injective, then so is* f.

       4. *If* $g \circ f$ *is surjective, then so is* g.

We prove (2) and (3), with (1) and (4) left as exercises. To prove (2), we must show that the function $g \circ f$ is onto. Recall that to prove a function is surjective, you show that an arbitrary element of its co-domain is mapped to by some element of its domain. In this case, the co-domain of $g \circ f$ is Z, which is also the co-domain of g. Take $z \in Z$. Because we assume g is onto, there is an element $y \in Y$ that maps via g to z. Since f is also surjective and its co-domain is Y, there is an element $x \in X$ so that $f(x) = y$. This is the key idea in proving (2).

The goal of (3) is to prove that f is injective. How do we go about proving this? Start by assuming $f(a) = f(b)$ and show that $a = b$. We can utilize the injectivity of $g \circ f$ to unwind from Z to X to prove that $a = b$.

Proof Suppose $f : X \to Y$ and $g : Y \to Z$ are functions. We first show that if both f and g are surjective, then $g \circ f$ must be as well.

Let $z \in Z$. Because g is surjective, there exists some $y \in Y$ so that $g(y) = z$. Similarly, because f is surjective, there must exist $x \in X$ so that $f(x) = y$. Consequently, then,

$$(g \circ f)(x) = g(f(x)) = g(y) = z,$$

proving the surjectivity of $g \circ f$.

Next, assume the composition $g \circ f$ is injective. We will show that f must also be injective.

To that end, let $a, b \in X$ with

$$f(a) = f(b).$$

It follows, then, that

$$g(f(a)) = g(f(b)),$$

or equivalently,

$$(g \circ f)(a) = (g \circ f)(b).$$

By the injectivity of $g \circ f$, it must be that $a = b$, showing that f, too, is injective. $\square$

Combining two of the results in Theorem 6.23 immediately yields the following.

Corollary 6.24. *If $f : X \to Y$ and $g : Y \to Z$ are bijections, then $g \circ f$ is a bijection.*

Proof Suppose $f : X \to Y$ and $g : Y \to Z$ are bijections. Parts (1) and (2) of Theorem 6.23 show that $g \circ f$ is bijective. $\square$

We shift gears and consider now a new function that *sometimes* exists, and it depends only on one already defined function, as opposed to the two

required in defining function composition. A function maps each element of
the domain to some element in the co-domain. It is natural to ask if that
mapping can be undone. It is not always the case that it can, but when it is
possible, we call such a function *invertible*.

Definition 6.25. A function $f : X \to Y$ is *invertible* if there exists a
function $g : Y \to X$ such that

$$g \circ f = \mathrm{Id}_X$$

and

$$f \circ g = \mathrm{Id}_Y.$$

The function g, denoted by $g = f^{-1}$, is called an *inverse function of f*.

According to the above definition, if g is an inverse function of $f : X \to
Y$, then $g(f(a)) = a$ for every $a \in X$, and for every $b \in Y$ it follows that
$f(g(b)) = b$. One interpretation of the inverse function is that g undoes f in
the sense that the input-output association established by f is reversed by g.
A simple consequence of the definition is that if g is an inverse function of f,
then f is an inverse function of g. See Exercise 16.

The next example demonstrates Definition 6.25 via real-valued functions.

Example 6.15. Consider the real-valued function

$$f(x) = \frac{2+x}{3-x}.$$

Then f is invertible, with

$$f^{-1}(x) = \frac{3x-2}{x+1}.$$

To verify this, note that

$$(f \circ f^{-1})(x) = f\left(f^{-1}(x)\right)$$
$$= f\left(\frac{3x-2}{x+1}\right)$$
$$= \frac{2 + \frac{3x-2}{x+1}}{3-x}$$
$$= x.$$

Likewise,

$$(f^{-1} \circ f)(x) = f^{-1}(f(x))$$

$$= f^{-1}\left(\frac{2+x}{3-x}\right)$$

$$= \frac{3\left(\frac{2+x}{3-x}\right) - 2}{\frac{2+x}{3-x} + 1}$$

$$= x.$$

Because both

$$f \circ f^{-1} = x$$

and

$$f^{-1} \circ f = x,$$

it follows from Definition 6.25 that f and f^{-1} are indeed inverses of one another.

Definition 6.25 defines *an* inverse function of another function rather than *the* inverse function. Recalling familiar examples from algebra, we might believe that invertible functions have one and only one inverse, yet nothing in the definition here prohibits a function from having two or more inverses. The following theorem proves that this cannot happen.

Theorem 6.26. *The inverse of an invertible function is unique.*

Recall that to prove the uniqueness of a particular mathematical object, assume that there are two of those objects and somehow show that they must actually be equal. In this case, it is the definition of a function's inverse that proves the two assumed inverses are equal.

Proof Let $f : X \to Y$ be an invertible function, with g and h inverses of f. Then, by Definition 6.25,

$$f \circ h = \mathrm{id}_Y$$

and

$$g \circ f = \mathrm{id}_X.$$

Then,

$$g = g \circ \text{id}_Y$$
$$= g \circ (f \circ h)$$
$$= (g \circ f) \circ h \text{ (by Theorem 6.22)}$$
$$= \text{id}_X \circ h$$
$$= h,$$

proving the uniqueness of the inverse function.   □

We used properties of functions to make this proof of uniqueness quite "clean." Alternatively, we could have shown that $g(y) = h(y)$ for all $y \in Y$. The tools used would have been identical, but we would have had to carry about the parenthetical notation "(y)" throughout the proof. The use of Theorem 6.22 made this unnecessary.

Definition 6.25 says that *if* a certain function exists, then a function is invertible. This hints that such a function may not exist and functions may not be invertible. Rather than looking for specific examples of such functions, can we determine a general property that guarantees invertibility? While the following theorem may be obvious when the domain and co-domain of the function are finite sets, it turns out that, in *all* cases, knowing the function is bijective is both necessary and sufficient in knowing that the function has an inverse.

Theorem 6.27. *A function* $f : X \to Y$ *is invertible if and only if f is bijective.*

Proof Consider the function $f : X \to Y$ and suppose that it is invertible with inverse function f^{-1}. We show that f is bijective.

To prove that f is injective, assume $f(a) = f(b)$ for $a, b \in X$. Then,

$$a = f^{-1}(f(a))$$
$$= f^{-1}(f(b))$$
$$= b.$$

Hence, f is injective. Note that for $y \in Y$, since $f^{-1}(y) \in X$, we have

$$f\left(f^{-1}(y)\right) = y.$$

This proves that f is surjective and consequently bijective.

Now assume that f is bijective. Define $g : Y \to X$ by $g(y) = x$ if and only if $f(x) = y$.

Showing that g is indeed a function is left as an exercise. It is only necessary then to show that g is the inverse of f. That is, we must show that

$$g \circ f = \mathrm{id}_X$$

and

$$f \circ g = \mathrm{id}_Y.$$

For this, let $x \in X$ with $f(x) = y$, so that $g(y) = x$. Thus,

$$\begin{aligned}(g \circ f)(x) &= g(f(x)) \\ &= g(y) \\ &= x,\end{aligned}$$

showing that

$$g \circ f = \mathrm{id}_X.$$

Similarly,

$$f \circ g = \mathrm{id}_Y.$$

Consequently, f is invertible with $g = f^{-1}$. □

As a result of the previous theorem, the proof of the Corollary 6.28 immediately falls in our laps.

Corollary 6.28. *If $f : X \to Y$ is a bijection, then $f^{-1} : Y \to X$ is bijective.*

Proof Because f is the inverse of f^{-1}, the result follows from Theorem 6.27. □

Having considered properties of both the newly defined function composition and invertibility, we move now to determine relationships between the two.

Theorem 6.29. *Let $f : X \to Y$ and $g : Y \to Z$ be bijections Then,*

$$(g \circ f)^{-1} = f^{-1} \circ g^{-1}.$$

What must we actually show to prove this result? Dissect the statement

$$(g \circ f)^{-1} = f^{-1} \circ g^{-1}.$$

This claims that the inverse of *some particular function* is $f^{-1} \circ g^{-1}$. How do we show two functions are inverses of one another? We show that when composed, the appropriate identity functions result. In this case, we must show that the compositions $g \circ f$ and $f^{-1} \circ g^{-1}$ are the required identity functions. Knowing that function composition is associative expedites the process.

Proof Let $f : X \to Y$ and $g : Y \to Z$ be bijections. To prove

$$(g \circ f)^{-1} = f^{-1} \circ g^{-1},$$

we must show both

$$(g \circ f) \circ (f^{-1} \circ g^{-1}) = \mathrm{id}_Z$$

and

$$(f^{-1} \circ g^{-1}) \circ (g \circ f) = \mathrm{id}_X.$$

Theorem 6.22 gives

$$(g \circ f) \circ (f^{-1} \circ g^{-1}) = g \circ (f \circ f^{-1}) \circ g^{-1}$$

and consequently,

$$\begin{aligned}
(g \circ f) \circ (f^{-1} \circ g^{-1}) &= g \circ (\mathrm{id}_Y) \circ g^{-1} \\
&= g \circ (\mathrm{id}_Y \circ g^{-1}) \\
&= g \circ g^{-1} \\
&= \mathrm{id}_Z,
\end{aligned}$$

as was to be shown. The second case is shown similarly and is left as an exercise.   □

We have barely touched upon the structures of certain sets of functions, yet you ought to notice parallels with particular sets of numbers under certain operations. The real numbers have a multiplicative identity, and every nonzero real number has a unique multiplicative inverse. Likewise, the identity function acts as an identity under function composition, and bijective (i.e., invertible) functions have unique inverses. Other parallels exist, but we leave those to the abstract algebraist to more thoroughly investigate. We have the necessary tools to create clear, concise proofs of results involving functions, their inverses and their compositions. With your proof writing skills undoubtedly becoming more personalized and succinct at this point, you can seamlessly integrate these new concepts.

Exercises

1. Let $f, g : \mathbb{R} \to \mathbb{R}$ be given by $f(x) = 3x^2 + 1$ and $g(x) = \dfrac{2}{x}$. Find

   (a) $(f \circ g)(1)$.
   (b) $(g \circ f)(1)$.
   (c) $(g \circ f)(0)$.
   (d) $(f \circ f)(1)$.
   (e) $(g \circ g)(1)$.

      (f) $(f \circ g)(0)$.

2. Let $f : \mathbb{Z}^+ \to \mathbb{Z}$ and $g : \mathbb{Z} \to \mathbb{R}$ be given by $f(x) = 5 - x$ and $g(x) = 2^x$. Find

      (a) $(g \circ f)(2)$.

      (b) $(f \circ g)(2)$.

      (c) $(g \circ f)(10)$.

      (d) $(g \circ g)(3)$.

      (e) All $x \in \mathbb{Z}$ such that $(f \circ g)(x)$ is undefined.

3. Let $f, g : \mathbb{R} \times \mathbb{R} \to \mathbb{R} \times \mathbb{R}$ be given by $f(x, y) = (-y, 2x)$ and $g(x, y) = (x + y, e^x)$. Find

      (a) $(g \circ f)(1, 0)$.

      (b) $(f \circ g)(2, 1)$.

      (c) $(g \circ f)(-1, -2)$.

      (d) $(f \circ f)(1, 0)$.

      (e) $(g \circ g)(1, 0)$.

4. Let $f : \mathbb{Z}^* \to \mathcal{P}(\mathbb{Z})$ be given by $f(n) = \{n, 2n, 3n, \ldots\}$ and π_2 the second projection function on $\mathbb{Z}^* \times \mathbb{Z}^* \times \mathbb{Z}^*$. Find

      (a) $(f \circ \pi_2)(1, 2, 3)$.

      (b) $(f \circ \pi_2)(-1, -2, -3)$.

5. If $f, g : \mathbb{R} \to \mathbb{R}$ are defined as follows, find $(f \circ g)(x)$ and $(g \circ f)(x)$.

      (a) $f(x) = x^3$; $g(x) = 2 - 3x$.

      (b) $f(x) = e^x$; $g(x) = e^x$.

      (c) $f(x) = \frac{3x+1}{2+x}$; $g(x) = \frac{4x+2}{5x}$ (Dom$(f) =$ Dom$(g) = \mathbb{R}^+$).

      (d) $f(x) = 9x$; $g(x) = x \bmod 3$.

      (e) $f(x) = \lfloor x \rfloor$; $g(x) = \lceil x \rceil$.

      (f) $f(x) = \sqrt{x + 3}$; $g(x) = x^4$ (Dom$(f) = \{x \in \mathbb{R} \mid x \geq -3\}$).

6. Show that the following functions are inverses of one another. Assume that their domains and co-domains are appropriately defined.

      (a) $f(x) = 2x - 5$; $g(x) = \frac{1}{2}(x + 5)$.

      (b) $f(x) = (x - 3)^3$; $g(x) = \sqrt[3]{x} + 3$.

      (c) $f(x) = \frac{9}{5}x + 32$; $g(x) = \frac{5}{9}(x - 32)$.

      (d) $f(x) = \frac{x+4}{3x-2}$; $g(x) = \frac{2x+4}{3x-1}$.

      (e) $f(x) = \tan(x)$; $g(x) = \arctan(x)$ (on $\left(-\frac{\pi}{2}, \frac{\pi}{2}\right)$).

7. Show that $f, g : \mathbb{R} \to \mathbb{R}$, as defined, are not inverses of one another.

      (a) $f(x) = x + 2$; $g(x) = -2x$

      (b) $f(x) = 4$; $g(x) = \frac{1}{4}$

      (c) $f(x) = \sin(x)$; $g(x) = \cos(x)$

8. Prove: $(\mathrm{Id}_X)^{-1} = \mathrm{Id}_X$.

9. Explain why $f : \mathbb{R} \to \mathbb{R}$ is not invertible.

   (a) $f(x) = x^2$.

   (b) $f(x) = \dfrac{1}{x}$ $(\mathrm{Dom}(x) = \mathbb{R}^*)$.

   (c) $f(x) = |x|$.

   (d) $f(x) = \cos(x)$.

10. Suppose $f : X \to Y$ and $g : Y \to Z$ are increasing (see Exercise 23 following 6.2).

    (a) Is $g \circ f$ increasing? Prove or disprove.

    (b) Is f^{-1} increasing, if f is invertible? Prove or disprove.

11. Prove that the restriction of a bijective function to a subset of its domain is still an invertible function.

12. Prove that every function has some restriction that is an invertible function.

13. Suppose $A \neq \varnothing$, with f and g two permutations on A.

    (a) Show that $f \circ g$ is a permutation on A.

    (b) Show that f is invertible and that f^{-1} is a permutation on A.

14. Complete the proof of Theorem 6.21 by showing that if $f : X \to Y$, then $\mathrm{Id}_Y \circ f = f$.

15. Prove the remaining two results of Theorem 6.23.

16. Suppose that $f : X \to Y$ and that the function $g : Y \to X$ is the inverse function of f. Prove that f is the inverse function of g.

17. Prove that the function g in Theorem 6.27 is indeed a function.

18. Prove the second case of Theorem 6.29:

$$(f^{-1} \circ g^{-1}) \circ (g \circ f) = \mathrm{id}_X.$$

19. Let $f, g, h : X \to X$ with $f \circ g$ and $g \circ h$ bijective. Prove that f, g and h are all bijective.

20. Suppose $\pi : A \times B \times C \to A \times B$ is given by $\pi(a, b, c) = (a, b)$. Prove that $\pi_1 \circ \pi = \pi_1'$, where π_1 and π_1' are the first projection functions on $A \times B$ and $A \times B \times C$, respectively.

21. Is the projection function $\pi_1 : A \times B \to A$ invertible? If not, under what hypotheses would it be invertible?

22. (a) Suppose $f : X \to Y$ and $g, h : Y \to Z$. If f is surjective and $g \circ f = h \circ f$, prove that $g = h$.

    (b) Suppose $f, g : X \to Y$ and $h : Y \to Z$. If h is injective and $h \circ f = h \circ h$, prove that $f = g$.

23. Prove that $f : \mathbb{R} \to \mathbb{R}$ given by $f(x) = x^3 + x$ is invertible.

24. For general function $f : X \to Y$ and $g : Y \to Z$, describe the domain and range of $(g \circ f)$.

25. Disprove: if $f^{-1}(x)$ is the inverse of $f(x)$, then

$$f^{-1}(x) = \frac{1}{f(x)}.$$

26. If A is a set with m elements and B is a set with n elements $(m \neq n)$, prove that any function $f : A \to B$ is not invertible.

27. A *left inverse* (resp. *right inverse*) function for $f : X \to Y$ is a function $g : Y \to X$ satisfying

$$g \circ f = \mathrm{Id}_X \ (\text{resp.} \ f \circ g = \mathrm{Id}_Y).$$

If g is a left inverse for $f : X \to Y$ and h is a right inverse for f, prove that $g = h$.

7

Cardinality

Suppose for a moment that you have a handful of candy in both your right and left hands. Each hand holds the same number of pieces. You decide to eat a few of the pieces from your right hand. Is the number of pieces of candy in your left hand still the same as the number in your right? Clearly not, you are probably thinking. The right hand must have fewer pieces than the left hand, since some of the pieces were removed from it.

Imagine now that you are holding all of the positive integers in each hand. From the right, remove all of the odd integers and ask the same question as before: do your hands hold the same number of things? While the situation seems the same as in the preceding paragraph (the hands held an equal quantity of things, and some were removed from the right hand), it is quite different in one significant way. There are *infinitely* many integers, whereas in the first scenario, you likely visualized holding a *finite* number of pieces of candy (though you may have wished to be holding an infinite amount of candy ...). Can we still determine which hand has more? Everything that is in your right hand is in your left hand, and on top of that, the left hand has all the odd integers as well. Intuitively, it seems like there *must* be more in the left than in the right. But our intuition is misled because of one particular question:

What does *more* mean in the infinite sense?

Consider the numbers in your hands from a different perspective. For each integer in your left, its double is in your right, and likewise, for each integer in your right, its half is in your left. Considering the situation this way, perhaps you are starting to question the initial feeling that the left hand has more than the right hand.

Rather than considering positive integers, what if we were to consider rational numbers, using the real number line as a visual reference? There are infinitely many positive integers on the real number line, each equally spaced one unit apart. Yet, between any two points on the number line, there are *infinitely* many rational numbers. For example, between 1.00001 and 1.000001, there are infinitely many rational numbers. Clearly, then, there must be many, many more rational numbers than positive integers. But as before, our intuition fails us. Surprisingly, it turns out that the positive integers are just as numerous, whatever that may mean, as the rational numbers! To determine *what that means*, we must have a precise definition.

These two examples support a major theme of this text: *definitions are important*. In the coming sections, we will solidify the concept of two sets having the same number of elements. It is an idea fairly new to mathematics (first appearing a little over 125 years ago), established by the aforementioned Georg Cantor. His investigation and theory of transfinite numbers (numbers that are greater than all finite numbers) stirred controversy amongst his contemporaries: Leopold Kronecker (referring to Cantor as a "corrupter of youth"), Henri Poincaré (who called the idea a "grave disease"), Hermann Weyl and Luitzen Brouwer [16] all were aghast at his work.[1] Yet, today, Cantor is held in high regard, both as a mathematician and a philosopher.

In this chapter we consider the concept of quantity, both in the finite and infinite cases. One simple question inspires us: what does it mean for two sets to have the same size? Once defined, the answer to this will be the driving force behind many of our key results.

7.1 The Finite

The development of number systems and enumeration is a fascinating journey, studied by mathematical historians and anthropologists for years. Indeed, entire books are devoted to this seemingly small piece in the history of mankind. As interesting as this is, our goal right now is not to investigate when, why or how today's number systems came about. Rather, we want to imagine ourselves *before* counting was ingrained in our consciousness.

Suppose you live in a time when counting was unknown and two early humans are comparing themselves to one another. They want to determine if the quantity of fingers on their hands are the *same* (you and I would say they want to determine if they have the same number of fingers). They cannot count their fingers, like you and I would. The concept of *number* is completely foreign to them. How can they compare?

Put yourself in their position and take away everything you know about *number* and *quantity*. Counting is not allowed; there is no quantification for the number of fingers you have or your friend has. Without this intellectual tool, is it still possible to determine if you have the same quantity of fingers on left as on the right? What would you do if you wanted to know if you had the same number of fingers on your left hand as on your right?

Right now, you are probably holding your hands close to one another (or imagining doing so). Your left and right pinky fingers touch, then your two ring fingers, followed by your middle fingers and pointer fingers and lastly your left and right thumbs. All your fingers on the left hand have partnering fingers

---

[1]Some attribute the resistance to Cantor's declaration that it was God himself who shared the ideas with him [16]. Regardless of *how* Cantor came about these findings, they are a critical part of today's mathematical framework.

on the right (or if one of your hands has fewer fingers than the other, you are left with at least one unpartnered finger), and consequently, you know that you have the same *quantity* (or lack thereof, in the case of the unpartnered finger) of fingers on your left hand as your right. To arrive at this conclusion, at no point did you need to understand how to enumerate. You just had to know how to pair up objects.

This is the key idea of *cardinality*. It provides a strict notion of how to compare the sizes of two sets in terms of the mathematical objects developed in the previous chapter: functions.

> **Definition 7.1.** A set A is said to have the *same cardinality* as a set B if there exists a bijection $f : A \to B$. If the set A is empty or has the same cardinality as $\{1, 2, \ldots, n\}$ for some $n \in \mathbb{Z}^+$, then A is called *finite* and we say it has *cardinality* n (or 0 for the empty set), written $|A| = n$. If A is not finite, it is called *infinite*.

Such a formal definition of a finite set may seem to be overkill. Why not simply say, "A set is finite if it contains a finite number of elements?" We have used the word "finite" in this latter context for years. Can we not simply state that a *set* is now called finite if it satisfies this simple rule? Such a definition lacks precision. Definition 7.1 eliminates all possible ambiguities about showing that a set is or is not finite (though the latter is a bit more involved and is the focus of Sections 7.2 and 7.3).

Throughout this section we prove results particular to finite sets using this exact definition. Many of the results will be obvious and you may find yourself asking, "Why do we need to prove something so simple?" By this point in the text you ought to be crafting a logical response to your own question. Mathematicians cannot just claim results to be true. They have to *prove* them to be true, no matter how obvious they seem!

Whether or not two sets have the same cardinality is a relation defined on sets. Say that a set A is related to set B if A has the same cardinality as B. This relation, as proven in the next theorem, is an equivalence relation on sets, allowing us to speak more generally about sets having the same cardinality rather than constantly comparing the cardinalities of two specific sets.

Theorem 7.2. *Let U be some universal set. Define a relation R on $\mathcal{P}(U)$ by saying A is related to B under R if and only if A and B have the same cardinality. Then, R is an equivalence relation.*

To prove this result, we must show R is reflexive, symmetric and transitive. Can we find a bijection from any set to itself? Of course: the identity function. This will show the reflexivity of R.

For R to be symmetric, if A is related to B, we must show that B is related to A. Assuming A is related to B gives us an assumed bijection from A to B. A past result tells us that the inverse to this bijection is also a bijection; this will show that B is related to A.

Lastly, to prove R is transitive, we will assume bijections exist from A to B and from B to C. We have shown that the composition of bijections yields a bijection, hence transitivity will follow.

We firm up this scratch work into the following proof.

Proof Suppose U is a universal set and define R to be a relation on the power set of U given by $(A, B) \in R$ if and only if A and B have the same cardinality. We proceed in showing R satisfies the three criteria for being an equivalence relation.

Reflexive: Let $A \subseteq U$. Because the identity function on A is a bijection (Lemma 6.18), A has the same cardinality as A. So, $(A, A) \in R$.

Symmetric: Suppose $A, B \subseteq U$, with A having the same cardinality as B. Thus, there exists a bijection

$$f : A \to B.$$

Corollary 6.28 tells us that

$$f^{-1} : B \to A$$

is also a bijection, meaning that B has the same cardinality as A. Hence, if $(A, B) \in R$, then $(B, A) \in R$, proving the symmetry of R.

Transitive: Suppose A has the same cardinality as B and B has the same cardinality as C, where $A, B, C \subseteq U$. Because the composition of bijections is a bijection (Corollary 6.24), the existence of bijections

$$f : A \to B$$

and

$$g : B \to C$$

yields a bijection

$$g \circ f : A \to C.$$

Thus, if $(A, B) \in R$ and $(B, C) \in R$, then $(A, C) \in R$, and consequently, we have shown that R is transitive.

Because R satisfies the three properties above, it is indeed an equivalence relation.   □

Lemma 7.3 begins our exploration of finite sets, serving as a tool for proving some of the stronger results of this section. As you work through these later lemmas, theorems and corollaries, notice how the *concept* that the results indicate is typically clear. The main purpose of including them here is not to develop a complicated theory but to emphasize the proofs. Their complexity and use of strong notation reinforces an ongoing theme of this text: definitions and precision must be mastered.

Lemma 7.3. *If $f : A \to \{1, 2, \ldots, m\}$ is an injective function, for $m \in \mathbb{Z}^+$, then $|A| \leq m$.*

This result can be visualized quite easily. Suppose a room has m chairs in it. People file into the room and sit in some chair. If there are enough chairs for everyone to sit down (only one person per chair is allowed), then there could not be more than m people in the room.

How can we go about proving this result? To show that $|A| \leq m$, Definition 7.1 tells us that we must find a bijection from A to $\{1, 2, \ldots, n\}$, where $n \leq m$. If f mapped the elements of A to the first n positive integers, our proof would be quite straightforward. But, we have no idea about how f actually maps the elements of A, just that it does so in an injective fashion.

The thought that, "if we knew how f mapped the elements of A" points our scratch work in the right direction. We can choose to notationally *define* how f maps the elements in a clever and helpful way:

$$\text{Ran}(f) = \{a_1, a_2, \ldots, a_n\}.$$

We have in a sense ordered the image of f. From there, we can then map this set in a natural way to $\{1, 2, \ldots, n\}$.

Proof Let $f : A \to \{1, 2, \ldots, m\}$ be injective, for some $m \in \mathbb{Z}^+$. Then,

$$f(A) \subseteq \{1, 2, \ldots, m\},$$

and for notation, say

$$f(A) = \{a_1, a_2, \ldots, a_n\}$$

where $n \leq m$. Define a new function

$$f' : A \to f(A)$$

by $f'(a) = f(a)$. We know from Exercise 6.2.20 that f' is a bijection. Thus, A and $f(A)$ have the same cardinality.

Now,

$$g : f(A) \to \{1, 2, \ldots, n\}$$

given by $g(a_i) = i$ is a bijection (the proof of which is left as an exercise). By definition, $|f(A)| = n$, and this shows that $|A| \leq m$, as required. $\qquad\square$

It is often beneficial to consider what most of the results in this chapter demonstrate in common language. Lemma 7.3 can be remembered with the aforementioned chairs metaphor; for a function between finite sets to be injective, the domain can be no bigger than the co-domain. Is there a similar metaphor for remembering the cardinality of a union of sets? We begin first in the case that the sets are disjoint, as in Theorem 7.4.

Consider the sets as bags containing objects. If we pour their contents out into a single pile, how many things will be in the pile? Clearly the number of things that were in the first bag plus the number of things that were in the second bag. That is precisely what this theorem tells us.

Theorem 7.4. *Let A and B be disjoint finite sets. Then,*

$$|A \cup B| = |A| + |B|.$$

Proof Let A and B be disjoint nonempty sets of cardinality m and n, respectively, for some $m, n \in \mathbb{Z}^+$ (the result clearly holds if either A or B is empty). By definition, there exist bijections

$$f : A \to \{1, 2, \ldots, m\}$$

and

$$g : B \to \{1, 2, \ldots, n\}.$$

Write $a_i = f^{-1}(i)$ and $b_j = g^{-1}(j)$ (for i and j appropriately defined positive integers). We can then write

$$A \cup B = \{a_1, a_2, \ldots, a_m, b_1, b_2, \ldots, b_n\}.$$

Define $h : A \cup B \to \{1, 2, \ldots, m, m+1, \ldots, m+n\}$ by

$$h(x) = \begin{cases} i & \text{if } x = a_i \\ m+j & \text{if } x = b_j \end{cases}.$$

We will proceed to show that h is both injective and surjective. Take a general $x \in \{1, 2, \ldots, m+n\}$ and consider two cases: $x \leq m$ or $x = m + j$ for some $j \in \{1, 2, \ldots, n\}$. If $x \leq m$, then $h(a_x) = x$. In the other case, we have that

$$h(b_j) = m + j = x.$$

Thus, f is surjective.

To prove the injectivity of h, suppose

$$h(x) = c = h(y).$$

We again consider two cases. If $c \leq m$, then

$$x = y = a_c.$$

The Finite   333

Otherwise, if $c = m + j$, then

$$x = y = b_j.$$

Thus, h is injective and consequently a bijection, proving the result. □

The proof of the following corollary, a natural generalization of Theorem 7.4, is left as an exercise.

Corollary 7.5. *Suppose* A_1, A_2, ..., A_n *are a collection of disjoint finite sets. Then,*

$$\left| \bigcup_{i=1}^{n} A_i \right| = \sum_{i=1}^{n} |A_i|.$$

Applying Theorem 7.4 is straightforward, so long as we know the sets A and B are disjoint. For example, consider a particular college's student body, of which 740 students are commuters and 2,230 are residential students living on campus. How many students attend this school? Since a student must be exactly one commuter or residential student, we simply sum the number of each type of student: 2,970. This is Theorem 7.4 in action; the set of commuter students and residential students is disjoint.

Two sets of objects need not always be disjoint, however. Suppose you only know the above number of residential students at a college, yet you also knew that 1,114 of the college students are varsity athletes. Would you have been able to determine the enrollment at this college? In this situation we cannot simply sum the number of each type of student because some residential students may also be athletes. If we did sum the quantities, we would be counting those residential athletes twice. To compute the actual enrollment, we would need to know how many of the residential students were also athletes and subtract that number from the sum (hence eliminating the double counting).

When we count how many students are residential *or* athletes, we are asking for the cardinality of a union. Subtracting the number that are both residential *and* athletes, we are taking away the cardinality of an intersection. This is precisely Theorem 7.6, the proof of which is left as an exercise.[2]

Theorem 7.6. *Let A and B be finite sets. Then,*

$$|A \cup B| = |A| + |B| - |A \cap B|.$$

When counting, the number of things that are *this* or *that* is no more than the number of things that are *this* plus the number of things that are *that*. Why? Let us consider the college student example again. We know that 2,230 students are residential students and 1,114 are athletes. What is the *most*

---

[2]This result is often referred to as the Inclusion-Exclusion Principle.

number of students that could be classified as either residential or an athlete? It would be the sum of these two quantities: 3,344. This would be the scenario that there are no students that are both residential and an athlete, the same situation as Theorem 7.4. If there are students that are in both categories, those students are being double-counted in this sum and would have to be subtracted from the total, as in Theorem 7.6.

Corollary 7.7. *For finite sets A and B,*

$$|A \cup B| \leq |A| + |B|.$$

At first glance, it may seem that we need a constructive proof similar to that of Lemma 7.3 or Theorem 7.4. But if we refer back to the definition of cardinality, we know that $|A \cap B|$ must be non-negative. This together with the theorem from which this corollary stems immediately yields our desired result.

Proof For finite sets A and B, $|A \cap B| \geq 0$, so that

$$|A| + |B| - |A \cap B| \leq |A| + |B|,$$

proving the result.   □

From the union of two sets we move to the Cartesian product of two sets. The Cartesian product of two sets is a formalization of all possible pairs of the objects in one set with another. For example, if you were to pull a card from a standard 52-card deck and roll a single 6-sided die, how many possible outcomes would exist? For each of the 52 card choices, there are 6 possible numbers that could appear on the die. Thus, we multiply: $52 \cdot 6$ possible outcomes.[3] We validate this argumentative approach by using the explicit definition of finite sets to prove the result.

Theorem 7.8. *For finite sets A and B, $|A \times B| = |A| \cdot |B|$.*

Our approach follows from the thought process of the card-die example preceding the statement of this theorem. To count the possible outcomes in that example, we said that every choice of card yields 6 possible outcomes: one for each roll of the die. That is, we are counting 6 outcomes 52 times.

How can we generally create a similar approach? We will *partition* $A \times B$ by first coordinate. That is, group together all elements of $A \times B$ with a_1 as a first coordinate; there will be $|B|$ of these. Then, collect all elements of $A \times B$ with a_2 as a first coordinate; again, there are $|B|$ of these. Continuing this, we will partition $A \times B$ into $|A|$ partitioning sets, each of which has $|B|$ elements in it.

This is the thought process behind the proof. A little scratch work develops sound notation and a final draft of the proof appears below.

---

[3] This concept in elementary combinatorics is called the multiplication rule.

Proof If either A or B is empty, then $A \times B$ is empty and the result holds. Else, suppose A and B are finite sets with m and n elements, respectively. According to Definition 7.1, there exist bijections

$$f : A \to \{1, 2, \ldots, m\}$$

and

$$g : B \to \{1, 2, \ldots, n\}.$$

Thus, we can write $A = \{a_1, a_2, \ldots, a_m\}$ and $B = \{b_1, b_2, \ldots, b_n\}$ (as in the proof of Theorem 7.4).

For each $i \in \{1, 2, \ldots, m\}$, define

$$A_i = \{(a_i, b_k) \mid 1 \le k \le n, k \in \mathbb{Z}\}.$$

Because A and B are nonempty, each set A_i is nonempty. Moreover, for $i \ne j$, the sets A_i and A_j are disjoint, since $a_i \ne a_j$. Lastly,

$$\bigcup_{i=1}^{m} A_i = A \times B,$$

since $(a_i, b_j) \in A_i$ for all i and j. These three properties show that $A \times B$ is partitioned by the collection of sets A_i.

Note that $\pi_2 : A_i \to B$ given by $\pi_2(a_i, b_j) = b_j$ is a bijection, as it is a projection function (see Lemma 6.12 and Exercise 6.2.21), giving $|A_i| = |B|$ for all i. Thus,

$$\begin{aligned} |A \times B| &= \left| \sum_{i=1}^{m} A_i \right| \\ &= \sum_{i=1}^{m} |A_i| \ \text{(Corollary 7.5)} \\ &= \sum_{i=1}^{m} |B| \\ &= \sum_{i=1}^{m} n \\ &= m \cdot n \\ &= |A| \cdot |B|, \end{aligned}$$

as was to be shown. $\qquad\square$

Many corollaries appear to follow immediately from Theorem 7.8. Much like the theorem itself, the explicit proofs of them require a bit of work. The first two are left as exercises.

Corollary 7.9. *For finite sets A and B,*

$$|A \times B| = |B \times A|.$$

What is the intuition behind the above corollary? Think of it in terms of the previous card-die example. There are $52 \cdot 6$ possible outcomes for pulling a card and then rolling the die. Reversing the order of these actions yields $6 \cdot 52$ possible outcomes, an equal number.

Corollary 7.10. *Let A_1, A_2, ..., A_n be finite sets.[4] Then,*

$$|A_1 \times A_2 \times \cdots \times A_n| = |A_1| \cdot |A_2| \cdot \cdots \cdot |A_n|.$$

The third corollary considers two different Cartesian products: $A \times C$ and $B \times D$. What it says is that if A and B have the same number of elements (i.e., there are the same number of possible first coordinates for the two collections of ordered pairs), and, if C and D have the same number of elements (the same number of possible second coordinates for the ordered pairs), then there will be the same number of ordered pairs in both collections. While Corollary 7.11 is provable using the originally definition of cardinality (requiring the introduction and construction of numerous bijections between sets), calling upon the theorem for which it is a corollary makes the proof rather straightforward.

Corollary 7.11. *Let A, B, C and D be finite sets with $|A| = |B|$ and $|C| = |D|$. Then,*

$$|A \times C| = |B \times D|.$$

Proof Suppose for sets A, B, C and D that $|A| = |B|$ and $|C| = |D|$. Then,

$$|A \times C| = |A| \cdot |C| \text{ (by Theorem 7.8)}$$
$$= |B| \cdot |D|$$
$$= |B \times D| \text{ (by Theorem 7.8) },$$

as was to be shown. □

We continue our investigation of finite sets by exploring more that seems apparent. Theorem 7.13 is helpful. Suppose a box is holding n objects. You take some of the things from the box. How many things could you possibly be holding? Even a young child would tell you that you'd have no more than n, the number of things in the box to begin with. Though the concept is trivial, proving it requires effort. We begin with a lemma.

Lemma 7.12. *Any subset of $\{1, 2, \ldots, m\}$, where $m \in \mathbb{Z}^+$, is finite with cardinality less than or equal to m.*

---

[4]This result is sometimes referred to as the Generalized Multiplication Rule.

Proof Let $A \subseteq \{1, 2, \ldots, m\}$ for some $m \in \mathbb{Z}^+$. Define

$$f : A \to \{1, 2, \ldots, m\}$$

by $f(a) = a$. Clearly injective, Lemma 7.3 gives $|A| \leq m$. □

Armed with this lemma, we proceed to prove the aforementioned result.

Theorem 7.13. *If $A \subseteq B$ with B a finite set, then A is finite with $|A| \leq |B|$.*

Proof Let A be a subset of a finite set B, with $|B| = m$ for some non-negative integer m. If A is empty, then $|A| = 0$ and consequently, $|A| \leq |B|$. Else, suppose that A is nonempty.

Since B is finite, there exists a bijection

$$f : B \to \{1, 2, \ldots, m\}$$

for some $m \in \mathbb{Z}^+$. Then, by Exercise 20 following Section 6.2, $f|_A$ is also a bijection. Thus, $|A| = |f(A)|$. But,

$$f(A) \subseteq \{1, 2, \ldots, m\},$$

meaning that $|f(A)| \leq m$ by Lemma 7.12. Therefore, $|A| \leq |B|$. □

Chapter 2 introduced numerous ways for creating new sets from already existing ones. In this section, then, we investigate how the cardinality of the new sets relates to those of the original sets. So far we have considered the cardinality of unions, Cartesian products and subsets of sets. Set intersection and difference are investigated in various homework exercises. There is one other set that can be formed from a given set that we have yet to consider: the power set. We briefly hinted at this idea at the end of Section 6.2 when we asked if it is possible to find a bijection from A to $\mathcal{P}(A)$, stating that it does not seem possible when A is finite. Using the terminology of finite sets, we are claiming that

$$|A| \neq |\mathcal{P}(A)|.$$

The power set of A is the set of all subsets of A. When A is finite, we know that any subset of it is finite. But what can we say about the *number* of all such finite sets? A few basic examples are shown in the table below.

A	$|A|$	$\mathcal{P}(A)$	$|\mathcal{P}(A)|$
$\varnothing$	0	$\{\varnothing\}$	1
$\{a\}$	1	$\{\varnothing, \{a\}\}$	2
$\{a, b\}$	2	$\{\varnothing, \{a\}, \{b\}, \{a, b\}\}$	4
$\{a, b, c\}$	3	$\{\varnothing, \{a\}, \{b\}, \{c\}$ $\{a, b\}, \{a, c\}, \{b, c\}, \{a, b, c\}\}$	8

A pattern appears to be emerging just from these few examples: the cardinality of the power set of A (when A is finite) appears to be $2^{|A|}$. Theorem 7.14 shows us that this is indeed the case.

Theorem 7.14. *If A is a set with $|A| = n$, then $|\mathcal{P}(A)| = 2^n$.*

Our proof uses an idea presented in Example 6.13, what is referred to there as a natural bijection.

Proof Suppose A is a set with $|A| = n$, so that we write

$$A = \{a_1, a_2, \ldots, a_n\}.$$

Let B be the set of the first 2^n positive integers written in base 2 representation. Then, define $f : \mathcal{P}(A) \to B$ by

$$f(S) = b_n b_{n-1} \ldots b_1,$$

where

$$b_i = \begin{cases} 1 & \text{if } a_i \in S \\ 0 & \text{if } a_i \notin S \end{cases}.$$

It suffices to prove that f is a bijection.[5]
First, suppose

$$f(S_1) = f(S_2)$$

for $S_1, S_2 \subseteq \mathcal{P}(A)$. To show $S_1 = S_2$ we show that they are subsets of one another. Assume $a_i \in S_1$. Then, $b_i = 1$, and consequently, $a_i \in S_2$. Hence,

$$S_1 \subseteq S_2.$$

The claim that

$$S_2 \subseteq S_1$$

is shown similarly. Therefore f is injective.
Now, suppose

$$b_n b_{n-1} \ldots b_1 \in B.$$

Construct a subset S of A as

$$a_i \in S \text{ if and only if } b_i = 1.$$

By construction and definition of f, it follows that

$$f(S) = b_n b_{n-1} \ldots b_1,$$

---

[5]Images of some elements of $\mathcal{P}(A)$ may not be precisely in base 2 notation, as they may have leading zeros.

proving that f is surjective, and consequently, bijective. □

We conclude this section with a result concerning functions between finite sets. Suppose you roll two standard 6-sided dice. Are you guaranteed to roll two matching numbers? No, because you only have about a 16.7% chance of doing so. What if you were to roll three dice? Now can you say you'll roll at least two numbers that match? As before, we do not know it has to happen (though it is more likely than when rolling two dice, giving you a 41.7% chance of a pair in this case). How about four dice? Five? Six? With six dice, it is quite likely (98.5% likely) that you'll roll a pair but it's no guarantee; perhaps each die shows a different number. But if you were to roll seven dice? In this scenario, something is inherently different from rolling six or fewer dice.

In rolling seven die, why *must* at least two of the same number appear? In trying to construct the worst case scenario, the first six dice could all be different, but the seventh die must show some number that has already been rolled, creating a pair. That is, there are seven outcomes that must all be one of six possible choices. At least two of them must be the same. This is precisely the idea known as the *Pigeonhole Principle*.

Its name comes from a similar real-world parallel. Pigeons are known to roost in holes, known as pigeonholes. Suppose you look at a wall and there are m pigeonholes, yet on the ground there are greater than m pigeons. If you were to spook the birds, causing them all to fly into various pigeonholes, can you conclude anything about the number of birds in the holes? Yes; at least one hole holds two or more birds.

Theorem 7.15 formalizes this result. It is attributed to Peter Dirichlet (1805-1859), whose contributions to mathematics appear in number theory and analysis (in particular, the theory of Fourier series). Interestingly, he is also considered to be the first to propose the definition of a function appearing in Chapter 6 [33].

Theorem 7.15. (Pigeonhole Principle) *For finite sets A and B, no injective function $f : A \to B$ exists if $|B| < |A|$.*

Before proceeding with the proof of the Pigeonhole Principle, note that the statement of the result is not quite identical to the metaphor of pigeons roosting in holes. Here, we can consider the set A to be the pigeons and the set B to be the pigeonholes. If the number of holes is fewer than the number of pigeons, then, this result says there *cannot* exist an injective function mapping the pigeons to the holes. That is, there is no way to put exactly one bird in each hole. What does this mean? That at least one hole must contain two or more birds.

It is apt to mention that the Pigeonhole Principle is an existential result, much like the Intermediate Value and Mean Value Theorems of calculus.[6]

---

[6]These results state that a particular value exists within a certain interval, though there may be more than one such value. Where in the interval such values occur cannot be determined. All that can be concluded is the existence of one value satisfying a particular property.

There is no way to state *which* hole holds multiple pigeons or *how many* pigeons are in any particular hole (they all may flock to one, for example). All we know is what *some* hole holds more than one pigeon. Beyond that, we can make no conclusions.

Proof Let A and B be finite sets of cardinality m and n, respectively, with $n < m$. We can write

$$B = \{b_1, b_2, \ldots, b_n\}.$$

Let us assume for contradiction that $f : A \to B$ is injective. Then, by injectivity, we know

$$|f^{-1}(\{b_i\})| \leq 1$$

for each i. Moreover, we have that

$$\bigcup_{i=1}^{n} f^{-1}(\{b_i\}) = A,$$

since $f(a) \in \{b_1, b_2, \ldots, b_n\}$ for each $a \in A$. Because

$$f^{-1}(\{b_i\}) \cap f^{-1}(\{b_j\}) = \varnothing$$

for $i \neq j$, we have

$$
\begin{aligned}
m = |A| \\
= \left| \bigcup_{i=1}^{n} f^{-1}(\{b_i\}) \right| \\
= \bigcup_{i=1}^{n} |f^{-1}(\{b_i\})| \\
\leq n,
\end{aligned}
$$

a contradiction to the assumption proving such injective function f exists. $\square$

The Pigeonhole Principle is so straightforward it seems trivial, yet it can be used to make rather stunning conclusions. The next two examples exhibit just this, with the first being a fairly routine use of the principle while the second, Example 7.2, yields a somewhat surprising result.

> **Example 7.1.** Given any collection of 5 integers, show that there are two of them (different from one another) whose difference is divisible by 4.
>
> When looking at this, it is tempting to begin by considering all possible sums of pairs of the integers. However, this gets to be a bit overwhelming

quickly; there would be 10 different pairs to consider. Ideally we can use a transferable method, should we encounter a similar problem involving a larger collection of integers. We have previously seen that the Quotient-Remainder Theorem is a useful tool when dealing with divisibility.

To that end, call the integers x_1 through x_5, with the Quotient-Remainder Theorem yielding integers q_i and r_i $(0 \leq r_i < 4)$ so that

$$x_i = 4q_i + r_i.$$

Note that there are 5 remainders, each of which takes one of the 4 values 0, 1, 2 or 3. Thus, at least two of these remainders must be the same; assume WLOG that $r_1 = r_2$. Then,

$$\begin{aligned} x_1 - x_2 &= (4q_1 + r_1) - (4q_2 + r_2) \\ &= 4(q_1 - q_2), \end{aligned}$$

proving that $x_1 - x_2$ is divisible by 4.

Example 7.2. Prove that at any given Super Bowl, at least two people shake the same number of hands.

At first glance, this seems like it is rather straightforward. Maybe two people didn't shake anybody's hands. But we cannot definitively make this conclusion. Perhaps only one person shook no hands, and likewise, maybe only one person shook a single person's hands. The attendance at the 1980 Super Bowl was 103,985; that's a lot of possible hands people could or could not shake!

We are dealing with *any* Super Bowl, so let us say there are n people in attendance at a particular Super Bowl. If we wish to employ the Pigeonhole Principle, we must determine what plays the roles of the pigeons and the pigeonholes. In this case, the question being asked makes that clear. People are playing the role of the pigeons and handshakes the pigeonholes (since we wish to show two *people* have the same number of *handshakes*).

Then, how many possible handshakes are there amongst the n people? A person can shake anywhere from 0 hands to $n - 1$ hands (everybody except themselves). That's n possible handshakes; we cannot apply the Pigeonhole Principle to n people and n handshakes. However, if somebody were to shake no hands, would it be possible for someone else to shake everybody's hand? No, the person would have to shake the hand of the person who shakes no hands! Consequently, there are either 0 to $n - 2$ possible handshakes or 1 to $n - 1$ possible handshakes. In either case, there are $n-1$ possible numbers of handshakes amongst n people, so that the desired conclusion can be made.

As we have said throughout this section, the ideas encountered here come as no surprise. Thorough proofs require care and precision. As with most of the concepts introduced throughout this text, gaining comfort with the new definitions and results should be done at two levels: intuitive and verbatim. The intuitive level will lead towards strong investigative skills while knowing mathematical concepts precisely (the verbatim level) will help create pristine proofs.

Such an in-depth look at finite sets begs one particular question: can we generalize to the infinite? In the next two sections we attempt to do just this.

Exercises

1. Determine the cardinality of each of the following sets by finding a specific bijection between the set and a subset of the positive integers.

   (a) $\{\pm 1, \pm 2, \pm 3, \pm 4\}$.

   (b) $\{2, 4, 8, \ldots, 512\}$.

   (c) $\{x \in \mathbb{Z} \mid x = 4t + 2 \text{ for some } t \in \mathbb{Z}, 1 \le t \le 50\}$.

   (d) $\{x \in \mathbb{Z}^+ \mid x^2 + x \text{ is prime}\}$.

   (e) $\{1, 2, 3\} \times \{1, 2, 3\}$.

2. Prove that the function g of Lemma 7.3 is a bijection.

3. Prove directly using the definition of a finite set that if A and B are finite sets, then $A \cap B$ is finite.

4. (a) If A is a nonempty finite set and $a \in A$, prove that
   $$|A - \{a\}| = |A| - 1.$$

   (b) If A is a finite set with $|A| > n$, then if $a_1, a_2, \ldots, a_n \in A$, prove that
   $$|A - \{a_1, a_2, \ldots, a_n\}| = |A| - n.$$

5. Prove that the function g defined in the proof of Lemma 7.12 is a bijection.

6. Prove Corollary 7.5.

7. Prove Corollary 7.9.

8. Prove Corollary 7.10.

9. For finite sets A and B of cardinality m and n, respectively, find a bijection
   $$h : A \times B \to \{1, 2, \ldots, mn\}$$

and prove it is such.

10. Prove Theorem 7.6.

11. Provide a second proof of Theorem 7.14 utilizing Theorem 4.13 to more efficiently prove that the defined function f is a bijection.

12. Prove that the number of permutations of a finite set is finite.

13. If A and B are finite, show that the number of functions $f : A \to B$ is finite.

14. Prove each of the following, where A is a finite set.

    (a) $|A \times \{1\}| = |A|$.
    (b) $|A \times \{1, 2, \ldots, n\}| = n|A|$.

15. Prove each of the following, where A is a finite set of cardinality $n > 0$.

    (a) $|\mathcal{P}(A) \times \mathcal{P}(A)| = 2^{2n}$.
    (b) $|\mathcal{P}(A) \times \mathcal{P}(A) \times \cdots \times \mathcal{P}(A)| = 2^{mn}$, where m copies of $\mathcal{P}(A)$ appear in the Cartesian product.
    (c) $|\mathcal{P}(\mathcal{P}(A))| = 2^{2^n}$.

16. (a) If A is finite and $f : A \to B$ for any set B, prove that $f(A)$ is a finite set.

    (b) If B is finite and $f : A \to B$ is a function, must $f^{-1}(B)$ be finite? Prove or disprove.

17. A dice game is played by rolling a pair of standard 6-sided dice. With each roll, the sum of the upward facing sides is computed. Prove that if the pair is rolled 12 times, then then some sum will have occurred at least twice.

18. A bag contains 5 red balls, 10 blue balls, 1 yellow ball and 4 black balls. Balls are randomly selected from the bag.

    (a) How many balls must be selected to guarantee 2 of the same color are chosen?

    (b) How many balls must be selected to guarantee 2 blue balls are chosen?

19. A coding system assigns 4-digit codes to objects.

    (a) How many objects must be chosen to guarantee that 2 have the same code?

    (b) Items on a certain shelf are assigned codes from 4000 to 4199. If the shelf contains over 100 objects, none of which have the same code but are ordered from low-to-high codes, prove that at least two objects sitting next to each other on the shelf have consecutive codes.

20. How many people must be gathered to guarantee that two were born during the same hour (i.e., 12:00 AM - 12:59 AM) on the same day of the calendar year (disregarding the year they were born and disregarding leap years)?

21. (a) On the unit circle, 210 points are chosen, each a positive integer (up to 360) number of degrees, as measured in the standard method (counterclockwise from the positive x-axis). Prove that at least one pair of points are antipodal (that is, on opposite ends of a chord passing through the origin).

    (b) The integers 1 through 10 are written on a circle, in any order. Prove that there are three consecutive numbers on the circle whose sum is greater than 16.

22. If 55 elements from the set $\{1, 2, \ldots, 99, 100\}$ are chosen, prove that two of the elements will have a difference of 10.

23. If $m + 1$ elements of the set $\{1, 2, \ldots, 2m\}$ are chosen, prove that at least one of the chosen integers is odd.

24. The population of the United States was approximately 325 million in 2017. Prove that in 2017, there were

    (a) two people living in the United States who were born during the same minute on the same date (disregarding year of birth).

    (b) two people living in the United States with the same first and last initials who were born during the same hour (see Exercise 20) on the same day of the year in the same state.

25. What is the least number of integers between 1 and 50 one must choose to guarantee that one of them is divisible by 4?

26. If one selects 201 integers between 1 and 400, must there be a pair such that one number of the pair divides the other? Prove or disprove.

27. If 5 points are chosen on a square of side length 1, prove that at least 2 of them are at most $\sqrt{2}/2$ units apart.

28. A sequence of integers $a_1, a_2, \ldots, a_n$ is called *strictly increasing* if

$$a_1 < a_2 < \cdots < a_n;$$

the sequence is *strictly decreasing* if

$$a_1 > a_2 > \cdots > a_n.$$

A *subsequence* of the sequence is a sequence $a_{i_1}, a_{i_2}, \ldots, a_{i_m}$ where

$$1 \leq i_1 < i_2 < \cdots < i_m \leq n.$$

(a) Find a strictly increasing subsequence of length 4 in the sequence of integers
$$20,\ 47,\ 38,\ 71,\ 63,\ 55,\ 64,\ 81,\ 13,\ 8.$$

(b) Find a strictly decreasing subsequence of length 4 in the above sequence.

(c) Prove that a sequence of $n^2 + 1$ distinct real numbers must contain a subsequence of length $n + 1$ that is either strictly increasing or strictly decreasing.

7.2 The Infinite: Countable

In the previous section, a nonempty set A was defined to be *infinite* if no bijection between A and $\{1, 2, \ldots, n\}$, for any $n \in \mathbb{Z}^+$, exists. This is a simple concept, but how do we go about showing something *cannot* exist? Typically, a proof by contradiction is used. Assume that there *is* a bijection of the required form and arrive at some contradiction. The following lemma exhibits this.

Lemma 7.16. *The set of positive integers, $\mathbb{Z}^+$, is infinite.*

As mentioned prior to this lemma, trying to prove prove this result directly is fruitless. A direct proof would require us to prove that $\mathbb{Z}^+$ satisfies the precise definition of being infinite; we would show that $\mathbb{Z}^+$ is not finite. We have seen throughout this text that showing an object is *not* something is often challenging. Just because *we* could not show that it was the something does not mean that *nobody* could do so. So, we turn to indirect proof methods and consider a proof by contradiction. The assumption that $\mathbb{Z}^+$ is finite gives us a tool to work with: that a particular bijection from $\mathbb{Z}^+$ to $\{1, 2, \ldots, n\}$ exists.

What sort of contradiction might we aim to reach by assuming this? If we dissect our assumption a bit further, we may find something to investigate. The assumption is that a *bijection* exists, and bijections are both *injective* and *surjective*. Can either of these help us?

Consider injectivity. This means that every element of the co-domain, $\{1, 2, \ldots, n\}$, is mapped to by exactly one integer. For example, if $n = 4$, then there are only four integers that can possibly map to $\{1, 2, 3, 4\}$. But $\mathbb{Z}^+$ contains a whole lot more than four integers! In particular, if we take any five integers, could such a function into $\{1, 2, 3, 4\}$ possibly be injective? No!

This is the key idea, then. Take just *one more* integer than the n that map injectively into $\{1, 2, \ldots, n\}$ and show this violates our assumption of the function being injective. What tool tells us that it is impossible to injectively partner $n + 1$ objects with n things? The Pigeonhole Principle!

Proof Suppose for contradiction that $\mathbb{Z}^+$ is finite. By definition, then, there exists a bijection $f : \mathbb{Z}^+ \to \{1, 2, \ldots, n\}$, for some $n \in \mathbb{Z}^+$. Let

$$a_i = f^{-1}(\{i\});$$

that is, a_i is the positive integer mapped to i under f. Call a_j the greatest of these, so that $a_i \leq a_j$ for $1 \leq i \leq n$.

Consider then the set

$$B = \{a_1, a_2, \ldots, a_n, a_j + 1\},$$

a set of $n+1$ distinct positive integers. By the Pigeonhole Principle, then, $f|_B$, the restriction of f to B, cannot be injective, a contradiction since f itself is injective (Exercise 20a). Hence, $\mathbb{Z}^+$ is infinite. □

As an infinite set, $\mathbb{Z}^+$ plays a particularly important role. Any set with the same cardinality as $\mathbb{Z}^+$ is called *countably infinite*. Such sets are the focus of this section.

Definition 7.17. A set A is called *countably infinite* if it has the same cardinality as $\mathbb{Z}^+$, or equivalently, if there exists a bijection $f : A \to \mathbb{Z}^+$ (or $f : \mathbb{Z}^+ \to A$). A set is called *countable* if it is either finite or countably infinite.

It was Georg Cantor who introduced the world to this notion of cardinality and size of infinite sets. As the choice of language in Definition 7.17 suggests (i.e., *countably* infinite), not all infinite sets are countably infinite. Because of this, Cantor wished for some way to distinguish between these different sizes of infinity. To do so, he used the concept of *transfinite cardinals*. Just as we can say the cardinality of $\{a, b, c\}$ is 3, we can use transfinite cardinals to discuss the cardinality of infinite sets. Those infinite sets that are countably infinite are said to have cardinality $\aleph_0$ (pronounced "aleph nought," where $\aleph$ is the first letter of the Hebrew alphabet). Hence,

$$|\mathbb{Z}^+| = \aleph_0.$$

The choice of the word "countable" may seem awkward. Finite sets are clearly countable, as we can count the number of people in a classroom or the number of leaves on a tree. But how can something that is not finite be countable, since attempting to count it would be neverending? We choose to use the word "countable" not in the sense that "we can count all the elements" but rather that "each element is guaranteed to eventually be counted."

For example, imagine there is a roller coaster with a *really* long line of people waiting to ride it. If the line is finite, then there is a last person in line. However, if the line is countably infinite, there is no end to it; it stretches on forever. However, every person in line can figure out his or her position in line is. Veronica could be first in line, while Noah might be 237,114th in line (he must *really* want to ride this roller coaster). Everybody in line has some position. To determine what it is, start counting from the front of the line and know that at some point, you will eventually get around to counting that person, no matter how far back in line they are.

While we have mentioned that non-countable infinite sets (called *uncountable* sets) exist, we will have to wait until the next section to find a particular one. In the meantime, the following theorem tells us that *any* infinite set, countable or not, contains a countably infinite subset.

Theorem 7.18. *Every infinite set contains a countably infinite subset.*

Proof Suppose that A is an infinite set. Because A is nonempty, choose $a_1 \in A$. The set

$$A - \{a_1\}$$

is infinite (else $A = \{a_1\}$, a finite set), so we can choose a_2 as an element in it. Likewise, choose

$$a_3 \in A - \{a_1, a_2\},$$

which is possible since the set

$$A - \{a_1, a_2\},$$

as before, must be infinite. Continue this, choosing

$$a_i \in A - \{a_1, a_2, \ldots, a_{i-1}\}.$$

The set

$$\{a_i \mid i \in \mathbb{Z}^+\}$$

is the desired countable subset of A. The proof that it is indeed countably infinite is straightforward. Define f mapping this set to $\mathbb{Z}^+$ by $f(a_i) = i$. The proof that f is a bijection is left to the reader. $\square$

Corollary 7.19 follows immediately from Theorem 7.18. Its proof is left as an exercise.

Corollary 7.19. *Every nonempty set has a nonempty countable subset.*

While Theorem 7.18 tells us that *any* infinite set has *some* countably infinite subset, is there anything we can conclude about the subsets of a *countable* set? For example, consider the infinitely long line of people waiting for the roller coaster. We know that the entire line is countable. Is there anything we can say about any subcollection of those people? We may know nothing about the people in the line, but consider the set of all people in that line who were born on a Tuesday in March during a leap year. We have no way of knowing if such a person exists in the line, or alternatively, if every person in the line was born on such a day. But we do know one thing about this collection of people: they form a countable set.

If we simply proceed down the original line of people waiting to ride the roller coaster, every time we reach a person born on such a Tuesday we can count him or her. A person may be the hundredth person in the line but the first person born on a Tuesday in March during a leap year. We count that person. We continue until we find the second such person so we have a total of two. Continue this indefinitely and every person sharing this trait will eventually be counted.

Before proving Theorem 7.20, however, an important note should be made about countably infinite sets. If A is countably infinite, by definition there exists a bijection $f : A \to \mathbb{Z}^+$. Similar to what we did in the finite case, for each $i \in \mathbb{Z}^+$, let

$$a_i = f^{-1}(\{i\})$$

(that is, $f(a_i) = i$). In doing so, we now can write

$$A = \{a_1, a_2, a_3, \dots\}.$$

From this point forward, references to a countably infinite set will use this notation.

Theorem 7.20. *Every subset of a countable set is countable.*

When assuming that a set is countable, we must always remember that this really means we know one of *two* things: the set is finite or the set is countably infinite. We have to handle the two cases separately (and often using very different techniques).

If the set we assume to be countable is finite, our proof is over. Why? It is not because this is obvious but rather because this was a previous result (Theorem 7.13); no need to reinvent, or in this case reprove, the wheel.

Otherwise we can assume the set we wish to take a subset of takes the form

$$A = \{a_1, a_2, a_3, \dots\}.$$

That is, our subset of A consists of some number of the a_i, perhaps something like

$$\{a_5, a_7, a_{13}, a_{321}, \dots\}.$$

If you find yourself stuck, even with this example, step back from the work and ask yourself, "Did I assume anything helpful when I constructed my example?"

In this case, we *ordered* the elements of our subset in the same order they appeared in our exposition of A. In particular, a_5 is playing the role of the *first* element of the subset, a_7 the second, and so on. Considering the set of these subscripts as a nonempty subset of the positive integers, we can utilize the Well-Ordering Principle. With that in mind, we move to the proof of the result.

Proof Suppose that A is a countable set. If A is finite, then any subset of A is finite (Theorem 7.13). Otherwise, suppose that

$$A = \{a_1, a_2, a_3, \dots\}$$

is countably infinite and let $B \subseteq A$.

If $B = \varnothing$, then the result holds. Otherwise, employ the Well-Ordering Principle to choose $b_1 \in B$ so that b_1 has the smallest subscript when considered as an element of the set A.

Consider the set $B - \{b_1\}$. If this set is empty, then $B = \{b_1\}$ is a finite subset of A. Otherwise, choose b_2 to be the element of $B - \{b_1\}$ with the least subscript, as an element of A, which is possible again by the Well-Ordering Principle.

Continue this process. If at any step

$$B - \{b_1, b_2, \dots, b_i\}$$

is empty, then as before, B is finite and consequently countable. Otherwise

$$B = \{b_1, b_2, \dots\}$$

is an infinite subset of A (since each $b_i = a_j$ for some $a_j \in A$). Since the function $f : B \to \mathbb{Z}^+$ given by $f(b_i) = i$ is clearly a bijection, it follows that B is indeed countably infinite and thus countable. □

Having established these results about the existence of countably infinite sets, we proceed to show that various specific sets are countably infinite. The first result, showing that the set of *all* integers is countable, may be a bit surprising, similar to the discussion introducing this section. Our gut tells us that *clearly* there are more integers than simply positive integers. Thinking about the two sets on a number line, it seems that there are *half* as many positive integers as there are integers. It is the definition of being countably infinite that we must not push aside in favor of our intuition, however.

Theorem 7.21. *The set of integers, $\mathbb{Z}$, is countable.*

Before proving Theorem 7.21, this is an appropriate place to discuss a method for constructing such a proof. It is inspired by the infinitely long line for the roller coaster, mentioned previously. To demonstrate that $\mathbb{Z}$ is countable, we must be able to line up the integers in such a way that *every* integer will at some point appear in the line. Thus, trying to count the integers as in Figure 7.1 would not suffice.

The top row, representing $\mathbb{Z}^+$, of Figure 7.1 serves as the place in line, while the second row attempts to list all of the elements of the set we are counting, $\mathbb{Z}$. Why does this approach *not* work? Though we can continue the process indefinitely, only the non-negative integers will appear in line. At no point will -1 or any other negative integer appear; it simply does not have a place in the infinitely long line.

To count every integer, we proceed as in Figure 7.2. With this approach, one can definitively say that *every* integer will eventually be counted.

$$\mathbb{Z}^+: \quad 1 \quad 2 \quad 3 \quad 4 \quad 5 \quad \cdots$$
$$\downarrow \quad \downarrow \quad \downarrow \quad \downarrow \quad \downarrow$$
$$\mathbb{Z}: \quad 0 \quad 1 \quad 2 \quad 3 \quad 4 \quad \cdots$$

FIGURE 7.1
A failed attempt at counting $\mathbb{Z}$

$$\mathbb{Z}^+: \quad 1 \quad 2 \quad 3 \quad 4 \quad 5 \quad \cdots$$
$$\downarrow \quad \downarrow \quad \downarrow \quad \downarrow \quad \downarrow$$
$$\mathbb{Z}: \quad 0 \quad 1 \quad -1 \quad 2 \quad -2 \quad \cdots$$

FIGURE 7.2
A method for counting $\mathbb{Z}$

If one were asked to *show* that the integers are countable, Figure 7.2 would suffice. It only serves as our inspiration for a proof, though. An explicit proof requires fulfilling Definition 7.17 by defining and proving that a specific function is indeed a bijection.

Proof To prove that $\mathbb{Z}$ is countable, define $f : \mathbb{Z}^+ \to \mathbb{Z}$ by

$$f(n) = \begin{cases} \dfrac{n}{2}, & \text{if } n \text{ is even} \\[2mm] \dfrac{1-n}{2}, & \text{if } n \text{ is odd.} \end{cases}$$

First, to show that f is injective, assume $f(a) = f(b)$ for some $a, b \in \mathbb{Z}^+$. Note that a and b must be of the same parity, else $f(a)$ and $f(b)$ have opposite signs (and are nonsero, since $f(n) = 0$ if and only if $n = 0$). If both are even, then

$$\frac{a}{2} = \frac{b}{2}$$

gives $a = b$. Likewise, if both are odd, we see that $a = b$. Hence, f is injective.
To prove that f is surjective, take $m \in \mathbb{Z}$. If $m \geq 0$, then

$$f(2m) = m,$$

and if $m < 0$,

$$f(1 - 2m) = m,$$

proving that f is surjective and consequently bijective. □

Having established that the set of of all integers is countable, Corollary 7.22 should come as no surprise.

Corollary 7.22. *The set of all even integers is countable.*

Proof Because the set of all even integers is a subset of $\mathbb{Z}$, a countable set, it is countable, by Theorem 7.20. □

We can prove that the odd integers are countable in precisely the same fashion. Had these results come *before* proving that $\mathbb{Z}$ is countable, it would be natural to ask if combining these two results would force $\mathbb{Z}$ to be countable. If one can count the even integers and one can count the odd integers, one ought to be able to count the two sets collectively. Theorem 7.23 is a generalization from these two specific sets to any two countable sets.

Theorem 7.23. *If A and B are countably infinite sets, then $A \cup B$ is countable.*

A classic misstep in trying to prove that $A \cup B$ is countable is to say, "First count A and then count B. That will result in counting all of $A \cup B$." The flaw here is similar to the flaw appearing in Figure 7.1 when attempting to count $\mathbb{Z}$. In saying, "Count A first and then count B," we assume that we will eventually get around to counting the elements of B. But if A is infinite, if we start by only counting the elements of A, we will find ourselves *only* counting the elements of A; we will never count a single element of B!

The solution to this problem? Alternate counting elements of A and elements of B. The function f in the proof of this result does just that.

Proof Let $A = \{a_1, a_2, \dots\}$ and $B = \{b_1, b_2, \dots\}$ be countable sets. Define $f : \mathbb{Z}^+ \to A \cup B$ by

$$
f(n) = \begin{cases} a_k, & \text{if } n = 2k \text{ for some } k \in \mathbb{Z} \\ \\ b_k, & \text{if } n = 2k + 1 \text{ for some } k \in \mathbb{Z}. \end{cases}
$$

The proof that f is a bijection is left as an exercise. □

A natural generalization of Theorem 7.23 is that a finite union of countably infinite sets is countable, and it is left to the reader in Exercise 11. We generalize further, then, and consider a countably infinite number of countably infinite sets. Is it possible to count such a colletion? Theorem 7.24, presented without proof,[7] affirms it so.

An interesting exhibition of Theorem 7.24 is known as Hilbert's Grand Hotel, an idea put forth by David Hilbert (1862-1943) in the early 20th century [19]. This example is but one minor contribution of Hilbert's to modern

---

[7]The proof is omitted not because of its difficulty but because of its reliance on the *Axiom of Choice*.

mathematics. Geometry, functional analysis and invariant theory were all significantly advanced through his work, but no introductory proof text would be complete without mention of Hilbert's influence on logic. It is in this field that proofs are considered to be mathematical objects, adhering to the axioms and laws placed upon them. This philosophical approach to logic continues to give rise to research in language theory (both in computer science and in linguistics) as well active research in mathematics.

In regards to our work here, we describe Hilbert's Grand Hotel. If you were to ask anyone, "Can a hotel that has all its rooms occupied make room for another few guests," you would likely receive a response similar to, "Not unless some were to sleep in a closet." But Hilbert's hotel can accommodate more than just a few. He explains how his hotel can house *infinitely* many new guests.

Hilbert's Grand Hotel is imaginary in nature, because it contains a countably infinite number of rooms. Let us suppose each of them is occupied when at the front desk of the hotel arrive a countably infinite number of new guests requiring accommodations. To house them, the front desk calls up the current guests and asks them to switch rooms. Each guest is asked to move to the room number that is double their current room number, from Room M to room $2M$. In doing so, every odd numbered room is freed up, a countably infinite number of rooms to house the new guests.

Theorem 7.24. *A countably infinite union of countably infinite sets is countable.*

Figure 7.3 illustrates how to count the elements of a countably infinite number of countably infinite sets and can be thought of as a visual proof of Theorem 7.24. The countable union of sets can be enumerated $A_1, A_2, A_3, \ldots$ using the notation previously discussed. Each individual set, then, can also be enumerated using this same notation. So, we consider $A_i = \{a_{i1}, a_{i2}, a_{i3}, \ldots\}$, for each positive integer i. Now that a notation for all the elements of the countably infinite union of countably infinite sets is available, we proceed to count them.

As with counting the integers, however, we cannot just count the elements of A_1; no elements of A_2 will ever be counted. Likewise, we cannot count just the first elements of each set (that is, a_{11}, a_{21}, a_{31}, etc.). Because there are infinitely many sets, the process would not end and elements such as a_{12} would never be counted. Instead, we proceed "diagonally" through the enumerated sets. The method, known as the *diagonalization argument*, will prove to be quite useful in proving another well known set is countable.

Though it may seem unnecessary, we state the following corollary to Theorem 7.24. Its proof is left as an exercise.

Corollary 7.25. *A countable union of countable sets is countable.*

If we are searching for larger cardinalities, we have not gotten very far. Unions, finite or countably infinite, of countable sets are themselves countable.

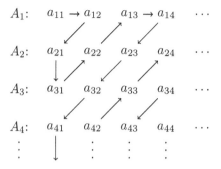

FIGURE 7.3
Counting a countably infinite collection of countably infinite sets

In other words, our sets have not increased in size. What other sets might we consider? A natural next place to look is towards Cartesian products of countable sets. Since we know that the Cartesian product of finite sets is finite, we focus on the Cartesian products of countably infinite sets. While the diagonalization argument of Figure 7.3 could be directly applied to show that $\mathbb{Z}^+ \times \mathbb{Z}^+$ is countable, we use Theorem 7.24 to show that $\mathbb{Z}^+ \times \mathbb{Z}^+$ is countable.

Corollary 7.26. $\mathbb{Z}^+ \times \mathbb{Z}^+$ *is countable.*

Proof To prove that $\mathbb{Z}^+ \times \mathbb{Z}^+$ is countable, define

$$A_i = \{(i, n) \mid n \in \mathbb{Z}^+\}.$$

Each A_i is countable, as $f : A_i \to \mathbb{Z}^+$ given by $f(i, n) = n$ is a bijection (details are left to the reader). Then,

$$\mathbb{Z}^+ \times \mathbb{Z}^+ = \bigcup_{i=1}^{\infty} A_i$$

is a countably infinite union of countably infinite sets. By Theorem 7.24, the result holds. □

As you may suspect, there is nothing special about the choice of $\mathbb{Z}^+$ in Corollary 7.26. The argument used in the proof is applicable to arbitrary countably infinite sets, as given in Corollary 7.27. Its proof is left as an exercise.

Corollary 7.27. *The Cartesian product of two countably infinite sets is itself countably infinite.*

If A, B and C are countably infinite sets, then what can we say about $A \times B \times C$? Corollary 7.27 gives that $A \times B$ is countably infinite. If we consider it as a single set, then, the Cartesian product of it with C would also be countably infinite (though we must take care with notation when explicitly proving this result). This argument can be applied to not just the Cartesian product of three sets but to any finite number of countably infinite sets.

Theorem 7.28. *Let A_1, A_2, ..., A_n be countably infinite sets. Then,*

$$A_1 \times A_2 \times \cdots \times A_n$$

is countably infinite.

Proof We prove that

$$A_1 \times A_2 \times \cdots \times A_n,$$

where each A_i is a countably infinite set, is countably infinite via induction on n.

Corollary 7.27 is the base case for the inductive argument. Assume then for the inductive hypothesis that for some general $n \in \mathbb{Z}^+$ that

$$A_1 \times A_2 \times \cdots \times A_n$$

is countably infinite.

Take A_{n+1} to be another countably infinite set. The set

$$(A_1 \times \cdots \times A_n) \times A_{n+1}$$

is, by Corollary 7.27, countably infinite. To prove that

$$A_1 \times \cdots \times A_n \times A_{n+1}$$

is countably infinite, consider the function

$$f : (A_1 \times \cdots \times A_n) \times A_{n+1} \to A_1 \times \cdots \times A_n \times A_{n+1}$$

given by

$$f((a_1, \ldots, a_n), a_{n+1}) = (a_1, \ldots, a_n, a_{n+1}).$$

The function f is a bijection (left to the reader), showing that

$$|(A_1 \times \cdots \times A_n) \times A_{n+1}| = |A_1 \times \cdots \times A_n \times A_{n+1}|.$$

Hence, the inductive step, and consequently the result, holds. $\square$

Therefore no matter how many copies of $\mathbb{Z}$ are included, the Cartesian product $\mathbb{Z} \times \mathbb{Z} \times \cdots \times \mathbb{Z}$ is countable. In our search for larger cardinalities, then, we turn our attention away from $\mathbb{Z}$ and to another specific collection of real numbers: the rational numbers.

In terms of considering *how many* rational numbers there are, the number line visualization mentioned in the introduction to this chapter makes it seem like their cardinality must be larger than that of the integers. How could one possibly find a bijection from $\mathbb{Z}$ to $\mathbb{Q}$? Theorem 7.29 tells us that this intuition is wrong, however. The rational numbers are indeed countable.

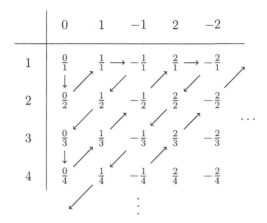

FIGURE 7.4
Counting the rational numbers

Theorem 7.29. *The rational numbers, $\mathbb{Q}$, are countable.*

Rather than prove Theorem 7.29 explicitly, we exhibit how to count $\mathbb{Q}$ in Figure 7.4. The table, which extends indefinitely downward and to the right, is a systematic method for displaying all rational numbers. The column represents the numerator of the rational number and the row the denominator. The method, then, to count the rational numbers is shown via the arrows and is often referred to as *Cantor's diagonalization process* and follows the approach of Figure 7.3. Following the path of the arrows, we align the rational numbers into an infinite list:

$$\frac{0}{1}, \frac{0}{2}, \frac{1}{1}, -\frac{1}{1}, \frac{1}{2}, \frac{0}{3}, \frac{0}{4}, \frac{1}{3}, -\frac{1}{2}, \frac{2}{1}, -\frac{2}{1}, \frac{2}{2}, -\frac{1}{3}, \frac{1}{4}, \ldots$$

Considering the entries of this list as elements of a set,

$$S = \{\frac{0}{1}, \frac{0}{2}, \frac{1}{1}, -\frac{1}{1}, \frac{1}{2}, \frac{0}{3}, \frac{0}{4}, \frac{1}{3}, -\frac{1}{2}, \frac{2}{1}, -\frac{2}{1}, \frac{2}{2}, -\frac{1}{3}, \frac{1}{4}, \ldots\},$$

shows that S is countable. However, S is not equal to the set of rational numbers. If its elements were put into lowest terms, it would contain repeated elements. However, $\mathbb{Q}$ *is* a subset of S and consequently by Theorem 7.20 it is countable.

Our search for a cardinality larger than that of being countably infinite, thus far, has come up empty handed. Forming new sets from old, such as unions and Cartesian products of countably infinite sets, does not yield a larger cardinality. Considering sets larger than $\mathbb{Z}^+$, such as $\mathbb{Z} \times \mathbb{Z} \times \mathbb{Z}$ and $\mathbb{Q}$, brought us right back to where we began; these sets actually are no bigger than $\mathbb{Z}^+$!

All these sets have the same cardinality: $\aleph_0$. This begs the question: are there infinite sets that are *not* countable? The natural set to consider next is $\mathbb{R}$, having already considered particular subsets of the real numbers such as $\mathbb{Z}^+$, $\mathbb{Z}$ and $\mathbb{Q}$. In the coming section we will show that $\mathbb{R}$ is indeed not a countably infinite set.

Exercises

1. In your own words, explain the notion of a *countably infinite set* as if you were describing it to a non-mathematician.

2. Show sets A and B have the same cardinality by finding a bijection between them.

    (a) $A =$ the set of even integers, $B =$ the set of odd integers
    (b) $A = \{4m \mid m \in \mathbb{Z}\}$, $B = \{5n \mid n \in \mathbb{Z}\}$
    (c) $A = \{1, 2\} \times \mathbb{Z}^+$, $B = \mathbb{Z}^+$
    (d) $A = \mathbb{Z}^+ \times \mathbb{Z}^+$, $B = \{(x, y) \in \mathbb{Z}^+ \times \mathbb{Z}^+ \mid x \leq y\}$

3. Show that each of the following sets is countable by defining an appropriate bijection.

    (a) $\{1, 10, 100, 1000, \ldots\}$
    (b) $\{e^x \mid x \in \mathbb{Z}^+\}$
    (c) $\{\ldots, \frac{1}{8}, \frac{1}{4}, \frac{1}{2}, 1, 2, 4, 8, \ldots\}$
    (d) $\{(2n, -5n) \mid n \in \mathbb{Z}\}$.

4. Prove that the following sets are countable.

    (a) $\mathcal{P}(A) \times \mathcal{P}(B)$, where A and B are finite sets
    (b) The set of all prime numbers
    (c) The set of all *lattice points* in $\mathbb{R}^3$: $\{(x, y, z) \mid x, y, z \in \mathbb{Z}\}$
    (d) The set of all lattice points in $\mathbb{R}^3$ lying on the x-, y- or z-axis
    (e) $P \times (\mathbb{Z} - P)$, where P is the set of all prime numbers
    (f) $\mathbb{Z}^+ \times \mathbb{Z} \times \mathbb{Q}$

5. (a) Partition $\mathbb{Z}$ into four partitioning sets, each of which is countably infinite.

    (b) Partition $\mathbb{Z}$ into four partitioning sets, exactly two of which are countably infinite.

    (c) Is it possible to partition $\mathbb{Z}$ into four partitioning sets, none of which is countably infinite? Prove or disprove.

6. If A is finite and B is countably infinite, prove that $A \cup B$ is countably infinite.

7. The bijection in the proof of Theorem 7.21 mapped $\mathbb{Z}^+$ to $\mathbb{Z}$. Provide an alternative proof of the result by defining a function from $\mathbb{Z}$ to $\mathbb{Z}^+$ and showing it is a bijection.

8. Prove Corollary 7.19.

9. Prove that an arbitrary intersection of countable sets is countable.

10. Prove that the function f in the proof of Theorem 7.23 is a bijection.

11. If A_i is a countably infinite set for all i, prove that

$$\bigcup_{i=1}^{n} A_i$$

is countably infinite.

12. Provide the definition of the function illustrated in Figure 7.3. That is, if

$$f : \bigcup_{i=1}^{\infty} A_i \to \mathbb{Z}^+,$$

with $A_i = \{a_{ij} \mid j \in \mathbb{Z}^+\}$, then $f(a_{ij}) = \ldots$.

13. Prove Corollary 7.25.

14. If A and B are finite with $|A| = |B|$.

   (a) Prove that any injection $f : A \to B$ is also a surjection.
   (b) Prove that any surjective function $g : A \to B$ is also injective.
   (c) Show that the previous two results are not necessarily true if A and B are infinite.

15. Prove Corollary 7.27.

16. Prove each of the following is countable.

   (a) The set P_n of all polynomials of degree less than or equal to a non-negative integer n with integer coefficients.
   (b) The set of all polynomials with integer coefficients.

17. If A is a finite set and B is a countably infinite set, prove that the set of all functions $f : A \to B$ is countable.

18. If A is countably infinite and $a \in A$, prove that

$$|A - \{a\}| = |A|.$$

19. Prove or disprove.

   (a) The set of all finite subsets of $\mathbb{Z}^+$ is countable.
   (b) Suppose that a set A is finite and a set B is countably infinite. Prove that the set of all functions $f : A \to B$ is countable.

7.3 The Infinite: Uncountable

Recall that a set A is called countably infinite *if* there exists a bijection $f : A \to \mathbb{Z}^+$. This choice of language would lead us to believe that there are sets for which such a bijection does *not* exist. Intuitively, what would such a set "look like?"

Countably infinite sets can, in some way, be "lined up" so that every element of the set has some place in line (that is, we can "count" all the elements of the set). This could not be done to an infinite set that is *not* countably infinite. Some of the elements of the set could get placed in line, but there would be others that were "left out." If we attempted to place them into the line, we would in turn leave others out or displace some that were already in line. No matter how we tried to rearrange it, such a set could not be "lined up and counted." We give sets like this a special name: uncountable.

> **Definition 7.30.** A set that is not countable is called *uncountable*.

Without hesitation we work towards proving that the set $\mathbb{R}$ is uncountable. Before proceeding, however, we first recall a property of the real numbers. Any real number can be written in a unique decimal form so long as we do not allow the decimal expansion to end in an infinite string of 9s. That is,

$$.73999\ldots = .74000\ldots .$$

To see that these are equal, one can use a geometric series representation of the term ending in 9s, or, one can make a clever observation. We know that the real numbers are dense (between any two distinct real numbers there is another real number). Thus, since .73999 and .74 are different real numbers, then we can find another real number between them, such as .739992. Is this possible for .73999... and .74?

The first few terms of such a real number between the two would be .73999..., but what about the remaining digits? If at any point in the expansion we change a 9 to another digit, the resulting decimal will be less than .73999.... For example,

$$.739998 < .739999\ldots .$$

In other words, there is *no* real number lying between .73999... and .74. Consequently, they must be *equal*.

With this, we proceed to proving a specific set of real numbers is uncountable (which in turn will show that $\mathbb{R}$ is uncountable). What proof technique seems appropriate? A set is uncountable if no particular bijection exists. Directly showing the non-existence of something, as we have seen is difficult. However, if we assume such a bijection *does* exist, then, perhaps a contradiction will arise.

Theorem 7.31. *The interval* $(0,1)$ *is uncountable.*

If we assume that $(0,1)$ is countably infinite, then we are assuming that *every* real number between 0 and 1 can be lined up and counted. What sort of contradiction would come from this assumption? We will find a real number between 0 and 1 that appears nowhere in the number sequence we are counting – a contradiction! How can we go about finding such a real number? We will construct it, from the assumed sequence, to be different from every real number in the image of the bijection.

Proof To prove $(0,1)$ is uncountable, assume for contradiction that it is countably infinite. Let $f : \mathbb{Z}^+ \to (0,1)$ be a bijection with

$$f(n) = .d_{n1}d_{n2}d_{n3}\ldots,$$

with the decimal representation not ending in an infinite string of 9s. We will find an element of $(0,1)$ that is not in the image of f, a contradiction to f being a bijection.

To that end, construct a real number

$$a = .a_1a_2a_3\ldots$$

by letting a_i be any digit except 0, 9 or d_{ii}. Then, since $a \neq .000\ldots$ and $a \neq .999\ldots$, we know $a \in (0,1)$. Moreover, $a \notin \text{Ran}(f)$ because at least one digit of a differs from $f(i)$ for every i, meaning $a_i \neq d_{ii}$.

Because of this, f is not a surjective function, contradicting the assumption that f was bijective. Hence, no such bijection exists and the interval $(0,1)$ is indeed uncountable. $\qquad\square$

To highlight how the proof of Theorem 7.31 works, let us consider a specific example and the construction of the number a in the proof. Suppose the following exhibits the first five mappings of a potential bijection f between $\mathbb{Z}^+$ and $(0,1)$:

$\mathbb{Z}^+$		$(0,1)$
1	$\longrightarrow$	.58451...
2	$\longrightarrow$	.33333...
3	$\longrightarrow$	.50000...
4	$\longrightarrow$	.99851...
5	$\longrightarrow$	.63627...

The real number a is constructed one term at a time. The first digit of $f(1)$ is 5. Thus, choose the first digit of a to be any digit except 0, 9 or 5. Let us choose 1. The second digit of $f(2)$ is 3. We choose the second digit of a to be any digit except 0, 9 or 3, say 6. Likewise, the third digit cannot be 0 or 9 (since 0 is the third digit of $f(3)$); choose the third digit of a to be 7. Similarly, we pick the fourth and fifth decimal terms to be 3 and 5, respectively. So far, then, we have constructed a as

$$a = .16735.$$

Continuing, a will differ from every single element of the range of f by at least one digit in its decimal expansion. Because two real numbers are the same if and only if all of their digits are the same (with respect to the unique representation mentioned previously), this is all it takes to show that a is not possibly in the range of f.

Having established that $(0, 1)$ is uncountable, recall Theorem 7.20: every subset of a countable set is countable. Alternatively, this could be stated as, "if a set is countable, then any subset of it is countable." What is the contrapositive of this statement? If a set has an uncountable subset, then that set itself is uncountable. This is exactly what we need to achieve our desired conclusion, that the set of real numbers is uncountable.

Corollary 7.32. *The set $\mathbb{R}$ is uncountable.*

Proof The interval $(0, 1)$ is an uncountable subset of $\mathbb{R}$, by Theorem 7.20, $\mathbb{R}$ must be uncountable. □

There is another set of real numbers that we have not mentioned to this point: the irrational numbers. It has the same density property as the rationals (between any two real numbers there is an irrational number (infinitely many of them, in fact)). Is it, too, countable? Because every real number is either rational or irrational, we have that $\mathbb{R}$ is the union of the rational numbers $\mathbb{Q}$ and the irrational numbers $\mathbb{R} - \mathbb{Q}$:

$$\mathbb{R} = \mathbb{Q} \cup (\mathbb{R} - \mathbb{Q}).$$

Hence, Corollary 7.33 follows.

Corollary 7.33. *The set of irrational numbers, $\mathbb{R} - \mathbb{Q}$, is uncountable.*

Proof If the set of irrational numbers, $\mathbb{R} - \mathbb{Q}$, were countable, then

$$\mathbb{R} = \mathbb{Q} \cup (\mathbb{R} - \mathbb{Q})$$

would be countable (Theorem 7.23). However, $\mathbb{R}$ is uncountable, meaning $\mathbb{R} - \mathbb{Q}$ must be uncountable. □

What have these results given us? Even though the sets $\mathbb{Z}^+$ and $\mathbb{R}$ are infinite in size, they are different sizes of infinity. In other words, there are more real numbers than there are integers or rational numbers. Similarly, even though both the rational and irrational numbers are dense in $\mathbb{R}$, there are more irrational numbers than there are rational numbers. The cardinality of the positive integers is denoted $\aleph_0$, and we call $|\mathbb{R}| = \mathbf{c}$, the cardinality of the *continuum*.

We are tempted to write $\aleph_0 < \mathbf{c}$, but we must be careful. The relation $<$ is defined on real numbers, not transfinite numbers or cardinalities. Definition 7.34 takes care of this.

Definition 7.34. Let A and B be infinite sets. We say the *cardinality of A is less than that of B*, denoted $|A| < |B|$, if there exists an injection $f : A \to B$ but no bijection between the sets A and B.

Now that we can say $\aleph_0 < \mathbf{c}$, two questions arise immediately. First, is there a set S for which $\mathbf{c} < |S|$? That is, is there a set larger than the real numbers? The second question is related. Is there a set smaller than the real numbers yet larger than the integers? In particular, does a set S exist satisfying $\aleph_0 < |S| < \mathbf{c}$? We discuss the latter question first.

Cantor believed that there was no uncountable set whose cardinality was less than that of $\mathbb{R}$. In other words, any infinite subset of $\mathbb{R}$ could be put into bijective correspondence with either $\mathbb{Z}$ or with $\mathbb{R}$. The claim sat unproven for years and came to be known as the *Continuum Hypothesis*. In 1940, Kurl Gödel (1906-1978) stunned the mathematics community by proving that the Continuum Hypothesis could not be *disproven* using our standard set theory axioms [21]. This, however, did not prove that the result *could* be proven, only that one could not prove it was false.

Then, nearly a quarter century later, Paul Cohen (1934-2007) provided another stunner. He showed that the standard axioms of set theory could not *prove* the Continuum Hypothesis [14, 15]. Together, Gödel and Cohen's results showed that the Continuum Hypothesis would never be proven true nor shown false. What does this mean? Basically, it means the Continuum Hypothesis is not true and it is not false, yet we can assume it to be true or we can assume it to be false. As mathematicians, we are able to assume it to hold or not hold and that assumption will lead to no logical contradiction with our assumed set theory axioms.

Thus, investigating the second of our two questions is a moot point. We move then to our other question. Is there a set S whose cardinality is greater than that of $\mathbb{R}$? A natural first set to look at would be $\mathbb{R} \times \mathbb{R}$. The Cartesian plane is visually much larger than just the real number line. Hence, its cardinality must be greater than $\mathbf{c}$. Once again, our intuition has led us astray. It turns out that

$$|\mathbb{R} \times \mathbb{R}| = |\mathbb{R}|.$$

If asked to prove that $\mathbb{R}$ and $\mathbb{R} \times \mathbb{R}$ have the same cardinality, it would be natural to try and find a bijection between the two sets. Such a search turns out to be quite difficult, however. Instead, we call upon Theorem 7.35, the Schröder-Bernstein Theorem.[8] It provides an often simpler way to show the *existence* of a bijection between two sets than to actually find the bijection itself. The theorem tells us that in place of finding a specific bijection, we need only find an injection from one set to the other and vice versa!

The proof of Theorem 7.35, while quite involved and heavily reliant on

---

[8]This theorem is also known as the Cantor-Schröder-Bernstein Theorem.

both definitions and strong notation, serves as a bit of a climax for this text. Upon working through and understanding the proof, you, the reader, are encouraged to spend a moment thinking about how far your proof constructing and writing skills have come. You have polished your style and developed *your* mathematical voice, whether it be through word choice, conciseness, structure or a bevy of other mathematics writing techniques. Your skills will continue to develop and improve, but you have created a solid foundation for the study of theoretical mathematics.

Theorem 7.35. (Schröder-Bernstein Theorem) *Let A and B be sets. If there exist injective functions $f : A \to B$ and $g : B \to A$, then there exists a bijective function $h : A \to B$. In particular, if the injective functions f and g exist, then*

$$|A| = |B|.$$

Proof Let A and B be sets with $f : A \to B$ and $g : B \to A$ injective functions. Note that if A and B are finite (see Exercise 14 following Section 7.2) or if either f or g is surjective, then the result holds. Thus, assume A and B are both infinite with neither f nor g surjective.

While we assume g is not surjective, if we restrict its co-domain to just $\text{Ran}(g) = g(B)$, then the function $g : B \to g(B)$ is in fact surjective, and consequently bijective. We consider then g to be this new function,[9] meaning that g is invertible.

For notation, define inductively the function $(g \circ f)^n$ (where n is a non-negative integer) by:

$$(g \circ f)^n = (g \circ f) \circ (g \circ f)^{n-1},$$

with

$$(g \circ f)^1 = g \circ f$$

and

$$(g \circ f)^0 = \text{id}_A.$$

With this, define a new set C as

$$C = \bigcup_{n=0}^{\infty} (g \circ f)^n (A - g(B)).$$

Claim 1: $C \subseteq A$.

Since $(g \circ f) : A \to A$, then

---

[9]Though we are technically defining a new function here, introducing a new function name, such as g_1, serves to muddy the proof with unnecessary notation. The two functions operate identically: $g(b) = g_1(b)$ for all $b \in B$. They simply have different co-domains (though their ranges are equal).

$$(g \circ f)^n : A \to A$$

for all $n \in \mathbb{Z}$, $n \geq 0$. Thus, Claim 1 holds.

Then, A is partitioned into C and $D = A - C$. Now, define a new function $h : A \to B$, which we will show is the desired bijection, by

$$h(a) = \begin{cases} f(a), & \text{if } a \in C \\ g^{-1}(a), & \text{if } a \in D. \end{cases}$$

Claim 2: h is indeed a function.

For $a \in A$, either $a \in C$ or $a \in D$. When $a \in C$, then

$$h(a) = f(a)$$

is defined. For $a \in D$, then

$$a \notin (g \circ f)^n (A - g(B))$$

for any $n \in \mathbb{Z}$, $n \geq 0$. Since $(g \circ f)^0 = \text{id}_A$, it follows that $a \notin A - g(B)$. Thus, $a \in g(B)$, meaning $g^{-1}(a)$ is defined.

Next, we show that h is indeed bijective.

Claim 3: h is injective.

Assume $h(a) = h(b)$ for $a, b \in A$. To prove $a = b$, we consider three possibilities:

1. $a, b \in C$
2. $a, b \in D$
3. $a \in C$, $b \in D$

Case 1: $a, b \in C$.

Since $a, b \in C$, we have

$$\begin{aligned} f(a) &= h(a) \\ &= h(b) \\ &= f(b), \end{aligned}$$

but f is assumed to be injective. Hence, $f(a) = f(b)$ implies the desired result: $a = b$.

Case 2: $a, b \in D$.

Since $a, b \in D$, we have

$$\begin{aligned} a &= g(g^{-1}(a)) \\ &= g(h(a)) \\ &= g(h(b)) \\ &= g(g^{-1}(b)) \\ &= b, \end{aligned}$$

showing that $a = b$.

Case 3: $a \in C$, $b \in D$.

Rather than show that h is injective, we show that this case cannot occur. Since

$$a \in \bigcup_{n=0}^{\infty} (g \circ f)^n (A - g(B)),$$

it must be that

$$a \in (g \circ f)^n (A - g(B))$$

for some $n \in \mathbb{Z}$, $n \geq 0$. Thus, $a = (g \circ f)^n(x)$ for some $x \in A - g(B)$. Because $a \in C$ and $b \in D$,

$$\begin{aligned} g^{-1}(b) &= h(b) \\ &= h(a) \\ &= f(a) \\ &= f((g \circ f)^n(x)). \end{aligned}$$

Thus,

$$\begin{aligned} b &= g(g^{-1}(b)) \\ &= g(f((g \circ f)^n(x))) \\ &= (g \circ f)^{n+1}(x). \end{aligned}$$

But if

$$b = (g \circ f)^{n+1}(x),$$

we must have that

$$b \in ((g \circ f)^{n+1}(A - g(B)) \subseteq C,$$

a contradiction to $b \in D$. Thus, it is not possible to have $a \in C$ and $b \in D$, as was shown in the two possible cases, h is injective.

Claim 4: h is surjective.
    Let $b \in B$. Since $g(B) \subseteq A$ and $A = C \cup D$, we have that $g(b) \in C$ or $g(b) \in D$.

Case 1: $g(b) \in C$
By the definition of C, it follows that

$$g(b) \in (g \circ f)^n (A - g(B))$$

for some $n \in \mathbb{Z}$, $n \geq 0$. For some $x \in A - g(B)$, then,

$$g(b) = (g \circ f)^n(x)$$
$$= g(f((g \circ f)^{n-1}(x))),$$

and by the injectivity of g, we have that

$$b = f((g \circ f)^{n-1}(x)).$$

Then,

$$(g \circ f)^{n-1}(x) \in C,$$

and consequently,

$$h((g \circ f)^{n-1}(x)) = f((g \circ f)^{n-1}(x))$$
$$= b,$$

proving that h is indeed surjective.

Case 2: $g(b) \in D$.
Because $g(b) \in D$, the definition of h yields

$$h(g(b)) = g^{-1}(g(b))$$
$$= b,$$

proving that h is surjective in this case.

Having shown that the function $h : A \to B$ is injective and surjective, the result follows, and consequently, by the definition of cardinality, $|A| = |B|$. $\square$

Armed with this new tool, we show that the cardinality of $\mathbb{R}$ and $\mathbb{R} \times \mathbb{R}$ are the same.

Theorem 7.36. *The sets $\mathbb{R}$ and $\mathbb{R} \times \mathbb{R}$ have the same cardinality.*

To use Theorem 7.35, we need only find two injective functions. But, rather than find a function from $\mathbb{R}$ to $\mathbb{R} \times \mathbb{R}$, we will call upon the fact that $\mathbb{R}$ and $(0, 1)$ have the same cardinality. This latter set is much more manageable.

Proof To prove $|\mathbb{R}| = |\mathbb{R} \times \mathbb{R}|$, we show instead that $(0, 1)$ and $(0, 1) \times (0, 1)$ have the same cardinality. Because $|(0, 1)| = |\mathbb{R}|$, the result will follow.
We employ the Schröder-Bernstein Theorem. The function

$$f : (0, 1) \to (0, 1) \times (0, 1)$$

given by

$$f(x) = (x, \frac{1}{2})$$

is an obvious injection. Thus, it suffices to find an injection from $(0,1) \times (0,1)$ to $(0,1)$.

To that end, define

$$g : (0,1) \times (0,1) \to (0,1)$$

by

$$g(x,y) = .x_1 y_1 x_2 y_2 \ldots,$$

where $x = .x_1 x_2 x_3 \ldots$ and $y = .y_1 y_2 y_3 \ldots$ are the decimal representations of x and y that do not terminate.[10]

If

$$g(x,y) = g(a,b),$$

then

$$.x_1 y_1 x_2 y_2 \ldots = .a_1 b_1 a_2 b_2 \ldots,$$

and consequently, $x_i = a_i$ and $y_i = b_i$ for all $i \in \mathbb{Z}^+$, proving that $x = a$ and $y = b$. Thus, g is injective.

By Theorem 7.35, because an injection from $(0,1)$ to $(0,1) \times (0,1)$, and vice versa,[11] exists, there must exist a bijection between $(0,1)$ and $(0,1) \times (0,1)$, proving the result. $\square$

Thus, our natural guess that

$$|\mathbb{R}| < |\mathbb{R} \times \mathbb{R}|$$

is incorrect. It turns out that $|\mathbb{R}^n| = \mathbf{c}$ for *every* positive integer n. What is the interpretation of this? Since cardinality is a count of how many elements are in a set, this result says that there are an *equal number* of points on the real number line as there are in $\mathbb{R}^3$, $\mathbb{R}^{30}$ or $\mathbb{R}^{3000}$! The proof of this, Corollary 7.37, is left as an exercise.

Corollary 7.37. *For every $n \in \mathbb{Z}^+$, $|\mathbb{R}^n| = |\mathbb{R}|$.*

Where then do we look for larger cardinalities? Perhaps a result that held for finite sets, that $|A| < |\mathcal{P}(A)|$ for any finite set A (Theorem 7.14), extends to infinite sets. In particular, can we say anything about the size of the set of all subsets of $\mathbb{R}$? This is quite the set to wrap your head around. It is the set of *all* subsets of $\mathbb{R}$. Just some of the elements $\mathcal{P}(\mathbb{R})$ contains are

1. $\{x\}$, where $x \in \mathbb{R}$ (all single-element subsets of $\mathbb{R}$),

---

[10] We choose to represent terminating decimals with an infinite string of 9s, such as choosing $.53999\ldots$ in place of $.54$.

[11] Note that neither of the injective functions defined in the proof is actually a bijection.

2. $\mathbb{R} - \{x\}$, where $x \in \mathbb{R}$ (the set of *every* real number except one),

3. $\mathbb{R} - \{x, x^2, x^3, \dots\}$, where $x \in \mathbb{R}$, and

4. $\{\dots \pi q, \frac{q}{\pi}, \pi^2 q, \frac{q}{\pi^2}, \dots \mid q \in \mathbb{Q}\}$.

Cantor was the first to show that in the infinite case, just as in the finite case, the cardinality of the power set of an infinite set is larger than the cardinality of the set itself. Often called Cantor's Theorem, it is considered one of his crowning achievements. Its proof is constructive in the same sense that Theorem 7.32 was, and this approach of "constructing" a counterexample to the proof's assumed claim has come to be known as Cantor's diagonalization argument (not to be confused with his diagonalization approach, used to enumerate the rational numbers).

Theorem 7.38. (Cantor's Theorem) *For any set A, $|A| < |\mathcal{P}(A)|$.*

Proof If A is finite, the result holds by Theorem 7.14. Otherwise, suppose A is infinite. The function $f : A \to \mathcal{P}(A)$ given by $f(a) = \{a\}$ is an obvious injection. Then, to show

$$|A| < |\mathcal{P}(A)|,$$

we must show that no bijection exists between A and $\mathcal{P}(A)$.

For contradiction, assume then that there does exist such a bijection:

$$g : A \to \mathcal{P}(A).$$

Define a particular subset S of A as

$$S = \{a \in A \mid a \notin g(a)\}.$$

This set S is clearly a subset of A, as it is defined to be elements of A that satisfy particular properties. Thus, $S \in \mathcal{P}(A)$. Since g is assumed to be a bijection whose range is $\mathcal{P}(A)$, it must be that S is in the range of g. Let $s \in A$ be the element mapped to S under g. That is, $g(s) = S$. Because S is a subset of A and $s \in A$, we have that either $s \notin S$ or $s \in S$. Consider both cases.

Suppose $s \notin S$. Then we have that s is an element of A but not an element of its image under g. By the definition of the set S, this means that $s \in S$, a contradiction to the supposition that $s \notin S$. Thus, this case cannot hold.

Then, we must have that $s \in S$. By the definition of S, we have that $s \notin g(s)$, or equivalently, $s \notin S$. Again, we have reached a contradiction.

Since both cases yield a contradiction, we must have that no bijection exists between A and $\mathcal{P}(A)$, proving that $|A| < |\mathcal{P}(A)|$. $\qquad \square$

Before discussing the implications of Cantor's Theorem, let us illustrate the proof via an example, much like we did for Theorem 7.31. Let $A = \{a, b, c, \dots\}$

and suppose we have a bijection $g : A \to \mathcal{P}(A)$, illustrated below.

A		$\mathcal{P}(A)$
a	$\longrightarrow$	$\{a, d, m\}$
b	$\longrightarrow$	$\{x\}$
c	$\longrightarrow$	$\{a, c, d, q, v\}$
d	$\longrightarrow$	A
e	$\longrightarrow$	$\varnothing$
$\vdots$	$\vdots$	$\vdots$

To construct the set S in the proof of Theorem 7.38, consider each element of the domain. Such an element is placed in S if and only if it is *not* an element of its image. That is, look at the elements in the left column. Are they elements of their images in the right column? The only way they are placed in S is if they are not.

In our example, since $a \in \{a, d, m\}$, we do not place a in S. But $b \notin \{x\}$, so then b becomes an element in S. Looking at the next three rows, we have that $c \notin S$ (since $c \in \{a, c, d, q, v\}$), $d \notin S$ (since $d \in A$), and $e \in S$ (since $e \notin \varnothing$).

What does all this mean in terms of cardinality? We knew that there were two different sizes of infinity: $\aleph_0$ and **c**. But now, Cantor's Theorem shows us that there are *infinitely many* sizes of infinity:

$$|\mathbb{Z}| < |\mathbb{R}| < |\mathcal{P}(\mathbb{R})| < |\mathcal{P}(\mathcal{P}(\mathbb{R}))| < \dots .$$

In spite of our intuition failing us, we were able to prove results throughout this section (in fact, throughout the entire text) by relying on precise definitions. They are *the main tools* a mathematician wields when crafting proofs. Regardless of how the proof's author has chosen to personalize the proof, there is often one and only one way to show that a certain definition is fulfilled. Know your definitions thoroughly and you will be on your way to crafting quality proofs.

Exercises

1. Give examples of uncountable sets A and B satisfying

    (a) $A \cap B = \varnothing$.
    (b) $A \cap B$ is finite and nonempty.
    (c) $A \cap B$ is countably infinite.
    (d) $A \cap B$ is uncountable.
    (e) $A - B$ is finite.
    (f) $A - B$ is countable.
    (g) $A - B$ is uncountable.

2. Prove or disprove.

(a) There exists a bijection $f : \mathbb{Q} \to \mathbb{R}$.

(b) There exists a bijection $g : \mathcal{P}(\mathbb{Q}) \to \mathbb{Z} \times \mathbb{Z} \times \mathbb{Z}$.

(c) $\{0,1\} \times \mathbb{R}$ is uncountable.

(d) If A is an uncountable set, then $|A| = |\mathbb{R}|$.

(e) Every infinite set is a subset of some countably infinite set.

(f) Every infinite set is a subset of some uncountable set.

(g) If an uncountable set is partitioned by two sets, one of which is uncountable, then the other partitioning set must be countable.

3. Prove that there do not exist countable sets A and B such that $A \cap B$ is uncountable.

4. Prove that if $A \subseteq B$ and A is uncountable, then B is uncountable.

5. Find a bijection between $[0,1]$ and

(a) $[0,8]$.

(b) $[1,3]$.

(c) $[-2,6]$.

6. Show that $(0,1)$ and $[0,1]$ have the same cardinality by finding a bijection between them.

7. Show that $|(0,1)| = |\mathbb{R}|$ by finding a bijection between $(0,1)$ and $\mathbb{R}$.

8. Use the Schröder-Bernstein Theorem to show

(a) $|[0,1]| = |[0,1)|$.

(b) $|[2,4]| = |(2,10]|$.

(c) $|(-1,3)| = |[6,14]|$.

9. Show that the functions f and g in the proof of Theorem 7.36 are not bijective.

10. Prove Corollary 7.37.

11. Let F be the subset of $(0,1)$ consisting of all those real numbers whose decimal expansion contains a 5.

(a) Show that $|F| = |\mathbb{R}|$.

(b) Show that $(0,1) - F$ is uncountable.

12. Show that the cardinality of the unit disk $(\{(x,y) \mid x^2 + y^2 \leq 1\})$ and the solid unit square $(\{(x,y) \mid |x| \leq 1, |y| \leq 1\})$ are the same.

13. If $S = \{A \mid A$ is a set and $|A| = |\mathbb{R}|\}$ and $|S| = |\mathbb{R}|$, prove that

$$\left| \bigcup_{A \in S} A \right| = |\mathbb{R}|.$$

14. Prove that the set of all functions $f : \mathbb{Z}^+ \to \mathbb{Z}^+$ is uncountable.

15. Suppose A and B are disjoint nonempty sets satisfying

$$|A| = |\mathbb{R}| = |B|.$$

Prove that $|A \cup B| = |\mathbb{R}|$.

16. Prove that there does not exist a infinite set A with $|A| < \aleph_0$.

17. Use the Schröder-Bernstein Theorem to prove that $|\mathcal{P}(\mathbb{Z}^+)| = |\mathbb{R}|$.

8

Introduction to Topology

Three mathematical fields are often considered the "main" areas of pure mathematics: analysis, algebra and topology. Any would be a worthy inclusion as a final chapter to this text, but the choice to introduce the reader to the basics of topology is purposeful (and not simply because it is the author's favorite amongst the three areas listed). It is grounded in basic set theory, so that its fundamental principles are considered common knowledge by this point. Yet those basic ideas require the utmost attention to detail in each and every definition and result. Thus, beyond the fun material and interesting results is a reinforcement of one of this text's main themes:

Know your definitions, specifically and intuitively.

So what *is* topology? Perhaps it is best described as the study of geometric and spatial properties that are preserved by continuous deformations of objects. For example, imagine you are handed clay in the shape of a cube and are told to mold it into something else. You would begin to press down on the sides of the clay, maybe rolling it around in your hands to smooth out corners. These are continuous deformations; two points on the clay that are close to each other prior to some deformation remain close to one another after the deformation.

However if you were to take your clay and tear it into two pieces, you have just performed a non-continuous deformation. Two points may have been close to one another prior to tearing the clay, but then after the ripping occurs, those two points are far apart.

To the topologist, two objects that can be continuously deformed into one another are considered the same object.[1] Determining properties that are preserved under continuous deformations, then, yields a method for telling when two objects are *not* the same. If their respective property values are different, yet that property has been proven to be preserved under continuous deformations, then there is no way that one of those objects could be continuously deformed into the other.

These concepts of closeness and continuous deformations are defined in terms of objects we have studied: sets and functions. The latter, functions, seems like a natural way to define a continuous deformation of an object. Yet

---

[1]The topologist considers a coffee cup and a donut (the standard donut shape, with a hole in the middle) the same shape, but both are very different from a single slice of toast.

why would a set be the thing needed to define closeness? It stems from the ideas that inspired the development of the topology we introduce here, called *point-set topology* [36].

Topology is assumed to share its roots with graph theory and Euler's 18th century investigation of the bridges of Königsberg. Graph theory's beginnings are founded in the mathematical tools Euler used to attack the question (and the further development of those tools). Topology got its start in a concept Euler described as the "geometry of position" [1].

Throughout the remainder of the 18th century through the 20th century, many of history's most prominent mathematicians had a hand in shaping topology, including the aforementioned Gauss, Poincaré, Cantor, Weyl and Brouwer. Some of these mathematicians based their thinking on how the real-number line behaves. In particular, properties of collections of open sets on the real-number line were generalized to properties of subsets of given sets. It is here that closeness finds its abstract definition.

This chapter barely touches upon the most basic ideas of point-set topology, introducing just a few definitions and their properties. Think of point-set topology's role in all of topology as that of algebra's role in all of calculus. In that sense, what you are about to see in the coming four sections parallels the learning about number systems and just a few properties of them. There is much more to learn before heading off into a more specific area of topology (such as algebraic, geometric or differential topology, just to name a few broad areas).

Beyond the pure mathematics, topology has established itself as an important tool for using mathematics to address real-world problems. Biology (how particular enzymes affect DNA), data analysis (understanding the structure of huge sets of data), physics (quantum field theory) and cosmology (describing the shape of the universe) are just some of the areas in which topology is applied. To get to those applications, however, we must understand the basics. And with that thought in mind, we proceed to the rest of this chapter.

8.1   Topologies and Topological Spaces

As mentioned in the introduction, point-set topology has its roots partially on the real-number line. There, open sets play the important role of determining closeness. Before we introduce this idea in Section 8.4, let us generalize collections of open sets in $\mathbb{R}$ to collections of subsets in any given set. Such a collection of subsets is called a *topology* on a set.

Definition 8.1. Let S be any set. A collection $\mathcal{T}$ of subsets of S is called a *topology* if

   (1) $\varnothing, S \in \mathcal{T}$,
   (2) if $U_1, U_2, \ldots, U_n \in \mathcal{T}$, for some $n \in \mathbb{Z}^+$, then

$$\bigcap_{i=1}^{n} U_i \in \mathcal{T}, \text{ and}$$

   (3) if $U_i \in \mathcal{T}$ for $i \in I$, where I is any index set, then

$$\bigcup_{i \in I} U_i \in \mathcal{T}.$$

Any element of $\mathcal{T}$ is called an *open set* in S. A set S together with a topology $\mathcal{T}$ is called a *topological space*.

How is this a generalization of open sets in $\mathbb{R}$? The first property, that the empty set and the entire set are open, is an obvious generalization, as both are open in $\mathbb{R}$. The other two become just as obvious when the term "open sets" is changed to "open intervals." The intersection of finitely many open intervals in $\mathbb{R}$ is an open interval (it is not true for infinitely many open intervals in $\mathbb{R}$; see Exercise 4). Likewise, the union of arbitrarily many open intervals in $\mathbb{R}$ is an open interval.

The following examples exhibit Definition 8.1.

Example 8.1. Let $S = \{1, 2, 3\}$ with

$$\mathcal{T}_1 = \{\varnothing, \{1, 2, 3\}, \{2\}, \{3\}, \{1, 2\}, \{2, 3\}\}$$

and

$$\mathcal{T}_2 = \{\varnothing, \{3\}, \{1, 2\}, \{1, 3\}, \{1, 2, 3\}\}.$$

Though both sets contain $\varnothing$ and S, only one of them is a topology on S. Set $\mathcal{T}_1$ is a topology on S because it is closed under both intersections and unions (because it is a finite set, we need not be concerned with finite intersections versus arbitrary unions).

Set $\mathcal{T}_2$, however, is not a topology on S. To justify this, we need only find one counterexample to one of the criteria of Definition 8.1. We have $\{1, 2\}, \{1, 3\} \in \mathcal{T}_2$ but

$$\{1, 2\} \cap \{1, 3\} \notin \mathcal{T}_2.$$

Topologies need not be defined only on finite sets. Indeed, every nonempty set has a topology defined on it (and in most cases at least two topologies on defined on it). Examples 8.2 and 8.3 define these.

Example 8.2. The collection of all subsets (i.e., the power set) of S, where S is any set, is called the *discrete topology*. The proof that it is indeed a topology on S is left as an exercise.

Example 8.3. For any nonempty set S, the set

$$\mathcal{T} = \{\varnothing, S\}$$

is a topology on S called the *trivial topology*. The proof that this is actually a topology on S requires almost no work; call on the Universal Bound identity of Theorem 2.8.

There are many ways to define a topology on $\mathbb{R}$; in the two previous examples we saw the trivial and discrete topologies are two such ways. Another follows from the discussion on open intervals in $\mathbb{R}$ at the start of this chapter.

Definition 8.2. The collection of all open intervals in $\mathbb{R}$ (intervals of the form (a, b), where $a < b$) and disjoint unions of open intervals, along with $\varnothing$ and $\mathbb{R}$, is called the *standard topology* on $\mathbb{R}$.

The fact that the standard topology on $\mathbb{R}$ is actually a topology follows from the fact that if (a, b) and (c, d) are non-disjoint intervals in $\mathbb{R}$ (with $a < c < b < d$), then

$$(a, b) \cap (c, d) = (c, b)$$

and

$$(a, b) \cup (c, d) = (a, d).$$

It is important to note that the standard topology on $\mathbb{R}$ is not the set of *single* open intervals on the real number line. For example,

$$(1, 2) \cup (4, 6) \cup (8, 10)$$

is not an open interval in $\mathbb{R}$, but it is indeed open in the standard topology. One might anticipate that the collection of single open intervals, though, plays an important role for this topology. Any nonempty element of the standard topology can be built by taking unions of open intervals. It turns out that this set is called a *basis* for the standard topology. But, before introducing this notion, an important parallel between open sets on the real number line and general open sets in a topology is given in Theorem 8.3.

Theorem 8.3. *Let S be a topological space. Then, $A \subseteq S$ is open if and only if for every $a \in A$, there exists an open set U in S with $a \in U$ and $U \subseteq A$.*

Proof Let S be a topological space. If A is open in S, then for any $a \in A$, the result holds since $A \subseteq A$.

Conversely, suppose for every $a \in A$ there exists an open set U_a in S with $a \in U_a$ and

$$U_a \subseteq A.$$

Then,

$$A = \bigcup_{a \in A} U_a,$$

an arbitrary union of open sets, which is necessarily open by Definition 8.1. Thus, the result holds. $\square$

What is the intution behind Theorem 8.3? Consider the two sets $[2, 4)$ and $(2, 4)$ in $\mathbb{R}$. Notice that if any element of the latter, which is open, is selected, it is contained in some interval *entirely contained* in $(2, 4)$. This is not true for every element of $[2, 4)$. Every open interval containing 2 will contain some elements less than 2, thus making it *not* contained in $[2, 4)$. See Exercise 5.

We proceed now to define the aforementioned *basis*.

Definition 8.4. Let S be a set. A collection $\mathcal{B}$ of subsets of S is called a *basis* (for a topology on S) if both properties below hold.

1. For all $s \in S$, there exists $B \in \mathcal{B}$ such that $s \in B$.
2. For B_1, $B_2 \in \mathcal{B}$, if $s \in B_1 \cap B_2$, then there exists $B_3 \in \mathcal{B}$ with $s \in B_3$ and

$$B_3 \subseteq B_1 \cap B_2.$$

The definition for a basis includes the phrase "for a topology on S," yet the definition does not describe *which* topology on S (and we have seen that sets can have numerous different topologies on them). How the basis constructs a topology is hidden in the discussion prior to Theorem 8.3. Open intervals in $\mathbb{R}$ do not constitute the entire standard topology on $\mathbb{R}$; however, the standard topology *does* consist of all possible unions of open intervals in $\mathbb{R}$, precisely the idea behind the next definition.

Definition 8.5. Let $\mathcal{B}$ be a basis on a set S. Then, $\mathcal{B}$ *generates a topology* $\mathcal{T}$ on S by defining $\mathcal{T}$ to have as elements $\varnothing$ and all unions of elements of $\mathcal{B}$.

By this point in the text, you are surely thinking, "We define this thing to be a topology, but we still must *show* it satisfies the definition of being a topology." Indeed you are correct. Before doing that in Theorem 8.7, we have the following necessary lemma. The proof is left as an exercise.

Lemma 8.6. *If $\mathcal{B}$ is a basis and $B_1, B_2, \ldots, B_n \in \mathcal{B}$ with*

$$s \in \bigcap_{i=1}^{n} B_i,$$

then there exists $B \in \mathcal{B}$ with $s \in B$ such that

$$B \subseteq \bigcap_{i=1}^{n} B_i.$$

We proceed now to prove that the topology generated by a basis is indeed a topology. To do so, we must show it satisfies the criteria of Definition 8.1.

Theorem 8.7. *The topology generated by a basis $\mathcal{B}$ on a set S is a topology.*

Proof Let $\mathcal{B}$ be a basis on a set S and $\mathcal{T}$ the topology generated by $\mathcal{B}$. In order for $\mathcal{T}$ to be a topology, we must show that the three properties of Definition 8.1 hold.

First, we have that $\varnothing \in \mathcal{T}$, directly from Definition 8.5. Then, by the definition of a basis, for each $s \in S$, there exists $B \in \mathcal{B}$ so that $s \in B$. Hence,

$$\bigcup_{B \in \mathcal{B}} B = S,$$

proving that $S \in \mathcal{T}$.

Next, take $U_1, U_2, \ldots, U_n \in \mathcal{T}$. Note that each U_i, by Definition 8.5, is a union of elements of $\mathcal{B}$ (or empty or all of S). We must show that

$$\bigcap_{i=1}^{n} U_i \in \mathcal{T}.$$

For this to hold, we must show that

$$\bigcap_{i=1}^{n} U_i = \bigcup_{j \in J} B_j,$$

where J is some index set and $B_j \in \mathcal{B}$. Let us assume U_i is nonempty, else we have that

$$\bigcap_{i=1}^{n} U_i = \varnothing,$$

an element of $\mathcal{T}$. Thus, take

$$s \in \bigcap_{i=1}^{n} U_i.$$

For each i, $s \in U_i$ implies that there exists some $B_i \in \mathcal{B}$ so that $s \in B_i$ and $B_i \subseteq U_i$. So,

$$s \in \bigcap_{i=1}^{n} B_i,$$

and by Lemma 8.6, $s \in B$ for some $B \in \mathcal{B}$. Therefore,

$$B \subseteq \bigcap_{i=1}^{n} B_i.$$

Thus,

$$\bigcap_{i=1}^{n} U_i = \cup_s B_s,$$

where the union is taken over all

$$s \in \bigcap_{i=1}^{n} U_i.$$

This proves the second property of Definition 8.1 holds. To show that $\mathcal{T}$ is closed under arbitrary unions, let $U_i \in \mathcal{T}$, where $i \in I$, some index set. Each U_i is either empty or acts as a union of elements of $\mathcal{B}$. Thus,

$$\bigcup_{i \in I} U_i$$

is the union of empty sets and elements of $\mathcal{B}$, which by Definition 8.5 is an element of $\mathcal{T}$. □

Though somewhat lengthy and seemingly complicated at first glance, the proof of Theorem 8.7 is straightforward, calling only upon the definitions of a topology, basis and topology generated by a basis (along with one small lemma). Understanding the "nuts and bolts" of this proof requires a firm understanding of the definitions. Keep in mind that all of these definitions are founded in simple set theoretic concepts: subsets, unions and intersections. Keep track of all the components and you can easily follow the thought progression of the proof.

The concept of a basis can be overwhelming and confusing. The definition is somewhat awkward; what does it really say? Moreover, *why* would we need a basis for a topology? Think of basis elements similar to prime numbers for the integers. They serve the role of building blocks. If we can understand how a basis operates or behaves in regard to new ideas, we often can understand what the new idea means relative to the entire topology.

It is important to note that you can define a set with a topology, or, you can define a basis on a set and generate a topology from that basis. Because sets can have many different topologies defined on them (we have seen a handful for $\mathbb{R}$ already), immediately we arrive at a natural question. Suppose we begin with a topological space (that is, a set with a pre-defined topology). Can we determine the basis that generates *that* topology? This is the goal of the next theorem. We leave the proof as an exercise.

Theorem 8.8. *Let S be a set with a topology $\mathcal{T}$. If a collection $\mathcal{C}$ of elements of $\mathcal{T}$ satifies the following property, then $\mathcal{C}$ is a basis of S generating $\mathcal{T}$:*

> *For each open set $U \in \mathcal{T}$ and for each $s \in U$, there exists $C \in \mathcal{C}$ such that $x \in C$ and $C \subseteq U$.*

From this theorem, we can determine a basis for various topologies we have already seen. In particular, we begin by finding a basis for the discrete topology on any set.

Example 8.4. Let S be a nonempty set and $\mathcal{T}$ the discrete topology on S. The following set $\mathcal{B}$ is a basis generating $\mathcal{T}$:

$$\mathcal{B} = \{\{s\} \mid s \in S\}.$$

To justify this claim, let $U \in \mathcal{T}$ with $s \in U$ (assuming U is nonempty, for if it were empty, the result holds vacuously). Then, $s \in \{s\}$, and since $s \in U$, we have that

$$\{s\} \subseteq U,$$

satisfying the hypotheses of Theorem 8.8.

Such a claim about the discrete topology comes as no surprise. Why? Intuitively, a basis is a collection of sets that can be combined to construct the topology. In the case of the discrete topology, the collection of all single-element sets plays this role. Any subset of S can be write as a union of single-element sets.

Consider now the standard topology on $\mathbb{R}$. Its elements, in addition to the empty set, are open intervals (a, b) in $\mathbb{R}$ and disjoint unions of open intervals. We might guess, then, that a basis generating this topology would just be *all* open intervals. A union of two open intervals is either an open interval or a disjoint union of open intervals. Thus, we can hope to build the standard topology from this collection.

Example 8.5. The standard topology $\mathcal{T}$ on $\mathbb{R}$ is generated by the basis

$$\mathcal{B} = \{(a, b) \mid a, b \in \mathbb{R} \text{ and } a < b\}.$$

To prove that $\mathcal{B}$ generates $\mathcal{T}$, let $U \in \mathcal{T}$ (assuming U is nonempty) with $s \in U$. Because U is either a single open interval in $\mathbb{R}$ or a disjoint union of such open intervals, it follows that $x \in (a, b)$, where (a, b) is one of the open intervals comprising U. But $(a, b) \in \mathcal{B}$, as $\mathcal{B}$ consists of all open intervals in $\mathbb{R}$. The claim holds by Theorem 8.8.

With just these few definitions and examples, we have laid the foundation for a thorough investigation into the basics of point-set topology. Because they are grounded in simple set-theoretic ideas from earlier in this text, we are prepared to move on to looking at particular topologies stemming from further set-theoretic concepts. In particular, in the next section, we consider the how the notions of *subset* and *Cartesian product* give rise to new topologies from existing ones.

Exercises

1. Determine all topologies on $\{1, 2\}$.

2. Determine if each of the following is a topology on $\{1, 2, 3, 4\}$.

    (a) $\{\varnothing, \{1, 2, 3, 4\}\}$
    (b) $\{\varnothing, \{1\}, \{2\}, \{3\}, \{4\}, \{1, 2, 3, 4\}\}$
    (c) $\mathcal{P}(\{1, 2, 3, 4\})$
    (d) $\{\varnothing, \{1, 3\}, \{1, 3, 4\}, \{1, 2, 3, 4\}\}$
    (e) $\{\varnothing, \{1, 2\}, \{3, 4\}, \{1, 2, 3, 4\}\}$
    (f) $\{\varnothing, \{1\}, \{3\}, \{2, 3\}, \{3, 4\}, \{2, 3, 4\}, \{1, 2, 3, 4\}\}$

3. Find a topology on $\{1, 2, 3\}$ with the given number of elements or explain why no such topology exists.

    (a) 0
    (b) 1
    (c) 4
    (d) 5
    (e) 8
    (f) 9

4. Find a collection of infinitely many open intervals in $\mathbb{R}$ such that their intersection is neither an open interval nor empty.

5. Let (a, b) be an open interval in $\mathbb{R}$, where $a < b$. If $x \in (a, b)$, find an explicit interval (c, d) containing x so that $(c, d) \subset (a, b)$.

6. Prove that a topology $\mathcal{T}$ on a set S is the discrete topology if and only if $\{s\} \in \mathcal{T}$ for all $s \in S$.

7. If S is a nonempty set, prove that the discrete topology on S is the same as the trivial topology on S if and only if $|S| = 1$.

8. On $\mathbb{R}$, the *finite complement topology* is defined to be the topology whose open sets are $\varnothing$ and all subsets S of $\mathbb{R}$ such that $\mathbb{R} - S$ is a finite set. Prove that the finite complement topology is indeed a topology.

9. If S is a finite set, prove that the finite complement topology (see Exercise 8) on S is the same as the discrete topology on S.

10. The set $\mathcal{T}$ containing $\varnothing$ and $\mathbb{R}$ and every interval of the form $(-\infty, x)$ (where $x \in \mathbb{R}$) is a topology on $\mathbb{R}$. Show that $\mathcal{T}$ satisfies the first and last criteria of Definition 8.1.

11. Let S be an infinite set. Show that

$$\mathcal{T} = \{U \mid S - U \text{ is infinite or } S - U = \varnothing \text{ or } S - U = S\}$$

is a topology on S.

12. Prove Lemma 8.6.

13. Find a basis for the discrete topology on $\{1, 2, 3, 4, 5, 6\}$.

14. (a) Prove that $\mathcal{B} = \{[x, y) \mid x, y \in \mathbb{R} \text{ and } x < y\}$ is a basis on $\mathbb{R}$.
    (b) Show that $[0, 1)$ and $(0, 1)$ are open in the topology generated by $\mathcal{B}$.

15. Determine if each of the following is a basis for $\mathbb{R}$.
    (a) $\{(n - 1, n + 1) \mid n \in \mathbb{Z}\}$
    (b) $\{[x_1, x_2] \mid x_1, x_2 \in \mathbb{R} \text{ with } x_1 \leq x_2\}$
    (c) $\{(-2x, 2x) \mid x \in \mathbb{R}\}$
    (d) $\{(x_1, x_2) \cup \{x_2 + 2\} \mid x_1, x_2 \in \mathbb{R} \text{ with } x_1 < x_2\}$

16. Show that $\mathcal{T} = \{(x, \infty) \mid x \in \mathbb{Q}\} \cup \{\varnothing, \mathbb{R}\}$ is not a topology on $\mathbb{R}$.

17. Prove Theorem 8.8.

8.2   Subspace and Product Topologies

In our development of sets in Chapter 2, once the basics of sets were defined, we set out to create new sets from old. The first method for doing so was to take subcollections of a set: *subsets*. A topological space is a set S with a topology $\mathcal{T}$ (which is simply a set of subsets of S) defined on it. Thus, it seems natural to ask: given a subset of S, does it inherit in some way the topology defined on the larger set?

Think about this in terms of our inspiration for this chapter: open sets on the real number line. The number line $\mathbb{R}$ plays the role of S in this parallel discussion, with the open intervals the elements of $\mathcal{T}$. A subset of S is then simply a small piece of the number line, say, the interval $[0, 10]$. If an open interval is in the topology on $\mathbb{R}$, how might $[0, 10]$ naturally inherit that open interval into this subset topology? A first response might be to take the intersection (that is, the overlap) of the open interval with $[0, 10]$. Whatever "part" of the open set that lies in this subset will constitute an open set in the subset. This intuitive reasoning ends up being mathematically sound and it gives rise to the definition of the *subspace topology*.

Definition 8.9. Let S be a topological space with topology $\mathcal{T}$. If $A \subseteq S$, then

$$\mathcal{T}_A = \{A \cap U \mid U \in \mathcal{T}\}$$

is a topology on A called the *subspace topology*.

As before, we must show that this claimed topology is actually a topology. We present first two necessary set-theoretic results that are used in the proof of this claim. The proofs of these are left as exercises.

Lemma 8.10. *For sets A_1, A_2, ..., A_n, and a set S,*

$$(A_1 \cap S) \cap \cdots \cap (A_n \cap S) = (A_1 \cap \cdots \cap A_n) \cap S.$$

Lemma 8.11. *For sets A_i, $i \in I$, where I is some index set, and any set S,*

$$\bigcup_{i \in I}(A_i \cap S) = \left(\bigcup_{i \in I} A_i\right) \cap S.$$

Think of Lemmas 8.10 and 8.11 as generalized De Morgan's Laws. Armed with them, we are prepared to show that the subspace topology is actually a topology.

Theorem 8.12. *If S is a topological space with topology $\mathcal{T}$ and $A \subseteq S$, then the subspace topology $\mathcal{T}_A$ is a topology on A.*

Proof Let S be a topological space with topology $\mathcal{T}$ and $A \subseteq S$. We proceed to show that the subspace topology $\mathcal{T}_A$ is a topology on A.

First, since

$$\varnothing \cap A = \varnothing$$

and

$$S \cap A = A,$$

it follows that $\varnothing$ and S are elements of $\mathcal{T}_A$.

Next, take $V_1, V_2, \ldots, V_n \in \mathcal{T}_A$. Then, $V_i = A \cap U_i$ for some $U_i \in \mathcal{T}$. By Lemma 8.10,

$$V_1 \cap V_2 \cap \cdots \cap V_n = (A \cap U_1) \cap (A \cap U_2) \cap \cdots \cap (A \cap U_n)$$

$$= A \cap \left(\bigcap_{i=1}^{n} U_i \right).$$

Since $\mathcal{T}$ is closed under arbitrary intersections, it follows that

$$\bigcap_{i=1}^{n} U_i \in \mathcal{T},$$

proving that

$$V_1 \cap V_2 \cap \cdots \cap V_n \in \mathcal{T}_A.$$

The proof that $\mathcal{T}_A$ is closed under arbitrary unions is left as an exercise. □

A subset of a topological space inherits the topology to form the subspace topology. A natural next question to ask is, if $\mathcal{B}$ is a basis for a topology on S and $A \subseteq S$, does the subspace topology $\mathcal{T}_A$ inherit this basis naturally? That is, would the intersection of the elements of $\mathcal{B}$ with A form a basis generating $\mathcal{T}_A$? Indeed so, as seen in the next result.

Theorem 8.13. *If $\mathcal{B}$ is a basis for a topology $\mathcal{T}$ on a set S and $A \subseteq S$, then*

$$\mathcal{B}_A = \{B \cap A \mid B \in \mathcal{B}\}$$

is a basis for the subspace topology $\mathcal{T}_A$ on A.

Proof Suppose $\mathcal{B}$ is a basis for a topology $\mathcal{T}$ on a set S and $A \subseteq S$. To show that the set

$$\mathcal{B}_A = \{B \cap A \mid B \in \mathcal{B}\}$$

is a basis for $\mathcal{T}_A$, the subspace topology on A, we call on Theorem 8.8.

Let $V \in \mathcal{T}_A$ and $a \in V$. Then, by the definition of the subspace topology,

$$V = U \cap A$$

for some $U \in \mathcal{T}$. By Theorem 8.8, because $a \in U$, an open set in S, there exists $B \in \mathcal{B}$ so that $a \in B$ and $B \subseteq U$.

Thus, $a \in B \cap A$, and

$$B \cap A \subseteq U \cap A,$$

or equivalently,

$$B \cap A \subseteq V.$$

Since $B \cap A \in \mathcal{B}_A$, we have shown that $\mathcal{B}_A$ satisfies the necessary criteria of Theorem 8.8 and is indeed a basis for $\mathcal{T}_A$. □

Example 8.6 demonstrates how the basis for a topological space can yield a basis for the subspace topology on a certain subset. It simply involves considering set intersections.[2] In particular, consider the closed interval $[a, b]$ in $\mathbb{R}$, supposing $\mathbb{R}$ to have the standard topology. In $\mathbb{R}$, open sets consist of open intervals and disjoint unions of open intervals. Then, the open sets in $[a, b]$, under the subspace topology, are the intersection of $[a, b]$ with such sets.

Example 8.6. In the subspace topology on $[0, 2]$, in $\mathbb{R}$ with the standard topology, the following sets are open.

(1) $(0, 1)$, since

$$(0, 1) = (0, 1) \cap [0, 2].$$

(2) $(\frac{1}{2}, 1) \cup (\frac{3}{2}, 2)$, since

$$(\tfrac{1}{2}, 1) \cup (\tfrac{3}{2}, 2) = ((\tfrac{1}{2}, 1) \cup (\tfrac{3}{2}, 2)) \cap [0, 2].$$

(3) $[0, 1)$, since

$$[0, 1) = (-1, 1) \cap [0, 2].$$

(4) $(1, 2]$, since

$$(1, 2] = (1, 3) \cap [0, 2].$$

(5) $[0, 2]$, since

$$[0, 2] = (-1, 3) \cap [0, 2].$$

Shifting gears, we consider another method we developed for creating new sets from previously defined sets: the Cartesian product. This concept can be extended to define a topology on the Cartesian product of two topological spaces.[3]

---

[2] An intuitive way to remember the subspace topology is to think of them simply as intersections with the bigger set. This is true for the topology as well as with the basis.

[3] As a topological space is simply a set with an accompanying topology defined on it, this phrasing, "Cartesian product of topological spaces" is defined, even though the Cartesian product was defined simply on sets.

Definition 8.14. Let S_1 and S_2 be topological spaces with topologies $\mathcal{T}_1$ and $\mathcal{T}_2$, respectively. Then, $S_1 \times S_2$ is a topological space with a topology, called the *product topology*, generated by the basis

$$\mathcal{B} = \{U \times V \mid U \in \mathcal{T}_1, V \in \mathcal{T}_2\}.$$

Note that the topology on $S_1 \times S_2$ is not simply $\mathcal{T}_1 \times \mathcal{T}_2$, nor are the elements of the topology all possible Cartesian products $U \times V$ where $U \in \mathcal{T}_1$ and $V \in \mathcal{T}_2$. The given set $\mathcal{B}$ is defined to be a *basis* for the product topology. And as with such a claim, we must show that $\mathcal{B}$ is actually a basis.

Theorem 8.15. *If S_1 and S_2 are topological spaces with topologies $\mathcal{T}_1$ and $\mathcal{T}_2$, respectively, then*

$$\mathcal{B} = \{U \times V \mid U \in \mathcal{T}_1, V \in \mathcal{T}_2\}$$

is a basis for a topology on $S_1 \times S_2$.

Proof Let S_1 and S_2 be topological spaces with respective topologies $\mathcal{T}_1$ and $\mathcal{T}_2$. We prove that

$$\mathcal{B} = \{U \times V \mid U \in \mathcal{T}_1, V \in \mathcal{T}_2\}$$

is a basis for $S_1 \times S_2$ directly from Definition 8.4. To that end, first note that $S_1 \times S_2 \in \mathcal{B}$, so that for all $(s_1, s_2) \in S_1 \times S_2$, there is an element B of $\mathcal{B}$ so that $(s_1, s_2) \in B$.

Next, take

$$(s_1, s_2) \in (U_1 \times V_1) \cap (U_2 \times V_2),$$

where $U_1, U_2 \in \mathcal{T}_1$ and $V_1, V_2 \in \mathcal{T}_2$. By Exercise 5d following Section 3.3,

$$(U_1 \times V_1) \cap (U_2 \times V_2) = (U_1 \cap U_2) \times (V_1 \cap V_2),$$

and because $\mathcal{T}_1$ and $\mathcal{T}_2$ are topologies, we have that

$$U_1 \cap U_2 \in \mathcal{T}_1$$

and

$$V_1 \cap V_2 \in \mathcal{T}_2.$$

Thus, we have that $(s_1, s_2) \in B$ for some $B \in \mathcal{B}$, as desired, and consequently proving that $\mathcal{B}$ is indeed a basis for a topology on $S_1 \times S_2$. $\qquad\square$

Just as with the subspace topology, it is natural to ask that if $\mathcal{B}_1$ is a basis for S_1 and $\mathcal{B}_2$ is a basis for S_2, do they together somehow form a basis for $S_1 \times S_2$? As suspected, they do; the proof of Theorem 8.16 is left as an exercise.

Theorem 8.16. *If $\mathcal{B}_1$ is a basis for a S_1 and $\mathcal{B}_2$ a basis for S_2, then*

$$\mathcal{B} = \{B_1 \times B_2 \mid B_1 \in \mathcal{B}_1,\ B_2 \in \mathcal{B}_2\}$$

is a basis that generates the product topology on $S_1 \times S_2$.

We can see Theorem 8.16 in action just be considering the standard topology on $\mathbb{R}$. It has as a basis the set of all open intervals (a, b), for $a, b \in \mathbb{R}$ with $a < b$. The Cartesian plane, $\mathbb{R}^2$, has a product topology defined on it, considering $\mathbb{R}^2$ as $\mathbb{R} \times \mathbb{R}$ (each with the standard topology). A basis generating this topology is

$$\mathcal{B} = \{(a, b) \times (c, d) \mid a, b, c, d \in \mathbb{R} \text{ with } a < b \text{ and } c < d\}.$$

We visualize $\mathcal{B}$ as the set of all *open rectangles* in the Cartesian plane.

We conclude this section posing and answering a significant question relating the two topologies discussed in this section. Suppose $A_1 \subseteq S_1$ and $A_2 \subseteq S_2$, where S_1 and S_2 are topological spaces with topologies $\mathcal{T}_1$ and $\mathcal{T}_2$, respectively.

Consider $A_1 \times A_2$. How can we construct a topology for this set? In lieu of the discussions and constructions from this section, there are two possible approaches. We could choose to view $A_1 \times A_2$ as a subset of $S_1 \times S_2$, a topological space. Or, we could choose to view $A_1 \times A_2$ as a Cartesian product of two topological spaces (each as a subspace).

In the first scenario, we obtain a topology on $S_1 \times S_2$ via the product topology, and then, obtain a topology on $A_1 \times A_2$ as a subspace of this topological space. In the latter situation, A_1 and A_2 become topological spaces with a topology defined via the subspace topology, and then we form their product and defining a topology on them via the product topology. In an ideal world, these two topologies would be the same. Otherwise, we would have to always be aware of which action took place first and then develop a strong notation to distinguish the two different topologies. Luckily, Theorem 8.17 saves us the hassle of having to do this. Regardless of the approach we take, the two resulting topologies on $A_1 \times A_2$ are the *same topology*. The use of basis elements significantly streamlines the proof, exhibiting once again why bases are powerful tools.

Theorem 8.17. *Let S_1 and S_2 be topological spaces with subsets $A_1 \subseteq S_1$ and $A_2 \subseteq S_2$. The topology on $A_1 \times A_2$ as a subspace of the Cartesian product $S_1 \times S_2$ is the same as the product topology on $A_1 \times A_2$, where A_i has the subspace topology obtained from S_i.*

Proof Suppose S_1 and S_2 are topological spaces with $A_1 \subseteq S_1$ and $A_2 \subseteq S_2$. According to Definition 8.14, a general basis element for $S_1 \times S_2$ is of the form $U_1 \times U_2$, where U_1 is open in S_1 and U_2 is open in S_2. Then, Definition 8.9 gives that a general basis element for $A_1 \times A_2$, as a subspace of $S_1 \times S_2$, is of the form

$$(U_1 \times U_2) \cap (A_1 \times A_2).$$

However, by Exercise 5d following Section 3.3,

$$(U_1 \times U_2) \cap (A_1 \times A_2) = (U_1 \cap A_1) \times (U_2 \cap A_2).$$

In the subspace topology on A_1, we have that $U_1 \cap A_1$ is open, and likewise, $U_2 \cap A_2$ is open in the subspace topology on A_2. Thus, the set

$$(U_1 \cap A_1) \times (U_2 \cap A_2)$$

is in the basis for the product topology on $A_1 \times A_2$. Consequently, we have shown that the two bases generating the desired topologies are equal, so that the topologies must be equal.   □

In the next section, we introduce the idea of *closed sets* and *closure* of a set to lead us towards a notion for closeness. As with all of our topological definitions thus far, it is a natural generalization of the real-number line concept of a *closed set*.

Exercises

1. Consider $\mathbb{R}$ with the standard topology. Which sets are open in $[-2, 2]$ with the subspace topology?

   (a) $(-2, -1) \cup (1, 2]$

   (b) $(-2, -1] \cup [1, 2]$

   (c) $[-2, -2) \cup (1, 2)$

   (d) $\displaystyle\bigcup_{i=1}^{\infty} \left(-\frac{1}{i}, \frac{1}{i} \right)$

   (e) $\displaystyle\bigcap_{i=1}^{\infty} \left(-\frac{1}{i}, \frac{1}{i} \right)$

2. Consider $\mathbb{R}$ with the standard topology. Which of the sets are open in $[0, 2)$ with the subspace topology?

   (a) $\{0\}$

   (b) $[0, 1)$

   (c) $(0, 2]$

   (d) $(0, 1) \cup (1, 2)$

3. Consider $\mathbb{R}$ with the standard topology. Which of the sets are open in

$$\{-1\} \cup (0, 1] \cup \{2\} \cup \{3\} \cup \dots$$

   with the subspace topology?

(a) $\{-1\} \cup \{1\}$

(b) $\{-1\} \cup \{2\}$

(c) $\{-1\} \cup \{0, 1\} \cup \{2\}$

(d) $\mathbb{Z}^+$

4. Determine a basis for the product topology on $S_1 \times S_2$, where S_1 and S_2 are the given topological spaces.

   (a) $S_1 = \{1\}$ with the discrete topology
       $S_2 = \{1, 2\}$ with trivial topology

   (b) $S_1 = \{1, 2\}$ with the trivial topology
       $S_2 = \{1, 2\}$ with the trivial topology

   (c) $S_1 = \{1, 2\}$ with the topology $\mathcal{T}_1 = \{\varnothing, \{1\}, \{1, 2\}\}$
       $S_2 = \{1, 2\}$ with the topology $\mathcal{T}_2 = \{\varnothing, \{2\}, \{1, 2\}\}$

   (d) $S_1 = \{1, 2\}$ with topology $\mathcal{T}_1 = \{\varnothing, \{1\}, \{1, 2\}\}$
       $S_2 = \{1, 2, 3\}$ with topology $\mathcal{T}_2 = \{\varnothing, \{1, 3\}, \{1, 2, 3\}\}$

5. Describe a general basis element for the product topology on $S_1 \times S_2$, where S_1 and S_2 are the given topological spaces.

   (a) S_1: $\mathbb{R}$ with the discrete topology
       S_2: $\mathbb{R}$ with the standard topology

   (b) S_1: $\mathbb{R}$ with basis given by $\{[a, b) \mid a, b \in \mathbb{R} \text{ and } a < b\}$
       S_2: $\mathbb{R}$ with the standard topology

6. Let $S = \{1, 2, 3, 4\}$ be a topological space under the topology

$$\mathcal{T} = \{\varnothing, \{1\}, \{4\}, \{1, 4\}, \{1, 2, 3, 4\}\},$$

   and $A_1 = \{1, 2, 3\}$ and $A_2 = \{2, 3, 4\}$.

   (a) Determine the subspace topologies $\mathcal{T}_{A_1}$ and $\mathcal{T}_{A_2}$.

   (b) Determine a topology for $A_1 \times A_2$ as a product space of the subspace topologies A_1 and A_2 with $\mathcal{T}_{A_1}$ and $\mathcal{T}_{A_2}$, respectively

   (c) Determine a topology on $A_1 \times A_2$ as a subspace of $S \times S$ with the product topology.

7. Proof Lemma 8.10.

8. Prove Lemma 8.11.

9. Prove that $\mathcal{T}_A$, in the proof of Theorem 8.12, is closed under arbitrary unions.

10. Determine a basis for the topology that $[0, 2]$ inherits from $\mathbb{R}$ with the standard topology.

11. Prove Theorem 8.16.

12. Consider $\mathbb{Z}$ as a subset of $\mathbb{R}$ with the standard topology. Describe the subspace topology $\mathcal{T}_{\mathbb{Z}}$.

13. Let S be a topological space with topology $\mathcal{T}$ and $A \subseteq S$ and $X \subseteq A$. Two subspace topologies can be defined on X.

(a) Considering X as a subspace of the topological space S with topology $\mathcal{T}$.

(b) Considering X as a subset of the topological space A with topology $\mathcal{T}_A$.

Prove that these two topologies are the same.

8.3   Closed Sets and Closure

As discussed at the start of this chapter, the notion of a topology comes from a generalization of open intervals on the real number line. Thus, it should come as no surprise that there is a parallel to the idea of a closed interval, a concept called a *closed* set in a topology. Defined simply, the concept states generically why an interval on the real number line is closed.

Definition 8.18. A subset A of a topological space S is *closed* if $S - A$ is open.

How is this a generalization of the concept of closed we are familiar with on the real number line? Supposing we only understood open intervals (i.e., those of the form $(-\infty, a)$, (a, b) or (a, ∞)), how could we describe a *closed* set? We might be inclined to say $[a, b]$ is closed because its complement $(-\infty, a) \cup (b, \infty)$ is open. Transfer this idea to a general topological space and you have Definition 8.18.

The following two examples demonstrate closed sets in two different topological spaces, first in a finite space followed by those in $\mathbb{R}$ with the standard topology. In the latter, we verify that sets we would intuitively think of as *closed* on the real number line are indeed closed topologically.

Example 8.7. Define the topology $\mathcal{T}$ on $S = \{1, 2, 3, 4\}$ as

$$\mathcal{T} = \{\varnothing, \{2\}, \{3\}, \{1, 2\}, \{2, 3\}, \{1, 2, 3\}, \{1, 2, 3, 4\}\}.$$

The following sets are closed in S:

(1) $\{4\}$ since

$$S - \{4\} = \{1, 2, 3\}.$$

(2) $\{1, 4\}$ since

$$S - \{1, 4\} = \{2, 3\}.$$

The following sets are *not* closed in S:

(1) $\{1\}$ since

$$S - \{1\} = \{2, 3, 4\}.$$

(2) $\{2, 4\}$ since

$$S - \{2, 4\} = \{1, 3\}.$$

Example 8.8. In $\mathbb{R}$ with the standard topology, the following sets are closed.

(1) $[2, 4]$ since

$$\mathbb{R} - [2, 4] = (-\infty, 2) \cup (4, \infty)$$

(2) $[4, \infty)$ since

$$\mathbb{R} - [4, \infty) = (-\infty, 4).$$

(3) $\{4\}$ since

$$\{4\} = (-\infty, 4) \cup (4, \infty).$$

As with open sets, whether or not a subset is closed in a particular topological space depends on the defined topology. Had $\mathcal{T}$ been defined differently in Example 8.7, $\{4\}$ may not have been closed while $\{2, 4\}$ may have been. In $\mathbb{R}$, $[2, 4]$ is closed in the standard topology, but in other topologies (such as the finite complement topology (see Exercise 8 following Section 8.1)), it is not. In one particular topological space, however, *every* subset is closed. The proof of Lemma 8.19 is left as an exercise.

Lemma 8.19. *If S is any set with the discrete topology, then every subset of S is closed.*

Lemma 8.19 exhibits a particularly important fact about subsets of topological spaces. A subset being open has no bearing on whether or not that same subset is closed, and vice versa (Lemma 8.19 exhibits a topological space where every subset is *both* open and closed (sometimes referred to as being *clopen*)). The properties are not opposites of one another; a subset that is not open is not automatically closed. Likewise, a subset that is not closed need not be open. Both properties are entirely dependent upon the topology.[4] The following example illustrates the possibilities of a subset being open or closed.

Example 8.9. Consider the topological space $S = \{1, 2, 3, 4\}$ with topology

$$\mathcal{T} = \{\varnothing, \{1\}, \{2\}, \{1, 2\}, \{1, 2, 3, 4\}\}$$

Subsets of S may be open, closed, both or neither, as demonstrated below.

(1) $\{1, 2, 3, 4\}$ is both open and closed, since

$$\varnothing, \{1, 2, 3, 4\} \in \mathcal{T}.$$

(2) $\{3, 4\}$ is closed but not open, since

$$\{1, 2\} \in \mathcal{T} \text{ but } \{3, 4\} \notin \mathcal{T}.$$

(3) $\{1\}$ is open but not closed, since

$$\{1\} \in \mathcal{T} \text{ but } \{2, 3, 4\} \notin \mathcal{T}.$$

(4) $\{2, 4\}$ is neither open nor closed, since

$$\{2, 4\}, \{1, 3\} \notin \mathcal{T}.$$

Since closed sets are defined in terms of open sets, the three results of Theorem 8.20 come as no surprise.

Theorem 8.20. *Let S be any topological space. Then,*

   1. $\varnothing$, S are closed.

   2. If V_i is a closed set for all $i \in I$, where I is some index set, then

$$\bigcap_{i \in I} V_i$$

   is closed.

---

[4]Some remember this fact by thinking of a *topologist's door*. To most people, a door is either open or closed. A topologist's door, however, may be open, closed, both open and closed or neither open nor closed.

3. *If $V_1, V_2, \ldots, V_n$ are closed, then*

$$\bigcup_{i=1}^{n} V_i$$

is closed.

Proof Let S be any topological space with topology $\mathcal{T}$. Because $\varnothing$ and S are open, it follows that $S - \varnothing = S$ and $S - S = \varnothing$ are closed.

Next, suppose V_i is closed for all $i \in I$, where I is some index set. Then, $S - V_i$ is open for every i, and consequently, so is

$$\bigcup_{i \in I} (S - V_i).$$

Applying Exercise 16 following Section 3.3, we have that

$$S - \bigcap_{i \in I} V_i = \bigcup_{i \in I} (S - V_i),$$

an arbitrary union of open sets, which is open. Thus, the desired arbitrary intersection is closed.

The proof of the third claim is left as an exercise. $\square$

Let's shift our attention to relating this new idea of closed sets to other previously defined topological structures. We begin with the subspace topology. There is a language issue that needs clarification, though. Given a topological space S and a subspace A of S, saying that a $B \subseteq A$ is closed is vague, because B is both a subset of S and a subset of A. To address this, we define what it means to be closed *in a particular set*.

Definition 8.21. If A is a subspace of a topological space S, we say that $B \subseteq A$ is *closed in A* if B is closed in the subspace topology of A (that is, if $A - B$ is open in A).

If we know what the closed sets in a given topological space are and consider the subspace topology on some subset of that space, is there a way to determine what the closed sets are in that subspace? Just as our intuition led us to the correct way of determining basis elements of the subspace, our intuition leads in the right direction here as well. If a subset is closed in the larger set, then its intersection with the subspace yields a closed subset in the subspace, and vice versa.

Theorem 8.22. *Suppose that S is a topological space and A is a subspace of S. Then, $B \subseteq A$ is closed in A if and only if B equals the intersection of a closed set in S with A.*

Proof Suppose S is a topological space, A a subspace of S, and $B \subseteq A$. We begin by assuming B is closed in A. Definition 8.21 tells us that $A - B$ is open in A, which by the definition of the subspace topology means that

$$A - B = U \cap A,$$

where U is some open set in S. Then, by definition, $S - U$ is closed in S. Because of this, and the set-theoretic result that

$$B = A \cap (S - U),$$

we have shown that B equals the intersection of a closed set in S with A, as desired.

Next, we assume

$$B = V \cap A,$$

where V is closed in S. The definition of being closed yields that $S - V$ is open, and in the subspace topology then, $A \cap (S - V)$ is necessarily open. Notice that

$$(S - V) \cap A = A - B.$$

Thus, $A - B$ is open, meaning that B is closed. Hence, the result holds. □

Consider the interval $[2, 5)$ on the real-number line. If we were to ask of it, "Is there an 'inside' to this interval that is open," we might be inclined to say, "yes, $(2, 5)$ plays such a role." Generalizing this idea to topological spaces, we can talk about the open inside of any set, a term we call the *interior* of the set.

Likewise, if we were asked to turn the set $[2, 5)$ into a closed interval, whatever that may mean, our intuition would be to respond with the interval $[2, 5]$. We have taken the half-open interval and closed the half that was open. As with the interior of an interval, we will generalize this same concept to any set in a topological space, forming what is called the *closure* of a set.

Definition 8.23. Let $A \subseteq S$, a topological space. The *interior of A*, denoted Int(A), is the union of all open sets contained in A. The *closure of A*, denoted Cl(A), is the intersection of all closed sets containing A.

The following lemma follows immediately from the previous definition.

Lemma 8.24. *Let $A \subseteq S$, a topological space. Then,*

1. Int$(A) \subseteq A$, *and*

2. $A \subseteq$ Cl(A).

As our discussion leading into Definition 8.23 suggested, the interior of a set is open and the closure of a set is closed.

Lemma 8.25. *If A is a subset of a topological space S, then*

    1. $\text{Int}(A)$ *is open, and*

    2. $\text{Cl}(A)$ *is closed.*

We prove (1) and leave (2) as an exercise.

Proof Let A be a subset of a topological space, with

$$\{U_i \mid i \in I \text{ (some index set)}, U_i \text{ is open and } U_i \subseteq A\}$$

the collection of all open sets in S contained in A. Then,

$$\bigcup_{i \in I} U_i$$

is open (by the definition of a topology), and hence, $\text{Int}(A)$ is open. □

We solidify our intuitive understanding of the interior and closure of intervals in $\mathbb{R}$ via the following example. Our intuition tells us that we *know* what the interior and closure of the given interval are. We use the definitions of those terms to verify our notions.

Example 8.10. In $\mathbb{R}$ with the standard topology, determine the interior and closure of $(0, 2]$.

Determining the interior of any set requires considering all open sets contained in the set. In this topological space, that means taking the union of all open intervals fully contained in $(0, 2]$. However, any open interval (or disjoint union of open intervals) contained in $(0, 2]$ must also be contained in $(0, 2)$, itself an open interval contained in $(0, 2]$. Hence,

$$\text{Int}((0, 2]) = (0, 2).$$

By a similar reasoning, since any closed interval containing $(0, 2]$ must also contain $[0, 2]$, we have

$$\text{Cl}((0, 2]) = [0, 2].$$

Example 8.7 supports an intuitive understanding of Definition 8.23: the interior of a set is the largest open set contained inside the set, while the closure is the smallest closed set containing the set. In the standard topology on $\mathbb{R}$, these are somewhat simple to find. In other topologies, a firm grasp of the topology leads us to finding interiors and closures.

Example 8.11. In $\mathbb{Z}$ with the discrete topology, find both the interior and closure of the set $A = \{1, 2, 3, 4\}$.

To find $\text{Int}(A)$, we must determine all open sets in $\mathbb{Z}$ that are contained in A. In this case, *every* subset of $\mathbb{Z}$ is open, including A itself. Thus,

$$\text{Int}(A) = A.$$

Likewise, to determine $\text{Cl}(A)$, we must determine all closed sets containing A. Once again, because every subset of $\mathbb{Z}$ is open, we have that $\mathbb{Z} - A$ is open, or equivalently, A is closed. Consequently,

$$\text{Cl}(A) = A.$$

The previous example goes against our intuition. When we think of open subsets of real numbers, we think of *open intervals* and definitely not finite sets of points. Once again, a theme of this chapter is reinforced: much depends on how the topology of a space is defined.

The next example considers the real numbers under yet a different topology: the finite complement topology (see Exercise 8.1.8). Working with closed sets in this topology requires care, simply due to the way the topology is defined. A nonempty set A is open in this topological space if $\mathbb{R} - A$ is finite. Then, a set B is closed if $\mathbb{R} - B$ is open, or equivalently, $\mathbb{R} - (\mathbb{R} - B)$ is finite.

Example 8.12. Find both the interior and closure of the set $A = [2, 4)$ in $\mathbb{R}$ with the finite complement topology.

To find $\text{Int}(A)$, we consider all open sets contained in A. What constitutes an open set in $\mathbb{R}$ with the finite complement topology? Any such set whose complement is finite.

Suppose B is such a set, so that $\mathbb{R} - B$ consists only of finitely many real numbers. Let us also assume that $B \subseteq A$. Then, because $B \subseteq A$, we have that $\mathbb{R} - A \subseteq \mathbb{R} - B$ (see Exercise 4f following Section 3.3). But $\mathbb{R} - A$ is infinite, a contradiction to $\mathbb{R} - B$ being finite. Thus, no such B could exist, proving that

$$\text{Int}(A) = \varnothing.$$

Because A itself is infinite, any closed set containing A must be infinite. To determine the closure of A, we must consider all closed sets containing A; what, then, are the infinite closed sets in this topological space? If a set B is closed, then $\mathbb{R} - B$ must be open. For such a set to be open, B

(the complement of $\mathbb{R} - B$) must be finite. This leads us to think that perhaps no such B exists and that $\mathrm{Cl}(A) = \varnothing$. But then we realize that cannot be possible; Lemma 8.24 tells us that $A \subseteq \mathrm{Cl}(A)$.

Remember, both $\varnothing$ and $\mathbb{R}$ are in the topology on $\mathbb{R}$. Both of these sets are open and closed, and $[2, 4) \subseteq \mathbb{R}$. Thus, $\mathrm{Cl}(A) = \mathbb{R}$.

What the previous example illustrates is sort of a worst case scenario for the interior or closure of a set. That is, if A is a subset of a topological space S, then

$$\varnothing \subseteq \mathrm{Int}(A) \subseteq A \subseteq \mathrm{Cl}(A) \subseteq S.$$

We finish the section with a theorem relating the interior and closure to other set theoretic concepts. First, however, Lemma 8.26 is necessary for proving parts of Theorem 8.27. Its proof is left as an exercise.

Lemma 8.26. *If S is a topological space and $A \subseteq S$, with $s \in S$, then $s \in \mathrm{Cl}(A)$ if and only if every open set containing s intersects A.*

Theorem 8.27. *Let S be a topological space with $A, B \subseteq S$.*

1. *If U is open in S with $U \subseteq A$, then $U \subseteq \mathrm{Int}(A)$.*

2. *If V is closed in S with $B \subseteq V$, then $\mathrm{Cl}(B) \subseteq V$.*

3. *If $A \subseteq B$, then $\mathrm{Int}(A) \subseteq \mathrm{Int}(B)$ and $\mathrm{Cl}(A) \subseteq \mathrm{Cl}(B)$.*

4. *A is open if and only if $A = \mathrm{Int}(A)$.*

5. *B is closed if and only if $B = \mathrm{Cl}(B)$.*

6. *$S - A = S - \mathrm{Cl}(A)$.*

7. *$\mathrm{Cl}(S - B) = S - \mathrm{Int}(B)$.*

8. *$\mathrm{Int}(A) \cup \mathrm{Int}(B) \subseteq A \cup B$.*

9. *$\mathrm{Int}(A) \cap \mathrm{Int}(B) = A \cap B$.*

We prove here a few of the claims above, leaving the rest as exercises.

Proof Let S be a topological space with $A, B \subseteq S$.

(1) Suppose U is open in S with $U \subseteq A$. Because

$$U \in \{V \mid V \text{ is open in } S \text{ and } V \subseteq A\},$$

it follows immediately that

$$U \subseteq \bigcup_{i \in I} V_i,$$

where I is some index set and V_i is open in S, $V_i \subseteq A$. Thus,

$$U \subseteq \text{Int}(A).$$

(3) Suppose $A \subseteq B$. Because every open set contained in A must be contained in B (see Example 2.8),

$$\{U_i \mid U_i \text{ is open and } U_i \subseteq A\} \subseteq \{V_j \mid V_j \text{ is open and } V_j \subseteq B\}.$$

Consequently,

$$\text{Int}(A) \subseteq \text{Int}(B).$$

(5) If B is closed, then

$$B \in \{V_i \mid V_i \text{ is closed and } B \subseteq V_i\},$$

and consequently,

$$\bigcap_{i \in I} V_i = B,$$

proving that B is closed if and only if $B = \text{Cl}(B)$.

(6) We use the set element method to prove

$$\text{Int}(S - A) = S - \text{Cl}(A).$$

First, suppose $s \in \text{Int}(S - A)$. Because this set equals the union of all open sets contained in $S - A$, we know that $s \in U$ for some

$$U \subseteq S - A,$$

where U is open in S. By Lemma 8.26, $s \notin \text{Cl}(A)$ (since $U \cap A = \varnothing$). So, $s \in S - \text{Cl}(A)$, proving that

$$\text{Int}(S - A) \subseteq S - \text{Cl}(A).$$

Next, $\text{Cl}(A)$ is closed, by Lemma 8.25, and

$$A \subseteq \text{Cl}(A).$$

So, $S - \text{Cl}(A)$ is open, and,

$$S - \text{Cl}(A) \subseteq S - A.$$

By (1), then,

$$S - \text{Cl}(A) \subseteq \text{Int}(S - A).$$

proving that

$$S - \text{Cl}(A) \subseteq \text{Int}(S - A),$$

and ultimately proving the two sets are equal. □

The basic concepts of topological spaces have been developed through these first three sections of this chapter. In particular, we have created a structure for generalizing collections of open intervals on the real number line. In our final section of the book, we use these definitions to generalize a major concept from calculus that is grounded in open intervals: continuity.

Exercises

1. Let $S = \{1, 2, 3, 4\}$ with topology $\mathcal{T}$ as defined. Find $\text{Int}(A)$ and $\text{Cl}(A)$, where $A = \{1, 3\}$.

    (a) $\mathcal{T}$ is the trivial topology.
    (b) $\mathcal{T}$ is the discrete topology.
    (c) $\mathcal{T} = \{\varnothing, \{1\}, \{2\}, \{1, 2\}, \{1, 2, 3, 4\}\}$.
    (d) $\mathcal{T} = \{\varnothing, \{1\}, \{3\}, \{1, 3\}, \{1, 2, 3, 4\}\}$.
    (e) $\mathcal{T} = \{\varnothing, \{1\}, \{3\}, \{4\}, \{1, 3\}, \{1, 4\}, \{3, 4\}, \{1, 3, 4\}, \{1, 2, 3, 4\}\}$.
    (f) $\mathcal{T} = \{\varnothing, , \{3\}, \{1, 3\}, \{2, 4\}, \{2, 3, 4\}, \{1, 2, 3, 4\}\}$.

2. Find $\text{Int}(A)$ and $\text{Cl}(A)$ when A is the subset of the given topology.

    (a) $A = [-1, 1)$ in $\mathbb{R}$ with the standard topology.
    (b) $A = (-\infty, -1]$ in $\mathbb{R}$ with the standard topology.
    (c) $A = (-4, 0] \cup \{1\} \cup (2, 4]$ in $\mathbb{R}$ with the standard topology.
    (d) $A = [-1, 1)$ in $\mathbb{R}$ with the discrete topology.
    (e) $A = \{1, 2, 3\}$ in $\mathbb{R}$ with the discrete topology.
    (f) $A = \{1, 2, 3\}$ in $\mathbb{R}$ with the finite complement topology (see Exercise 8 following Section 8.1).

3. Determine all closed sets in a nonempty topological space S with the trivial topology.

4. Prove the claim from Example 8.19: if S is any set with the discrete topology, then every subset of S is closed.

5. Determine all closed sets in $\mathbb{R}$ with the finite complement topology (see Exercise 8 following Section 8.1).

6. Prove that in any topological space, there exists a set that is both open and closed.

7. Prove (3) from Theorem 8.20.

8. In the proof of Theorem 8.22, two set theoretic claims were made. Prove them, referring to the proof for definitions of the appropriate sets.

    (a) $B = A \cap (S - U)$

(b) $(S - V) \cap A = A - B$

9. Prove (2) from Lemma 8.25.

10. Prove Lemma 8.26.

11. Prove that if A is any set in a topological space with the discrete topology, then

$$\text{Int}(A) = \text{Cl}(A) = A.$$

12. Prove the remaining parts of Theorem 8.27.

   (a) If V is closed in S with $B \subseteq V$, then $\text{Cl}(B) \subseteq V$.
   (b) If $A \subseteq B$, then $\text{Cl}(A) \subseteq \text{Cl}(B)$.
   (c) A is open if and only if $A = \text{Int}(A)$.
   (d) $\text{Cl}(S - B) = S - \text{Int}(B)$.
   (e) $\text{Int}(A) \cup \text{Int}(B) \subseteq A \cup B$.
   (f) $\text{Int}(A) \cap \text{Int}(B) = A \cap B$.

8.4   Continuous Functions

Recall from calculus that a real-valued function $f(x)$ is continuous at $x = a$ if

$$\lim_{x \to a} f(x) = f(a).$$

Stating this in terms of the definition of a limit, $f(x)$ is continuous at $x = 1$ if given any $\epsilon > 0$, there is a real number $\delta > 0$ so that if

$$x \in (a - \delta, a + \delta),$$

with $x \neq a$, then

$$f(x) \in (f(a) - \epsilon, f(a) + \epsilon).$$

What this means is that when x is close to a, then $f(x)$ is close to $f(a)$. This concept of closeness is precisely what is generalized when we define *continuity* on functions between topological spaces. Because the above definition involves open intervals, it is no surprise then that the definition of a general continuous function between topological spaces is in terms of open sets.

Definition 8.28. A function $f : S_1 \to S_2$ between topological spaces S_1 and S_2 is *continuous* if when V is open in S_2, then $f^{-1}(V)$ is open in S_1.

One of the primary goals of this section is to show that this new way of defining continuity on $\mathbb{R}$ (under the standard topology) is identical to how we have come to learn continuity since our days in calculus.

Definition 8.28 does more than generalize the concept of a continuous function. It gives us a method for talking about closeness in a topological space. Two elements of the space S are close if they are elements of the same open set. While this does not at first necessarily seem natural, if we recall that open sets are the parallel to open intervals, perhaps it is an adequate generalization of being close.

Example 8.13. Let $S_1 = \{1,2,3,4\}$ and $S_2 = \{1,2,3\}$ be topological spaces under the respective topologies

$$\mathcal{T}_1 = \{\varnothing, \{1\}, \{2\}, \{1,2\}, \{2,3\}, \{1,2,3\}, \{1,2,3,4\}\}$$

and

$$\mathcal{T}_2 = \{\varnothing, \{1,2\}, \{1,2,3\}\}.$$

Consider $f, g : S_1 \to S_2$ defined in the table below.

x	$f(x)$	$g(x)$
1	1	1
2	2	1
3	3	1
4	1	2

To determine if either f or g is continuous, we must consider the inverse image of *every* open set in S_2. Notice:

$$f^{-1}(\varnothing) = \varnothing$$
$$f^{-1}(\{1,2\}) = \{1,2,4\}$$
$$f^{-1}(\{1,2,3\}) = \{1,2,3,4\}.$$

Though two of these sets are open, $f^{-1}(\{1,2\})$ is not open in S_1. Hence, f is not continuous. The function g, however, is continuous, as $g^{-1}(V)$ is open for every $V \in \mathcal{T}_2$:

$$g^{-1}(\varnothing) = \varnothing$$
$$g^{-1}(\{1,2\}) = \{1,2,3,4\}$$
$$g^{-1}(\{1,2,3\}) = \{1,2,3,4\}.$$

Determining if the functions of the previous example were continuous amounted to simply computing inverse images of particular sets, since f and g were defined on a finite topological space. Such topological spaces are often simply conjured up examples to exhibit some definition or result (as was the case in Example 8.13). The more interesting examples involve functions whose domains and co-domains are certain infinite topological spaces, such as those we have encountered in past sections. Before looking at some of those situations, two particular functions are *always* continuous: the identity and constant functions.

Example 8.14. If S is any topological space, then $\mathrm{Id}_S : S \to S$ is continuous, since

$$\mathrm{Id}_S^{-1}(U) = U$$

for any $U \subseteq S$. Thus, if U is open in the co-domain, its inverse image (U) is open in the domain.

Example 8.15. If $f : S_1 \to S_2$ is given by $f(s) = s_2$ for some fixed $s_2 \in S_2$ (that is, f is a constant function), then f is continuous. The proof of this is left as an exercise.

Let us move to a specific function defined on $\mathbb{R}$ under the standard topology.

Example 8.16. Let $f : \mathbb{R} \to \mathbb{R}$ be given by

$$f(x) = x + 1,$$

where we consider the domain set of f having the standard topology and the co-domain having the finite complement topology (see Exercise 8 following Section 8.1). Show that f is continuous.

To show this, take an open set V in the co-domain and show that $f^{-1}(V)$ is open in the domain. There are three possibilities for V to consider.

(1) If $V = \varnothing$, then

$$f^{-1}(V) = \varnothing.$$

(2) $V = \mathbb{R}$, then

$$f^{-1}(V) = \mathbb{R}.$$

(3) V is a nonempty proper subset of $\mathbb{R}$, then

$$V = \mathbb{R} - \{v_1, v_2, \ldots, v_n\},$$

for some $v_i \in \mathbb{R}$. Then,

$$f^{-1}(V) = \mathbb{R} - \{v_1 - 1, v_2 - 2, \ldots, v_n - 1\},$$

a disjoint union of $n + 1$ open intervals.

In all cases, $f^{-1}(V)$ is open, proving that f is continuous.

Showing the function f of the previous example was continuous was rather straightforward. If f had been defined by

$$f(x) = x^2,$$

the proof of continuity becomes rather complex. Why? Consider the inverse image of *any* open set. Even if the co-domain was under the standard topology, this problem is significantly more involved than the linear function considered in Example 8.16.

However, we have developed an alternative way to investigate many topological spaces: a basis. If we understand how properties hold for a single basis element, we may be able to generalize that property to arbitrary unions of basis elements (that is, to open sets). The following theorem proves that this approach suffices when determining whether or not a function is continuous. If the inverse image of a basis element of the co-domain is open, then the function is continuous.

Theorem 8.29. *Let S_1 and S_2 be topological spaces with $\mathcal{B}_2$ a basis for S_2. Then, $f : S_1 \to S_2$ is continuous if and only if $f^{-1}(B)$ is open for every $B \in \mathcal{B}_2$.*

Proof Suppose S_1 and S_2 are topological spaces that that $\mathcal{B}_2$ is a basis for S_2. Let $f : S_1 \to S_2$. If f is continuous, then since B is open for every $B \in \mathcal{B}_2$, it follows from Definition 8.28 that $f^{-1}(B)$ is open.

Conversely, assume $f^{-1}(B)$ is open for every $B \in \mathcal{B}_2$ and take $V \in S_2$ to be an arbitrary open set. By Definition 8.4,

$$V = \bigcup_{i \in I} B_i$$

where $B_i \in \mathcal{B}_2$ and I is some index set. Then, by Exercise 21 following Section

6.1,

$$f^{-1}(V) = f^{-1}\left(\bigcup_{i \in I} B_i\right)$$

$$= \bigcup_{i \in I} f^{-1}(B_i),$$

a union of open sets in S_1. Thus, $f^{-1}(V)$ is open.   □

Now, we can show that $f : \mathbb{R} \to \mathbb{R}$ by $f(x) = x^2$, where both the domain and the co-domain are under the standard topology, is continuous. By the previous theorem, we need simply consider the inverse image of a general open interval (a, b). Notice that, in all cases, $f^{-1}((a, b))$ is open:

$$f^{-1}((a,b)) = \begin{cases} (-\sqrt{b}, \sqrt{a}) \cup (\sqrt{a}, \sqrt{b}) & \text{if } a \geq 0 \\ (-\sqrt{b}, \sqrt{b}) & \text{if } a < 0 \text{ and } b > 0 \\ \varnothing & \text{if } b \leq 0 \end{cases}.$$

A function is continuous if and only if the inverse image of an open set is open. Is there an alternative approach using closed sets? Because closed sets are defined in terms of their complements being open, we have the following.

Corollary 8.30. *A function between topological spaces is continuous if and only if the inverse image of a closed set is closed.*

To work towards our goal of showing that the standard calculus definition of continuity is identical to the topological definition of Definition 8.28, we must have a notion of closeness. Though we discussed this briefly at the start of the section, Theorem 8.31 accomplishes this a bit more than simply saying two elements of the same open set are close. If we consider the closure of a set to be the points close to it,[5] then a continuous function preserves this closeness. Two elements that are close in the domain (in the closure of a set) remain close after being mapped by the continuous function (they are in the closure of the image of the set).

Theorem 8.31. *Let $f : S_1 \to S_2$ be a continuous function between topological spaces, and $A \subseteq S_1$. If $s \in \text{Cl}(A)$, then $f(s) \in \text{Cl}(f(A))$.*

Proof Let $f : S_1 \to S_2$ be a continuous function between topological spaces and take $A \subseteq S_1$ and $s \in S_1$. We prove the result via the contrapositive, proving if $f(s) \notin \text{Cl}(f(a))$, then $s \notin \text{Cl}(A)$.

If $f(s) \notin \text{Cl}(f(a))$, then Lemma 8.26 yields the existence of an open set U with $f(s) \in U$ but

---

[5]What real number not in $[2, 4)$ is closest to $[2, 4)$? Our intuition says 4; indeed, 4 is in the closure of $[2, 4)$ but not in $[2, 4)$ itself.

$$U \cap f(A) = \varnothing.$$

Because U is open, we have that $f^{-1}(U)$ is open, and, $s \in U$. Moreover,

$$f^{-1}(U) \cap A = \varnothing,$$

otherwise U and $f(A)$ are not disjoint. Hence, $s \notin \text{Cl}(A)$ proving that the contrapositive of the desired result, and consequently the result itself, holds. □

Now that we have this result in our topological toolbox, we move on to showing that the calculus definition of continuity is identical to the topological definition. To translate the calculus definition, we have the following definition.

Definition 8.32. A function $f : S_1 \to S_2$ between topological spaces is ϵ-δ *continuous* if for every $s \in S_1$ and every open set $V \subseteq S_2$ such that $f(s) \in V$, then there exists an open set U in S_1 such that $s \in U$ and $f(U) \subseteq V$.

Why is this definition equivalent to the calculus definition continuity? See Exercise 10.

At last we are ready to show that our two definitions of continuity are equivalent. It is not necessary to discuss how the proof of this result is crafted. The proof is grounded in the definitions presented throughout this chapter; intuitive comfort with them and the "where did that step come from" questions are simply answered by applying those definitions at the appropriate times.

Theorem 8.33. *A function $f : S_1 \to S_2$ between topological spaces is ϵ-δ continuous if and only if it is continuous.*

Proof Let $f : S_1 \to S_2$ be a function between topological spaces. If we assume f is ϵ-δ continuous, take V to be open in S_2 with $s \in f^{-1}(V)$. By Definition 8.32, since $f(s) \in V$, there exists an open set U in S_1 with $s \in U$ and $f(U) \subseteq V$. Thus,

$$U \subseteq f^{-1}(V).$$

By Theorem 8.3, $f^{-1}(V)$ is open, as required.

Now, suppose f is continuous in the sense of Definition 8.28. Take $s \in S$ and $V \subseteq S_2$ such that $f(s) \in V$. Let

$$U = f^{-1}(V),$$

which is necessarily open and take $s \in U$. Since

$$f(U) \subseteq V,$$

we have that f is ϵ-δ continuous. $\qquad\square$

It is worth noting the directness of the previous proof. The proof is not shortened in any way. A few minor preliminary results and a couple definitions are the only tools necessary for it to hold. This is one of the nice aspects of point-set topology. Understanding fundamental concepts (sets and functions, in particular), and the theoretical concepts, though they appear high-level, are straightforward. It becomes an easy area to witness math happening, taking the specific and generalizing it.

You may have been thinking all along about the types of questions asked in calculus related to continuity. Typically, the standard questions used to reinforce the concept of continuity require determining if a piecewise-defined function is continuous, and perhaps you are wondering if the discussion of continuity here will consider such questions. We do but more generally. The result, known as the *Pasting Lemma*, reinforces how we approach such continuity questions in calculus itself.

Theorem 8.34. (Pasting Lemma) *Let S_1 be a topological space with A and B closed subsets of S_1 such that $A \cup B = S_1$. If $f : A \to S_2$ and $g : B \to S_2$ are continuous with $f(s) = g(s)$ for all $s \in A \cap B$, then $h : S_1 \to S_2$ given by*

$$h(s) = \begin{cases} f(s) & \text{if } s \in A \\ g(s) & \text{if } s \in B \end{cases}$$

is continuous.

Proof Take A and B to be closed subsets of a topological space S_1, with

$$A \cup B = S_1.$$

Suppose $f : A \to S_2$ and $g : B \to S_2$ are continuous functions satisfying

$$f(s) = g(s)$$

for all $s \in A \cap B$, and define $h : S_1 \to S_2$ by

$$h(s) = \begin{cases} f(s) & \text{if } s \in A \\ g(s) & \text{if } s \in B. \end{cases}$$

To show that h is continuous, it suffices by Lemma 8.30 to show that if V is closed in S_2, then $h^{-1}(V)$ is closed in S_1. To that end, take $V \subseteq S_2$ with V closed. Then, by the definition of h,

$$h^{-1}(V) = f^{-1}(V) \cup g^{-1}(V).$$

Since f and g are both continuous, by Lemma 8.30, we have that $f^{-1}(V)$ is closed in A and $g^{-1}(V)$ is closed in B. Thus, Theorem 8.22 gives

$$f^{-1}(V) = C_1 \cap A$$

and

$$g^{-1}(V) = C_2 \cap B,$$

where C_1 and C_2 are closed in S_1. Because the sets C_1 and A, along with C_2 and B, are closed in S_1, $f^{-1}(V)$ and $g^{-1}(V)$ are closed in S_1. Consequently, $h^{-1}(V)$ is closed in S_1, as desired. $\qquad\square$

Why is this result called the Pasting Lemma? Topologists often use a term called gluing when putting things together. Things that are glued together need not be physical or geometric; algebraic expressions, too, may be glued together. In this case, two functions are pasted together to form a third. Perhaps the choice of pasting over gluing is because it is the first time topology students experience gluing, so the result became commonly known as the Pasting Lemma (much like children learn to paste before glue, though the two actions really are the same thing). Regardless of what we call it, it is a crucial result for the construction of particular objects seen later on in topology and a fitting finish for our investigations of the subject.

Exercises

1. If $f : \{1,2,3\} \to \{1,2,3\}$ by $f(1) = 2$, $f(2) = 1$ and $f(3) = 3$, determine if f is continuous if the domain has topology T_1 and the co-domain has topology T_2.

   (a) $T_1 = T_2 = \{\varnothing, \{1\}, \{2\}, \{1,2\}, \{1,2,3\}\}$
   (b) $T_1 = \{\varnothing, \{1\}, \{1,2\}, \{1,2,3\}\}$
       $T_2 = \{\varnothing, \{1\}, \{3\}, \{1,3\}, \{1,2,3\}\}$
   (c) T_1 is the trivial topology
       T_2 is the discrete topology
   (d) T_1 is the discrete topology
       T_2 is the trivial topology
   (e) T_1 and T_2 are both the discrete topology

2. If $f : \{1,2,3,4\} \to \{1,2,3,4\}$ with $f(1) = 3$ and $f(2) = 1$, determine values of $f(3)$ and $f(4)$ so that f is continuous if the domain has topology T and the co-domain has the topology listed.

   $$T = \{\varnothing, \{3\}, \{4\}, \{1,2\}, \{3,4\}, \{1,2,3\}, \{1,2,4\}, \{1,2,3,4\}\}$$

   (a) trivial topology
   (b) discrete topology
   (c) $\{\varnothing, \{3\}, \{1,2,3,4\}\}$
   (d) $\{\varnothing, \{1\}, \{3\}, \{1,3\}, \{1,2,3,4\}\}$

3. Define a function $f : \mathbb{R} \to \mathbb{R}$, different from the identity function or any constant function, that is continuous if the domain and co-domain have the listed topologies.

(a) Domain: standard topology
Co-domain: standard topology

(b) Domain: standard topology
Co-domain: trivial topology

(c) Domain: trivial topology
Co-domain: standard topology

(d) Domain: finite complement topology
Co-domain: trivial topology

(e) Domain: finite complement topology
Co-domain: finite complement topology

4. (a) If S_1 is any set with the trivial topology, prove that any function with co-domain S_1 is continuous.

(b) If S_2 is any set with the discrete topology, then any function with domain S_2 is continuous.

5. Let $f : \mathbb{R} \to \mathbb{R}$, where both the domain and co-domain are topological spaces under the standard topology, be given by $f(x) = x^3$.

(a) Prove that if U is a basis element of the co-domain, then $f^{-1}(U)$ is open.

(b) Prove directly from Definition 8.32 that f is ϵ-δ continuous.

6. Let $f : S_1 \to S_2$ and $g : S_2 \to S_3$ be continuous functions between topological spaces. Prove that $g \circ f : S_1 \to S_3$ is continuous.

7. Let S_1 and S_2 be disjoint closed subspaces of a topological space S, with $f_1 : S_1 \to S_3$ and $f_2 : S_2 \to S_3$ continuous functions (where S_3 is some topological space), then prove that $f : S_1 \cup S_2 \to S_3$, defined below, is continuous.

$$f(s) = \begin{cases} f_1(s) & \text{if } s \in S_1 \\ f_2(s) & \text{if } s \in S_2. \end{cases}$$

8. If $f : S_1 \to S_2$ is a continuous function between topological spaces and A is a subspace of S_1, prove that $f|_A : A \to S_2$ is continuous.

9. Prove Corollary 8.30.

10. Prove that a function $f : \mathbb{R} \to \mathbb{R}$ is ϵ-δ continuous if and only if for every $x \in \mathbb{R}$ with $f(x) \in V$, where V an open set, there exists an open set U, $x \in U$, with $f(U) \subseteq V$.

11. Let S_1 and S_2 be topological spaces and $S_1 \times S_2$ their product space. Prove that $\pi_1 : S_1 \times S_2 \to S_1$ is continuous.

12. Give an example showing that the Pasting Lemma does not hold if A and B need not be closed.

13. Let $f_1 : S \to S_1$ and $f_2 : S \to S_2$ be continuous functions. Show that $h : S \to S_1 \times S_2$ given by

$$h(s) = (f_1(s), f_2(s))$$

is continuous, where $S_1 \times S_2$ has the product topology.

14. If S_1 and S_2 are topological spaces and $f : S_1 \to S_2$ is a bijection with both f and f^{-1} continuous, then f is called a *homeomorphism*.

   (a) Find a homeomorphism $f : S \to S$, with $f(1) = 3$, where $S = \{1, 2, 3\}$ where the domain has topology
   $$\{\varnothing, \{1\}, \{2\}, \{1, 2\}, \{2, 3\}, \{1, 2, 3\}\}$$
   and the co-domain has topology
   $$\{\varnothing, \{2\}, \{3\}, \{1, 2\}, \{2, 3\}, \{1, 2, 3\}\}.$$

   (b) Prove that the identity function is a homeomorphism.

   (c) Prove that if $f : S_1 \to S_2$ is a homeomorphism, then its inverse $f^{-1} : S_2 \to S_1$ is a homeomorphism.

   (d) Prove that the composition of homeomorphisms is a homeomorphism.

Appendix A: Properties of Real Number System

The basic axioms and properties of the real number system are assumed as prerequisite knowledge for this text. They are used in proofs in numerous sections and typically they are implemented without any sort of justification. This is actually good practice. We aim to write polished proofs that are not overburdened with too much information. Including justification for each and every use of an axiom of the real numbers will only serve to make a proof seem jumbled. Know your audience and trust that they can follow the basic laws of the real numbers, including those presented below.

Real Number Axioms

The real number axioms (commonly called *laws*) presented below fall into two categories. The first involves two operations: *addition* (denoted $a + b$) and *multiplication* (denoted ab). These rules establish guidelines for basic arithmetic, and because of this, they are known as **algebraic axioms** for the real numbers. Think of them as defining how equality works in the real number system.

The second set of laws establish an ordering for the real numbers, and hence are called **order axioms** for the real numbers. They establish a system to demonstrate how inequality works in the real number system.

Algebraic Axioms

- *Closure laws*: If $x, y \in \mathbb{R}$, then

  1. $x + y \in \mathbb{R}$, and
  2. $xy \in \mathbb{R}$.

- *Commutative laws*: If $x, y \in \mathbb{R}$, then

  1. $x + y = y + x$, and
  2. $xy = yx$.

- *Associative laws*: If $x, y, z \in \mathbb{R}$, then

  1. $(x + y) + z = x + (y + z)$, and

2. $(xy)z = x(yz)$.

- *Distributive laws*: If $x, y, z \in \mathbb{R}$, then

    1. $x(y + z) = xy + xz$, and
    2. $(x + y)z = xz + yz$.

- *Identity laws*: If $x \in \mathbb{R}$, then

    1. $0 + x = x = x + 0$, and
    2. $1x = x = x1$.

- *Inverse laws*: If $x \in \mathbb{R}$, then

    1. $x + (-x) = 0 = (-x) + x$, and
    2. if $x \neq 0$, then $x(\frac{1}{x}) = 1 = (\frac{1}{x})x$.

- *Cancellation laws*: If $x, y, z \in \mathbb{R}$, then

    1. if $x + y = x + z$, then $y = z$.
    2. if $xy = xz$ (for $x \neq 0$), then $y = z$.

Order Axioms

- *Translation invariance law*: If $x, y, z \in \mathbb{R}$ and $x < y$, then

$$x + z < y + z.$$

- *Transitivity of order*: If $x, y, z \in \mathbb{R}$ with $x < y$ and $y < z$, then

$$x < z.$$

- *Scaling of order law*: If $x, y, z \in \mathbb{R}$ with $x < y$ and $z > 0$, then

$$xz < yz.$$

- *Trichotomy law*: If $x, y \in \mathbb{R}$, then exactly one holds:

$$x < y, \ y < x, \text{ or } x = y.$$

Axioms of Integers

Various axioms or laws hold specifically for the integers, and as those with the real numbers, are often used in proofs throughout this text without mention.

- *Closure laws*: If $m, n \in \mathbb{Z}$, then $m + n \in \mathbb{Z}$ and $mn \in \mathbb{Z}$.

- *Archimedean law*: If $x \in \mathbb{R}$, then there exists $n \in \mathbb{Z}$ such that $x < n$.

Principle of Mathematical Induction

The proof technique known as *induction* is taken to be an axiom.[6]

A predicate $P(n)$ defined for $n \in \mathbb{Z}^+$ is proven true using the *Principle of Mathematical Induction* by showing

1. $P(1)$ is true, and
2. for integers $k \geq 1$, $P(k)$ being assumed true implies $P(k+1)$ is true.

---

[6]Because of the Principle of Mathematical Induction's equivalence with the Principle of Strong Mathematical Induction and the Well-Ordering Principle (see Secs. 4.2 and 4.3), we just as simply could have assumed one of these results as an axiom.

Appendix B: Proof Writing Tips

As with writing an expository paper, mathematical proofs can be written correctly but not *elegantly*. As you create more and more proofs, you will come to recognize not just what does *not* look right in a proof but also what makes the proof feel *better*. You will develop your likes and dislikes, your inclusions and exclusions and your structuring style. However, there are certain fundamental things, as discussed throughout this text, that add or detract from proofs. They are summarized below.

General Writing Tips

1. **Be consistent with your style choices**.

   Peruse some mathematical literature and you will find a multitude of proof styles. Some will be succinct and to-the-point, heavily utilizing mathematical symbols. Others will read with flow, making use of transitional language to connect thoughts. Regardless of how the author writes, her approach to the mathematics will be homogeneous throughout the paper (and in particular within every proof). The reader may never think, "There is consistent style here," but if the form changes, the reader will surely notice.

2. **Give your proof cohesion and flow by having it follow a natural train of thought**.

   A reader of a proof should never have to pause and think, "Where did *that* come from?" When writing your proofs, maintain a logical progression of ideas. This is often achieved through drafting your proof and then rereading it as a cohesive whole.

3. **Use writing techniques, such as transitions and word choice, to strengthen your proof and guide your audience**.

   A mathematical proof is an actual piece of writing. Because of this, the strategic use of transitional language, varied word choice, clause placement and other writing techniques will not only make your proof more elegant but will allow the reader to follow the progression of the proof. Be careful though. As with many things, too much of a good thing can be harmful. Being overly creative, metaphoric or unnecessarily flowery in your language can detract from your proof. Find the perfect balance for *your* style.

4. **Where necessary, describe what the goal of an upcoming section is or what organizational approach the proof will take.**

   If there is a rather long or complicated section of a proof, it is good practice to inform your reader what the purpose of that section is. Think of it as a roadmap. Rather than leave your reader wondering why you are doing a particular step, you tell them: "Next, we will show this because ... "

5. **Use correct grammar.**

   Mathematical writing is academic writing. As such, it must be grammatically correct. Complete, grammatically correct sentences must be used.

6. **Do not incorrectly use mathematical notation in place of written words.**

   How a symbol is pronounced does not necessarily allow that symbol to serve as shorthand for those written words. For example, the symbol $\mathbb{R}$ cannot always be used in place of "real numbers" and $\wedge$ cannot always substitute for "and." A poor sentence would be,

   $$\text{If } x \text{ is a } \mathbb{R}, \text{ then } x^2 \geq 0.$$

   But, it is appropriate to use the symbol in its set-theoretic capacity, such as

   $$\text{If } x \in \mathbb{R}, \text{ then } x^2 \geq 0.$$

   The same can be said about a multitude of mathematical symbols: logical connectives, operators, relations and others.

7. **Use white-space, linebreaks and alignment to improve a proof's presentation.**

   Utilizing a page properly (spacing, centering, justifying, etc.) can drastically improve the overall aesthetic appeal of a proof. Additionally, if a reader must return to parts of the proof to recall certain things, it is much easier to do so when it is properly spaced.

8. **Be clear and precise.**

   "Let x be a number." What type of number is x? An integer? A real number? Do not leave your readers wondering. Precision in mathematics is crucial.

9. **Do not be repetitive.**

   It is enough to state something once in a proof. Once it is stated, assume your reader will be aware of it throughout the entire proof.

Writing Mathematics

Due to the nature of the subject, there are certain things particular to writing mathematics that deserve mentioning.

1. **Use proper and uniform terminology for organizing results.**

    Words such as *conjecture, lemma, theorem* and *corollary* provide a hierarchy for results. Use them strategically and consistently throughout your writing.

2. **Proofs must be self contained. Precisely define each and every object appearing in your proof.**

    Mathematical proofs must be complete. Variables or mathematical objects inside the proofs must be properly defined within the proof itself; assumptions made outside the proof do not carry over into the proof.

3. **Know your audience.**

    Writing a proof for a beginning mathematician is much different from writing a proof for a student who has taken multiple upper-level mathematics courses. An algebraically strong audience can follow work that leaves out numerous algebra steps, but some readers need all of those steps to be present. Identify your intended audience and make sure the proof addresses their needs.

4. **Do not reuse a specific variable within its scope in a proof.**

    Once a variable is defined in a proof, that variable cannot be redefined. However, outside of the scope of that variable, it is not bad to redefine the variable (and sometimes it is actually good to reuse the variable). For example, in a proof by cases, a variable x defined within Case 1 is not defined outside of Case 1. However, within Case 1, if there were subcases, x would be as originally defined within each and every subcase.

    If the cases were dependent upon some property of x (such as Case 1: x is even, and, Case 2: x is odd), it is appropriate and good practice to reuse and redefine x in subsequent cases. Why? Because x plays the same role in every case of the proof, and this makes the proof easier to follow.

5. **Be consistent with what style of variables represent what type of mathematical object.**

    Making sure the same type of variable is used to represent similar mathematical objects is important. For example, if capital letters from the beginning of the English alphabet are being used to represent sets, a reader will automatically know, even if lost in the proof, "This thing B must be a set."

6. **When variables within a proof are dependent upon one another, define them in an appropriate order.**

   A phrase such as, "Let $x \in \mathbb{R}$," fixes x an arbitrary real number. If x is dependent upon another variable, introduce x at the appropriate moment using appropriate language. For example, saying, "Let $x, n \in mathbbZ$ with x even. Then, $x = 2n$," is incorrect. Both variables become fixed after the first sentence and it need not be the case that $x = 2n$.

7. **Appropriately justify your steps.**

   Readers should not scratch their heads and wonder how certain conclusions came to be. Justify steps appropriately by citing previous results and explaining complicated methods. As you develop more mathematical machinery, this tip becomes intimately related to knowing your audience. As you advance in your proof writing, you will find yourself justifying fewer of the simple or straightforward claims. If you find yourself in doubt about whether or not you should justify, however, it is best to include even just a brief explanation of why a step holds.

8. **Do not reinvent the wheel within a proof. If additional cases of a proof are proven identically to something already proven, use that previous part of the proof to streamline your work.**

   Many times in a proof there are multiple cases to consider and/or prove. If the proofs of the cases are identical, except for the choice of variable, state that these additional cases are proven *mutatis mutandis*. Be careful, however. If the additional cases have added assumptions or techniques in their proofs, they must be included.

9. **Be consistent with notational norms.**

   Be familiar with what symbols are typically used to represent particular things throughout books and articles in the area of math you are working on. For example, uppercase letters from the beginning of the Roman alphabet $(A, B, C, \dots)$ typically represent sets. Lowercase Greek letters $(\alpha, \beta, \theta, \dots)$ often represent angle measures. The lowercase Roman letters (i, j, m, n) typically play the role of index variables. Sticking with normal notation will help guide your reader.

10. **State when you are using an indirect proof technique.**

    You want your readers to know the direction of your proofs. If an indirect proof technique is being used, state so early on in your proof.

11. **Indicate the cases in a proof by exhaustion and present them clearly.**

Labelling cases is important. It denotes added hypotheses for that portion of the proof and assists the reader in following how the proof is organized. While using linebreaks is particularly helpful in this respect, the approach for denoting the cases is flexible. Section headings ("Case 1: $x < 0$") or transitional language ("Now, suppose $x < 0$.") are two possible approaches. Whichever is used, being consistent throughout the proof is a necessity.

12. **Using biconditional results, rather than two conditional results, can make a proof more efficient**.

    Some of the tools in your mathematical toolbox, such as various lemmas, theorems and definitions, are biconditional statements. Because of this, they can expedite arguments within a proof by replacing two justifications of conditional statements with one justification of a biconditional statement. If they can be used without hurting the integrity of your proof, do so to create proofs that are less redundant.

13. **When possible, make counterexamples specific**.

    Counterexamples that disprove universal statements ought to be specific elements of the statement's domain that make the predicate false. Using variables in place of specific domain elements is vague and requires your reader to connect unnecessary dots. Additionally, using variables to claim a predicate is false means that you are making a claim that the predicate is false for *all* possible variable assignments. This is often not the case.

14. **"Without loss of generality" can be used in a proof when particular cases of the proof can be applied and generalized to all others**.

    Use "WLOG" when making an assumption in a proof that narrows the proof to a special case without introducing additional hypotheses to the problem. That particular assumption, proof, and result can be applied trivially to all additional cases.

15. **Strive to keep readers focused on the result when proving that "the following are equivalent:" by organizing the proof in a natural way or informing the readers how the proof will proceed**.

    Prove the statements in the order they are presented or tell the reader the order in which you will prove them. This keeps the reader focused on the *result* and not confused about the arrangement of the proof.

16. **Eliminate unnecessary work by attempting to prove a list of statements are equivalent using the fewest number of implications possible**.

Proving $A \Leftrightarrow B$, $B \Leftrightarrow C$, $C \Leftrightarrow D$ can be done by proving six conditional statements, but it possible to prove using just four conditional statements: $A \Rightarrow B$, $B \Rightarrow C$, $C \Rightarrow D$, and $D \Rightarrow A$. Prove such claims as efficiently as possible.

17. **When proving via induction, be clear when you state and use the inductive hypothesis**.

    State your inductive hypothesis fully and clearly, and declare when it is used in the inductive step.

18. **Sometimes less is more.**

    Finding the proper balance between too little and too much explanation is important in creating elegant proofs. You want your readers to follow your train of thought. You do not want them to pause and think, "Of course that's what that means; that was unnecessary to have in the proof."

19. **Stay keen to what types of mathematical objects you are working with**.

    Knowing what types of mathematical objects are being dealt with may help you determine what tools may be useful in deriving a proof.

20. **Unnecessary variables congest a proof.**

    If variables are not necessary, do not include them. Readers will become confused when trying to keep straight an overabundance of variables.

21. **Do not use symbols to start sentences.**

    A sentences should start with a word rather than a mathematical symbol or number. Transitional language often makes this easy. The sentence,

    "x and y are nonzero integers,"

    is poorly written. However,

    "The terms x and y are nonzero integers,"

    is properly presented. Similarly, expressions such as $\ln(x)$ and $\tan(x)$ are considered mathematical symbols and should not be the first components of sentences:

    "$\ln(x)$ is an increasing function,"

    is poor writing, whereas

"The function $\ln(x)$ is increasing,"

is very well written.

22. **Write mathematics in an expository style.**

Beginning proof writers often resort to an approach seen in high school geometry or in logical deductions. Each step of a proof is given its own line or bullet point. This seems somewhat natural; proofs are very systematic and follow a train of thought: "from that I know *this*, then that gives me *this*, from which I know *this* ..." Bullets or line-by-line approaches are a great way to structure the inquiry process. They keep your thoughts organized. The final product, however, should be written in paragraph form.

23. **Place mathematical expressions appropriately.**

Longer expressions should be placed on their own lines, while shorter expressions can be placed in line with text. While there is no precise cut-off on what constitutes a longer expression, use your best judgment. Similarly, equations require their own line. Consider the following:

"The equation $x^2 + 5x + 10 = 0$ has no real solutions,"

is better presented as

"There are no real solutions to the equation
$$x^2 + 5x + 10 = 0.\text{"}$$

24. **Do not over- or under-use mathematical symbols.**

Too many mathematical symbols can make reading a mathematical proof quite difficult. Similarly, not using symbols when available can make for awkward writing. The following two examples exhibit this.

"There are many real number solutions to the equation x^2 plus y^2 equals z^2."

"Because $\int_1^3 f(x)\,dx = 4$, $.5\int_1^2 f(x)\,dx + .5\int_2^3 f(x)\,dx = 4.$"

25. **Keep scratch work and examples out of proofs.**

The proof is the final product of much work. It is the result of mistakes, observing examples and numerous drafts. That scratch work should never find its way into a proof. Additionally, statements of theorems or definitions should not be included within a proof. For example,

"Recall that $m \in \mathbb{Z}$ is even if $m = 2d$ for some $d \in \mathbb{Z}$,"

should never appear in a proof. It is part of the scratch work process. Referencing definitions or theorems, as justification for steps, as below, is appropriate.

"Thus, we have $m = 2a^3 + 1$, and since, by closure, $a^3 \in \mathbb{Z}$, we have shown, by Definition 3.7, that m is odd."

Appendix C: Selected Solutions and Hints

Chapter 1: Symbolic Logic

Section 1.1

1. Statements: (a), (c), (e), (f), (g)

3(a)i. $P \wedge Q$

3(a)iii. $(P \wedge Q) \wedge R$

3(b)ii. I didn't stay up past midnight or I didn't pass my exam.

5. $\sim(P \wedge Q)$: It is not true that the pen is out of ink and that I must complete this assignment by tomorrow morning.

$\sim P \vee \sim Q$: The pen is not out of ink, or, I do not need to complete this assignment by tomorrow morning.

6b. E: 7 is even; P: 7 is prime. (a): $P \vee Q$; (b) $\sim P \wedge \sim Q$; (c) 7 is not even and 7 is not prime.

10. Truth table:

P	$P \vee \sim(P \vee \sim P)$
T	T
F	F

11. Truth table:

P	Q	$Q \wedge (P \vee Q)$	$\sim P \vee \sim Q \vee (P \wedge Q)$
T	T	T	T
T	F	F	T
F	T	T	T
F	F	F	T

13. (a) neither; (b) tautology; (c) neither

15. neither associative $((P \times Q) \times R \not\equiv P \times (Q \times R))$ nor commutative $(P \times Q \not\equiv Q \times P)$

17. Show the logical statements have identical truth tables.

19. Show the logical statements have identical truth tables.

21. Show the logical statements have identical truth tables.

23. Logically equivalent statements: (c)

Section 1.2

1a. If your mother does not have to work, then we will go on vacation.

1c. If it snows more than 12 inches, then classes are canceled.

1e. If it thunders, then Linda's cat Rudy hides.

2b. If the plants are to survive, then they must be given water.

2d. If you take a direct flight from Chicago to St. Louis, then you will have traveled from Chicago to St. Louis.

3a. If I hear this song, then I remember my first kiss.

3d. If the video game is reset, then those buttons were hit simultaneously.

5. True statements: (a), (b), (c), (d), (f)

7a. Statement: If Harper apologizes, then she caused the accident.

Converse: If Harper caused the accident, then she apologized.

Contrapositive: If Harper did not cause the accident, then she will not apologize.

7c. Statement: If Brad watches this movie, then he cries.

Converse: If Brad cries, then he watched the movie.

Contrapositive: If Brad does not cry, then he did not watch the movie.

8. Truth table:

P	Q	R	$(P \Rightarrow Q) \Rightarrow R$	$P \Rightarrow (Q \Rightarrow R)$
T	T	T	T	T
T	T	F	F	F
T	F	T	T	T
T	F	F	T	T
F	T	T	T	T
F	T	F	F	T
F	F	T	T	T
F	F	F	F	T

9. Truth table:

P	Q	R	$(P \Leftrightarrow Q) \Leftrightarrow R$	$(P \Leftrightarrow R) \wedge (P \Leftrightarrow Q)$
T	T	T	T	T
T	T	F	F	F
T	F	T	F	F
T	F	F	T	T
F	T	T	F	T
F	T	F	T	F
F	F	T	T	F
F	F	F	F	T

11. $(P \vee \sim P) \Rightarrow (Q \wedge \sim Q)$

13. $(P \Leftrightarrow \sim P$

15a. The answer is negative and I was correct.

15c. Campus officials will attend when they are not invited, or, campus officials will be invited and they will not attend.

16b. $(P \wedge Q) \wedge (\sim Q \wedge \sim R)$

16c. $(\sim R \wedge P) \wedge (P \wedge \sim Q)$

18. Can only determine that (c) is true

20a. Neither

20c. Neither
22. Yes

Section 1.3
1a. Valid
1c. Valid
1e. Valid
2b. Invalid
2d. Valid
2f. Invalid
3b. Specialization
3d. Modus Tollens
3f. Cases
4b. Transitivity: If Flight 1345 leaves Seattle on time, then Sandra and I will do lunch together.
4d. Cases: We will go to the water park.
5a. Converse error
5c. Inverse error
7. Build a truth table.
9. Consider the definition of a valid argument.
11. Valid, since the conclusion is a tautology.

Section 1.4 **1a**. 4. Specialization 1; 5. Cases 2, 3, 4
1c. 6. Specialization 2; 7. Modus Ponens 3, 5; 8. Elimination 6, 7; 9. Theorem 1.17 1; 10. Specialization 9; 11. Modus Tollens 8, 10; 12. Conjunction 4, 11
2b.

1. $P \wedge (Q \wedge R)$	Assumption
2. $(Q \wedge R) \Rightarrow S$	Assumption
3. $Q \wedge R$	Specialization 1
4. S	Modus Ponens 2, 3

2d.

1. $(P \vee Q) \vee R$	Assumption
2. $\sim Q \wedge \sim R$	Assumption
3. $P \vee (Q \vee R)$	Associativity 1
4. $\sim(Q \vee R)$	Theorem 1.1 2
5. P	Elimination 3, 4

3b.

1. $R \wedge \sim Q$	Assumption
2. $P \Rightarrow Q$	Assumption
3. $S \Rightarrow \sim(R \wedge \sim P)$	Assumption
4. $\sim Q$	Specialization 1
5. $\sim P$	Modus Tollens 2, 4
6. R	Specialization 1
7. $R \wedge \sim P$	Conjunction 5, 6
8. $\sim(\sim(R \wedge \sim P))$	Double Negative 7
9. S	Modus Tollens 3, 8

3d.

1. $\sim P \vee \sim Q$	Assumption
2. $\sim T \Leftrightarrow (P \wedge Q)$	Assumption
3. $\sim T \vee S$	Assumption
4. $S \Rightarrow (W \wedge U)$	Assumption
5. $\sim(P \wedge Q)$	Theorem 1.1 1
6. $(\sim T \Rightarrow (P \wedge Q)) \wedge ((P \wedge Q) \Rightarrow \sim T)$	Theorem 1.17 2
7. $\sim T \Rightarrow (P \wedge Q)$	Specialization 6
8. $\sim(\sim T)$	Modus Tollens 5, 7
9. S	Elimination 3, 8
10. $W \wedge U$	Modus Ponens 4, 9
11. U	Specialization 10

5a. N: We left at noon. S: We arrived before the sun set.

$$N \Rightarrow \sim S$$
$$\underline{S \hphantom{xxxxx}}$$
$$\therefore \sim N$$

1. $N \Rightarrow \sim S$	Assumption
2. S	Assumption
3. $\sim(\sim S)$	Double Negative 2
4. $\sim N$	Modus Tollens 1, 3

5d. P: Paul is captain. Q: Quinn is alternate captain. R: Rory is alternate captain. T: Trish is in the starting lineup. W: Will is injured.

$$P \wedge (Q \wedge R)$$
$$T \Rightarrow \sim Q$$
$$\underline{W \Rightarrow T \hphantom{xx}}$$
$$\therefore \sim W$$

1. $P \wedge (Q \wedge R)$	Assumption
2. $T \Rightarrow \sim Q$	Assumption
3. $W \Rightarrow T$	Assumption
4. $Q \wedge R$	Specialization 1
5. Q	Specialization 4
6. $\sim(\sim Q)$	Double Negative 5
7. $\sim T$	Modus Tollens 2, 6
8. $\sim W$	Modus Tollens 3, 7

6b. P: A poem is popular among people with real taste. M: A poem is modern. I: A poem is interesting. Y: A poem is written by you. A: A poem has affection. S: A poem is on the subject of soap bubbles.

$$\sim P \Rightarrow \sim I$$
$$\sim A \Rightarrow \sim M$$
$$Y \Rightarrow S$$
$$P \Rightarrow \sim A$$
$$\underline{\sim M \Rightarrow S}$$
$$Y \Rightarrow \sim I.$$

Chapter 2: Sets

Section 2.1 **1a.** $\{1,2,3,4,5,6\}$

1c. $\{1,2,3,4\}$

2a. $\{x \in \mathbb{Z}^+ \mid 1 \leq x \leq 9 \text{ and } x = 2n+1 \text{ for some } n \in \mathbb{Z}\}$

2c. $\{3n\pi \mid n \in \mathbb{Z}^+ \text{ and } 1 \leq n \leq \pi\}$

3a. 1

3c. 2

3e. 2

5. True; (a), (c), (f), (g), (j), (l)

7. $\{x \in \mathbb{Z}^+ \mid x < 0\}$; $\{x \in \mathbb{R} \mid x^2 + 1 < 0\}$; $\{x \in \mathbb{Z} \mid x \neq \frac{x}{1}\}$

9a. $\varnothing$, $\{2\}$, $\{8\}$, $\{12\}$, $\{2,8\}$, $\{2,12\}$, $\{8,12\}$

9b. $\varnothing$

10b. $3 \in A$ but $3 \notin B$

10c. $5 \in A$ but $5 \notin B$

11. $\{x \in \mathbb{Z} \mid x = 4n \text{ for some } n \in \mathbb{Z}\}$; closed under both operations

13. False; $A = \{a\}$

15a. $A_1 = \{1\}$; $A_2 = \{2,4,8\}$; $A_{-3} = \{-3,9,-27\}$

16a. $D_2 = \{\ldots -4,-2,0,2,4,\ldots\}$; $D_3 = \{\ldots -6,-3,0,3,6,\ldots\}$; $D_4 = \{\ldots -8,-4,0,4,8,\ldots\}$

16c. D_n is the set of all integer multiples of n

17. $D \subseteq C$ but $C \nsubseteq D$; consider $f(x) = |x|$

19a. $S_1 = \{S \mid S \subset \{\varnothing, \{-1,1,\{-1,1\}\}\}\} = \{\varnothing, \{-1,1,\{-1,1\}\}\}$; $S_\pi = \{\varnothing, \{-\pi,\pi,\{-\pi,\pi\}\}\}$

19c. $\varnothing$, $\{-i,i,\{-i,i\}\}$, $\{\varnothing, \{-i,i,\{-i,i\}\}\}$

Section 2.2 **3b.** $\{a, e\}$

3d. $\{b, c, d, f, g\}$

4a. $U = \{-2, -1, 0, 1, 2\}$

4c. $\{-2, -1, 0\}$

4e. $\varnothing$

5b. $\{(\varnothing, 4), (\varnothing, 5), (1, 4), (1, 5), (4, 4), (4, 5)\}$

5d. $\{(\varnothing, \varnothing), (\varnothing, 1), (\varnothing, 4), (1, \varnothing), (1, 1), (1, 4), (4, \varnothing), (4, 1), (4, 4)\}$

5f. $\{\varnothing, \{\varnothing\}, \{1\}, \{4\}, \{\varnothing, 1\}, \{\varnothing, 4\}, \{1, 4\}, \{\varnothing, 1, 4\}\}$

6b. $\{((a, b), d), ((a, c), d)\}$

6c. $\{\varnothing, \{(a, b)\}, \{(a, c)\}, \{(a, b), (a, c)\}\}$

7a. $\{(1, 2, i), (1, 3, i)\}$

8a. $A \cap B = \{2, 3\} \subseteq \{2, 3, 5, 7\} = A$

8c. $A \cup A^C = \{2, 3, 5, 7\} \cup \{1, 4, 6, 8, 9, 10\} = \{1, 2, \ldots, 9, 10\} = U$

9b. Take $A = \{1\}$.

9d. In $U = \{1, 2, 3\}$, let $A = \{1\}$, $B = \{2\}$

10a. Suppose $A \subseteq B \cap D$. To show $A \subseteq B$, take $a \in A$. By definition of subset, $a \in B \cap D$, so that $a \in B$, showing $A \subseteq B$.

11a. True

11c. True

12a. $(-1, 1)$

12d. $(-\infty, \infty)$

13a. (a) $[.5, 1.5]$; (b) $[-1, 3]$; (c) $\{1\}$; (d) $[-1, 3]$

14b. $\mathbb{Z}$

15a. Partitions $\mathbb{Z}^+$

15c. Does not partition $\mathbb{Z}^+$

15e. Does not partition $\mathbb{Z}^+$

18. Let $a, b \in S$. $\{\{a\}, S - \{a\}\}$ and $\{\{b\}, S - \{b\}\}$ are partitions of S.

20a. $\{-2, -1\}, \{-1, 0, 1\}, \{1, 2\}$

20c. $\{-2, -1\}, \{0\}, \{1, 2\}$

22a. $\mathbb{R}^-, \mathbb{R}^+$

22c. $\mathbb{R} \times \mathbb{R} \times \mathbb{R}^-, \mathbb{R} \times \mathbb{R} \times \mathbb{R}^+, \mathbb{R} \times \mathbb{R} \times \{0\}$

23b. $\{b, c, d, f, g, i, o, u\}$

23d(i). $\varnothing$

23d(ii). A

Section 2.3 **1a.** T

1c. F

2a. T

2c. T

2e. F

3a. T

3c. T

4b. (a) $\varnothing$; (b) $\{(-1 \pm \sqrt{13})/2\}$

5b. If P is the set of all people, then $\forall x \in P$, x is treated a particular way.

6b. False; $(0 + 1)(0 - 1) < 0$

6e. Vacuously true; domain set is empty

7a. True; $\frac{2}{0} \notin \mathbb{Q}$

7d. True; $f(x) = |x|$ is continuous but not differentiable

8a. T

8c. F

8e. T

9b. True: $D = \mathbb{Z}^+$; False: $D = \mathbb{Z}$

10a. If $x \in \mathbb{R}$, then $\sqrt{x^2} = |x|$.

10c. If $x \in \mathbb{R}$ and $x^2 > 4$, then $|x| > 2$.

11b. $\exists\, x \in D \mid x = a$.

11d. $\exists\, x \in \{a, b, c\} \mid x$ has Property A and x does not have Property B.

12a. $\forall x \in S,\ x \leq 0$.

12c. $\forall x \in \mathbb{Z}^+,\ x < 0$ or $1 < x$.

12e. $\forall x \in \mathbb{Q},\ P(x)$ and $x \geq 3$.

13b. $\exists\, x, y \in S \mid x \neq y$ and both $(P(x) \wedge Q(x))$ and $(P(y) \wedge Q(y))$, or, $\forall x \in S$, $\sim P(x) \vee \sim Q(x)$,

14a. $x = \frac{1}{2}$

14c. $x = -1$

15b. $\mathbb{Z} - \mathbb{R} = \varnothing$

15d. $\mathbb{Z} \subseteq \mathbb{Q}$

17. (c), (e), (f) must be true

Section 2.4 **1a.** F

1c. F

2b. T

3a. F

3c. F

4a. $\forall x \in P,\ \exists\, y \in I \mid Q(x, y)$.

4c. $\forall x \in P,\ Q(x, \text{chocolate})$.

4e. $\exists\, x \in P \mid \forall y \in I,\ \sim Q(x, y)$.

5b. $\forall x, y \in \mathbb{R}$ with $x < y$, $\exists\, z \in \mathbb{Q} \mid x < z < y$.

5c. P: the set of all people; $L(x)$: length of hair of person x; $\exists\, x \in P \mid \forall y \in P - \{x\},\ L(y) < L(x)$.

6b. There is some integer that is an additive identity and its product with any integer is itself.

7a. $\exists\, x \in D \mid \forall y \in E,\ \sim P(x, y)$.

7c. $\exists\, m \in \mathbb{R} \mid \exists\, n \in \mathbb{Q},\ \sim P(m) \wedge Q(n)$.

8b. P: the set of all professors; C: the set of all classes; $L(x, y)$: Professor x likes teaching class y

(a) $\exists\, c \in C \mid \forall p \in P,\ \sim L(p, c)$.

(b) $\forall c \in C,\ \exists\, p \in P$ such that $L(p, c)$.

(c) Every class has some professor that likes to teach it.

9b. $x = 1,\ y = -1$

10. Only (e) is true.

11. True: (a), (b), (d)

14a. Universal Modus Tollens
14c. Universal Converse Error
14e. Universal Modus Tollens
15b. True

Chapter 3: Introduction to Proofs

Section 3.1 **1b.** A number is odd if it equals 2 times "something" plus 1
2a. $2(21)$
2c. $2(8n^4 + 6n^3 - 10n + 4)$
3a. $2(64) + 1$
3c. $2(36m - 3) + 1$
4a. Even
4c. Even
5. 1, 2, 3, 4, 6, 8, 12 are proper; 24 is a non-proper divisor
6a. $1122 = 6(187)$
6c. $125p + 50q = 5(25p + 10q)$
6e. $6^m = 3(2 \cdot 6^{m-1})$
7a. 8π is not even.
7c. $7 - 4$ is odd but 7 is odd and 4 is even.
7e. Consider 6.
8a. $p = 2m$ implies $p + 6 = 2(m + 3)$
8c. $(2m + 1) + (2n + 1) + (2p + 1) = 2(m + n + p + 1) + 1$
8e. $(2m + 1) + (2n + 1) = 2(m + n + 1)$
9. The proof improperly presents the algebraic steps, explaining what is to be done rather than simply doing it. Some variables are undefined.
11. The variables are undefined and the derivation need not have $n + 7$ on the left side throughout. Present derivations as a single string of equalities.
12b. Because $x = 2$ is a solution to $14 - x = 2x^2 + 4$, the existential statement "$\exists\, x \in \mathbb{R} \mid 14 - x = 2x^2 + 2$" is true.
13a. $x = 0$
13c. $x = 1$
14a. $\varnothing$ has no elements.
14c. $0 \notin \mathbb{Z}^*$, so the antecedent of the implication is always false.
17b. 6 is perfect, 12 is abundant, and all others are deficient
17d. $1 < p$
18a. $4 = \pi \cdot \frac{4}{\pi}$; $e = 5 \cdot \frac{5}{e}$

Section 3.2 **1a.** Assume $x, y \in \mathbb{Z}$ with x even and y odd. Show $x + y$ is odd.
1c. Assume $x, p \in \mathbb{Z}$ with x composite and p prime. Show xp is composite.
2a. Let $x, y \in \mathbb{Z}$ such that $x = 2m$ for some $m \in \mathbb{Z}$. Then, $xy = 2(my)$.
2c. Let $x, y \in \mathbb{Z}$ such that $x = 2m$ and $y = 2n + 1$ for some $m, n \in \mathbb{Z}$. Then, $x - y = 2(m + n - 1) + 1$.
3a. Let $x \in \mathbb{Z}^*$. Then, $x = x(1)$.
3c. Suppose $x, y \in \mathbb{Z}$ with $x \mid y$. Then, $y = xm$ for some $m \in \mathbb{Z}$, and

$y^2 = x^2(m^2)$.

5a. ($\to$) n even implies $n = 2m$. Then, $n + 1 = 2m + 1$. ($\leftarrow$) $n + 1$ odd implies $n + 1 = 2m + 1$, so that $n = 2m$.

5c. Suppose $a = bq + r$. If $d|a$ and $d|b$, then $a = dm$ and $b = dn$. Thus, $d(m - nq) = r$. If $d|b$ and $d|r$, then $b = dm$ and $r = dn$. Thus, $a = d(mq + n)$.

7a. $(-1)^2 = 1$

7c. $2^2 - 1 = 3$

8a. $a|b$ implies $b = am$. Then, $bc = a(mc)$, showing $a|bc$.

8c. False; consider ± 1.

11. Variable choice is inconsistent.

13. Consider if $n - 2 = 1$.

Section 3.3 **1a.** Assume that $x \in A - B$. Show that $x \in A$.

1c. Assume that $x \in (A - B) - C$. Show $x \in A - C$.

2a. $A = \{1\}$, $B = \{1, 2\}$, $C = \{1, 3\}$

2c. $A = \{1\}$, $B = \{2, 3\}$, $C = \{3, 4\}$

3a. Suppose A and B are sets and let $x \in A$. Then, $x \in A \cup B$ since we can conclude that $x \in A$ or $x \in B$.

3c. Let A, B and D be sets and let $x \in A \cap (B \cup D)$. Then, by definition of set intersection, $x \in A$ and $x \in B \cup D$, or equivalently, $x \in A$ and $x \in B$, or, $x \in A$ and $x \in D$. Thus, $x \in (A \cap B) \cup (A \cap D)$, proving $A \cap (B \cup D) \subseteq (A \cap B) \cup (A \cap D)$. To prove $(A \cap B) \cup (A \cap D) \subseteq A \cap (B \cup D)$, take $x \in (A \cap B) \cup (A \cap D)$. Then, by definition of set union, $x \in A \cap B$ or $x \in A \cap D$, or equivalently, $x \in A$ and $x \in B$, or, $x \in A$ and $x \in D$. Consequently, $x \in A$ and $x \in B \cup D$, showing $x \in A \cap (B \cup D)$, proving the result.

4c. Let A, B and C be sets with $A \subseteq B$. To prove $A \cap C \subseteq B \cap C$, take $x \in A \cap C$. Then, $x \in A$ (and hence $x \in B$, since $A \subseteq B$) and $x \in C$, by definition of set intersection. By the same definition, we have that $x \in B \cap C$, proving the result.

5a. Prove $A \cup (A \cap B) \cup (A \cap C) \subseteq A$ and $A \subseteq A \cup (A \cap B) \cup (A \cap C)$. The latter holds, because if $x \in A$, then $x \in A$ or $x \in A \cap B$ or $x \in A \cap C$. To prove the first, if $x \in A \cup (A \cap B) \cup (A \cap C)$, it follows by the definition of set union that $x \in A$ or $x \in A \cap B$ or $x \in A \cap C$. In all cases, $x \in A$, proving $A \cup (A \cap B) \cup (A \cap C) \subseteq A$ and consequently the desired set equality.

5c. Use the definition of set difference, repeatedly, to prove $(A-C)-(B-C) \subseteq (A - B) - C$ and $(A - B) - C \subseteq (A - C) - (B - C)$.

6d. Let A and B be sets with $A \subseteq B$. Take $S \in \mathcal{P}(A)$. Then, $S \subseteq A$. By Theorem 2.8, $S \subseteq B$, proving $\mathcal{P}(A) \subseteq \mathcal{P}(B)$.

6e. $A \subseteq A$, so $A \in \mathcal{P}(A)$.

6f. Suppose $\mathcal{P}(A) \subseteq \mathcal{P}(B)$. Since $A \in \mathcal{P}(A)$, so $A \in \mathcal{P}(B)$. Thus, $A \subseteq B$.

9. An additional step is needed from "$x \in A \cap B$ and $x \in D$" to "$x \in A$ and $x \in B \cap D$."

10. Which sets are A, B or C? How do they disprove the claim?

12. Does not hold.

13a. If $x \in B_6$, then $x = 6n = 2(3n)$.

15. To prove $(\mathbb{Z} \times \mathbb{Z}^+) \cap (\mathbb{Z}^+ \times \mathbb{Z}) \subseteq \mathbb{Z}^+ \times \mathbb{Z}^+$, take $(x, y) \in (\mathbb{Z} \times \mathbb{Z}^+) \cap (\mathbb{Z}^+ \times \mathbb{Z})$. Since $(x, y) \in \mathbb{Z} \times \mathbb{Z}^+$, it must be that $y \in \mathbb{Z}^+$. Likewise, $(x, y) \in \mathbb{Z}^+ \times \mathbb{Z}$ yields $x \in \mathbb{Z}^+$, showing $(x, y) \in \mathbb{Z}^+ \times \mathbb{Z}^+$.

Section 3.4 **1a.** Assume that $3|a$ or $3|b$. Show that $3|ab$.
1c. Assume that $p > 2$ is even. Show that p is not prime.
2a. Assume $m \in \mathbb{Z}$ and $5m + 4$ is odd. Assume for contradiction that m is even.
2c. Assume that there do not exist $x, y \in \mathbb{Z}^+$ such that $x^2 - y^2 = 1$.
5a. Suppose x is even. Then, $x = 2m$, so that $x^2 = 2(2m^2)$.
5c. Suppose $a < 0$. Then, $a - 1 < 0$ and since $a^2 > 0$, $a^3 - a^2 = a^2(a - 1) < 0$.
5e. Suppose x and y have different parity. If $x = 2m$ and $y = 2n + 1$, for some $m, n \in \mathbb{Z}$, then $x + y = 2(m + n) + 1$, so that $x + y$ is odd.
6b. Let $n|m$ and $n|(m + 1)$. Then, $m = na$ and $m + 1 = nb$ for $a, b \in \mathbb{Z}$. Then, $(m + 1) - m = n(b - a)$, showing $n|1$, a contradiction, since $n \geq 2$.
6d. Suppose $a \nmid b$ and $a|c$ and for contradiction that $a|(b + c)$. Then, $c = am$ and $b + c = an$ for some $m, n \in \mathbb{Z}$. This gives $b = (b + c) - c = a(n - m)$, so that $a|b$, a contradiction.
9b. Suppose $x \in (A - C) \cap (B - C)$. Then, $x \in A$, $x \in B$ and $x \notin C$, or, $x \in A \cap B$ and $x \notin C$. But, $A \cap B \subseteq C$, meaning $x \in C$, a contradiction.
10a. By contrapositive, suppose a is odd, so that $a = 2m + 1$ for some $m \in \mathbb{Z}$. Then, $a^3 = 2(4m^3 + 6m^2 + 3m) + 1$, which is odd.
11b. Mimic the proof of Theorem 3.30, using part (a).
12b. Suppose $a > b$ and use the fact that $a^2 + b^2 > 0$ to show $a^3 + ab^2 > b^3 + a^2b$.
12d. For contradiction, assume $a^2 - b^2 = c^2$ has such a solution. Note that c being prime means the only factors of c^2 are 1, c and c^2.
13b. Prove directly, using the definition of a rational number.
13d. Use a proof by contradiction, contradicting that $\sqrt{2} \notin \mathbb{Q}$.
13f. Prove directly, constructing the additive inverse.
14b. $2 = \sqrt{2} \cdot \sqrt{2}$
15. Mirror the proof of Theorem 3.30, arriving at the same contradiction.

Section 3.5 **1a.** Quotient: 2; remainder: 5
1c. Quotient: -20; remainder 6
2b. Must consider all cases guaranteed by the Quotient-Remainder Theorem, not just the three listed
3a. $n = 0, 1, 2$
3c. $n = 2k$, $n = 2k + 1$
3f. $n = 4k$, $n = 4k + 1$, $n = 4k + 2$, $n = 4k + 3$, though the cases $n = 2k$, $n = 2k + 1$ suffice
4. (b) Consider the cases $n = 3k$, $n = 3k + 1$ and $n = 3k + 2$, for some $k \in \mathbb{Z}$. If $n = 3k$, then $3|n$. If $n = 3k + 1$, then $3|(n + 2)$. If $n = 3k + 2$, then $3|(n + 1)$.
4. (e) Consider the cases $n = 2k$ and $n = 2k + 1$, for some $k \in \mathbb{Z}$. If $n = 2k$, then $n^3 + n = 2(4k^3 + k)$. If $n = 2k + 1$, then $n^3 + n = 2(4k^3 + 6k^2 + 4k + 1)$.
5a. Consider the 4 cases as in Theorem 3.21.

5b. Use Theorem 3.40.

8a. Suppose $n = 2m$. Consider the cases of m being even ($m = 2k$) and m being odd ($m = 2k + 1$).

10a. If n is odd, then $n = 4k + 1$ or $n = 4k + 3$ for some $k \in \mathbb{Z}$.

10c. Consider $n(n + 1)$ under 3 cases: $n = 3k$, $n = 3k + 1$, $n = 3k + 2$.

10e. Consider the seven cases guaranteed by the Quotient-Remainder Theorem upon division by 7.

11b. Consider $n(n + 1))(n + 2)(n + 3)$ under 4 cases: $n = 4k$, $n = 4k + 1$, $n = 4k + 2$, $n = 4k + 3$.

12b. Consider $n = 3k$, $n = 3k + 1$, $n = 3k + 2$.

Chapter 4: Mathematical Induction

Section 4.1 **1a.** Base case ($n = 0$): $1 = \frac{3^1 - 1}{2}$
Inductive hypothesis: Suppose for a general $k \in \mathbb{Z}$, $k \geq 0$, that $1 + 3 + 3^2 + \ldots + 3^k = \frac{3^{k+1} - 1}{2}$.

1c. Base case ($n = 1$): $1^4 = \frac{1(1+1)(2 \cdot 1 + 1)(3 \cdot 1 + 3 \cdot 1 - 1)}{30}$
Inductive hypothesis: Suppose for a general $k \in \mathbb{Z}^+$ that $\sum_{i=1}^{k} i^4 = \frac{k(k+1)(2k+1)(3k^2 + 3k - 1)}{30}$.

1e. Base case ($n = 1$): $\frac{1}{1 \cdot 3} = \frac{1}{2 \cdot 1 + 1}$
Inductive hypothesis: Suppose for a general $k \in \mathbb{Z}^+$, $\frac{1}{1 \cdot 3} + \frac{1}{3 \cdot 5} + \frac{1}{5 \cdot 7} + \ldots + \frac{1}{(2k-1) \cdot (2k+1)} = \frac{k}{2k+1}$

2a. Base case ($n = 1$): $1 = 1^2$
Inductive hypothesis: Suppose $1 + 3 + \ldots + (2k - 1) = k^2$ for a general $k \in \mathbb{Z}^+$.
Inductive step: $1 + 3 + \ldots + (2k - 1) + (2(k + 1) - 1) = k^2 + (2k + 1)$ (IH) $= (k + 1)^2$.

2c. Base case ($k = 1$): $1 = \frac{1(3-1)}{2}$
Inductive hypothesis: Suppose for a general $n \in \mathbb{Z}^+$ that $1 + 4 + 7 + \ldots + (3n - 2) = \frac{n(3n-1)}{2}$.
Inductive step: $1 + 4 + \ldots + (3n - 2) + (3(n + 1) - 2) = \frac{n(3n-1)}{2} + (3n + 1)$ (IH) $= \frac{3n^2 + 5n + 2}{2} = \frac{(n+1)(3n+2)}{2}$.

2f. Base case ($n = 1$): $1(1 + 1) = \frac{1(1+1)(1+2)}{3}$
Inductive hypothesis: Suppose for a general $k \in \mathbb{Z}^+$ that $\sum_{i=1}^{k} i(i + 1) = \frac{k(k+1)(k+2)}{3}$.
Inductive step: $\sum_{i=1}^{n+1} i(i + 1) = \frac{k(k+1)(k+2)}{3} + (k + 1)(k + 2)$ (IH) $= \frac{(k+1)(k+2)(k+3)}{3}$.

4a. Base case ($n = 1$): $5 | (6^1 - 1)$
Inductive hypothesis: Assume for a general $k \in \mathbb{Z}^+$ that $5 | (6^k - 1)$.
Inductive step: $6^{k+1} - 1 = 5 \cdot 6^{k+1} + (6^k - 1)$, which is divisible by 5 by Exercise 3.2.6 and the inductive hypothesis.

4c. Prove $3 | 4^n - 1$.

4e. Base case ($n = 0$): $5 | 0$

Appendix C: Selected Solutions and Hints

Inductive hypothesis: Assume $5|(2^{3k} - 3^k)$ for a general integer $k \geq 0$.
Inductive step: $2^{3(k+1)} - 3^{k+1} = 5 \cdot 2^{3k} + 3(2^{3k} - 3^k)$, and by the inductive hypothesis and Exercise 3.2.6, this is divisible by 5.
5b. Base case $(m = 4)$: $4! > 4^2$
Inductive hypothesis: Assume for $m \geq 4$, $m \in \mathbb{Z}$, that $m! > m^2$.
Inductive step: $(m + 1)! = (m + 1)m! > (m + 1)m^2$ (IH) $> (m + 1)(m + 1)$ (since $m \geq 4$).
5e. Base case $(m = 2)$: $4^2 + 7 < 5^2$
Inductive hypothesis: Assume for a general $m \geq 2$ that $4^m + 7 < 5^m$.
Inductive step: $4^{m+1}+7 = 3 \cdot 4^m+4^m+7 < 3 \cdot 4^m+5^m$ (IH) $< 4 \cdot 5^m+5^m = 5^{m+1}$.
6d. Base case $(n = 1)$: $\frac{1}{1\cdot4} = \frac{1}{3+1}$
Inductive hypothesis: Suppose for a general $k \in \mathbb{Z}^+$, $\frac{1}{1\cdot4} + \frac{1}{4\cdot7} + \frac{1}{7\cdot10} + \ldots + \frac{1}{(3k-2)\cdot(3k+1)} = \frac{k}{3k+1}$
Inductive step: $\frac{1}{1\cdot4} + \frac{1}{4\cdot7} + \ldots + \frac{1}{(3k-2)\cdot(3k+1)} + \frac{1}{(3(k+1)-2)\cdot(3(k+1)+1)} = \frac{k}{3k+1} + \frac{1}{(3k-1)\cdot(3k+4)}$ (IH) $= \frac{k+1}{3(k+1)+1}$.
8. Use the Triangle Inequality for both the base case and the inductive step.

Section 4.2 **1a.** 2, 3, 4, 5, 6
1c. 1, 1, 2, 4, 8
2a. b_0 is undefined
2c. Division by 0
3a. Suppose that b_k is odd for $k, n \in \mathbb{Z}$ with $1 \leq k \leq n$.
4a. No lower bound on k.
5a. Base cases $(n = 1, 2, 3)$: $1 \leq 2^1$; $1 \leq 2^2$; $3 \leq 2^3$
Inductive hypothesis: Suppose $a_k \leq 2^k$ for $n, k \in \mathbb{Z}$, $3 \leq k \leq n$.
Inductive step: $a_{n+1} = a_n + a_{n-1} + a_{n-2} \leq 2^{n-2}(2^2 + 2 + 1)$ (IH) $\leq 2^{n+1}$.
5e. Base cases $(n = 1, 2)$: $a_1 = 2(1) - 1$; $a_2 = 2(2) - 1$
Inductive hypothesis: Suppose $a_k = 2k - 1$ for $k, n \in \mathbb{Z}$ with $2 \leq k \leq n$.
Inductive step: $a_{n+1} = 2a_n - a_{n-1} = 2(2n - 1) - (2(n - 1) - 1)$ (IH) $= 2n - 1$
5f. Base cases $(n = 1, 2, 3)$: $0 < f_n \leq 1$ for $n = 1, 2, 3$
Inductive hypothesis: Suppose $0 < f_k \leq 1$ for $n, k \in \mathbb{Z}$, $3 \leq k \leq n$.
Inductive step: $f_{n+1} = f_n \cdot f_{n-2}$; $0 < f_n \leq 1$ and $0 \leq f_{n-2} \leq 1$ (IH), so $0 < f_n \cdot f_{n-2} \leq 1$.
6a. Base cases $(k = 0, 1)$: $F_0, F_1 \in \mathbb{Z}^+$
Inductive hypothesis: Assume $F_k \in \mathbb{Z}^+$ for $k, n \in \mathbb{Z}$, $1 \leq k \leq n$.
Inductive step: The inductive step holds via closure of $\mathbb{Z}^+$ under addition.
6c. Base cases $(n = 0, 1)$: $F_0^2 = 1 = F_0 F_1$; $F_0^2 + F_1^2 = 2 = F_1 F_2$
Inductive hypothesis: Suppose $\sum_{i=0}^k F_i^2 = F_k F_{k+1}$ for $k, n \in \mathbb{Z}$, $1 \leq k \leq n$.
Inductive step: $\sum_{i=0}^{n+1} F_i^2 = F_n F_{n+1} + F_{n+1}^2$ (IH) $= F_{n+1}(F_n + F_{n+1}) = F_{n+1}F_{n+2}$.
8a. Base 2: 10011; base 3: 201
8c. Base 2: 10000001; base 3: 11210
9b. (i) 11; (ii) 31
11. Mimic the proof of Theorem 4.13.

13. 36

15. $T_n = 2T_{n-1} + 2T_{n-2}$, $n \geq 2$, with $T_0 = 1$, $T_1 = 3$

17. Proof by strong induction, similar to Exercise 5a.

19. Use strong induction on a group of $n \geq 8$ people. Base cases consist of $n = 8, 9, 10, 11$ people. For the inductive step, create a group of 4 and consider the remaining people.

Section 4.3 **1a.** 21

1c. 1

3. $\binom{n-1}{r} + \binom{n-1}{r-1} = \frac{(n-1)!}{r!(n-1-r)!} + \frac{(n-1)!}{(r-1)!(n-r)!} =$
$(n-1)! \left(\frac{1}{r(r-1)!(n-r-1)!} + \frac{1}{(r-1)!(n-r)(n-r-1)!} \right)$

4a. $x^4 + 4x^3a + 6x^2a^2 + 4xa^3 + a^4$

4c. $1 + 3\sqrt{2} + 12 + 4\sqrt{2}$

5b. Prove via induction, using Pascal's Formula in the inductive step.

5d. Prove via induction, using Pascal's Formula in the inductive step.

6a. $\binom{n}{n-r} = \frac{n!}{(n-r)!(n-(n-r))!} = \binom{n}{r}$

6c. $0 = (1 + (-1))^n$; apply the Binomial Theorem

6e. $3^n = (2 + 1)^n$; apply the Binomial Theorem

7a. Apply Theorem 4.26.

8. If $a|b$ and a and b are relatively prime, then $a = 1$.

10a. Not a subset of $\mathbb{Z}^+$.

10c. Not nonempty.

10e. Not a subset of $\mathbb{Z}^+$.

11b. Consider $S = \{5^{2n} - 1 \mid n \in \mathbb{Z}^+, 8 \nmid (5^{2n} - 1)\}$. Suppose for contradiction that S is nonempty. Apply the Well-Ordering Principle and arrive at a contradiction to find an element of S smaller than that originally assumed to be the least element of S.

12a. 2

12b. 1

13b. Suppose c is a common divisor of x and y. Show cd is a common divisor of a and b.

13d. If $d|a$ and $d|(a + n)$, then $d|((a + n) - a)$.

13f. Suppose $d|a$ and $d|bc$ with $d > 1$. Let p be a prime divisor of d. Either $\gcd(a, b) \geq p$ or $\gcd(a, c) \geq p$, a contradiction.

15. $p_1^{4m_1} p_2^{4m_2} \cdots p_n^{4m_n}$

16. An integer ending in k 0s means it has k factors of 10; $10 = 2 \cdot 5$.

Chapter 5: Relations

Section 5.1 **1b.** $\varnothing$, $\{(1, m)\}$, $\{(2, m)\}$, $\{(1, m), (2, m)\}$

1d. $\varnothing$

2b. $\{(-1, 2), (1, 2)\}$; $\text{Dom}(S) = \{-1, 1\}$; $\text{Ran}(S) = \{2\}$

2d. $\{((3, 3), 0), ((4, 4), 0), ((5, 5), 0), ((6, 6), 0), ((6, 3), 3), ((7, 4), 3),$
$((8, 5), 3), ((9, 6), 3), ((0, 3), -3), ((1, 4), -3), ((2, 5), -3), ((3, 6), -3)\}$

3a. $\mathrm{Dom}(R) = \{1,3\}$; $\mathrm{Ran}(R) = \{2,3\}$

3c. $\mathrm{Dom}(f) = \{x \in \mathbb{R} \mid -2 < x < 2\}$; $\mathrm{Ran}(f) = \{y \in \mathbb{R} \mid 0 \leq y < 4\}$

6a. R from $\{1,2,3,4\}$ to $\mathbb{Z}$ given by xRy if and only if $y = 2x$.

6c. R from $\{0,1,2,\ldots,9\}$ to $\{0,1,2\}$ given by xRy if and only if y is the remainder when x is divided by 3.

7. $I_A \subseteq A \times A$. Its digraph consists only of loop at every vertex.

10. $(a,b) \in (R^{-1})^{-1}$ if and only if $(b,a) \in R^{-1}$; $(b,a) \in R^{-1}$ if and only if $(a,b) \in R$.

13. $\varnothing \subseteq A \times B$ for any sets A and B.

14. $A \times B$ has nm elements. $\mathcal{P}(A \times B)$ has 2^{nm} elements.

17a. Reflexive: (i), (v); irreflexive: (ii), (iv); symmetric: (i), (ii), (iv), (v); antisymmetric: none

18. Both claims are true. $A \times A \subseteq B \times B$ whenever $A \subseteq B$.

Section 5.2 **1a.** Transitive

1c. Symmetric, transitive

4a. Reflexive, not symmetric, transitive

4c. Reflexive, symmetric, transitive

4e. Reflexive, symmetric, transitive

5a. Not an equivalence relation

5d. Not an equivalence relation

6a. $[0] = \{0\}$; $[1] = \{-1,1\}$; $[2] = \{-\sqrt{2}, \sqrt{2}\}$

6c. $[3.1] = \{\ldots, -.9, .1, 1.1, 2.1, \ldots\}$; $[\pi] = \{\pi + n \mid n \in \mathbb{Z}\}$

6e. $[(2,3)]$ contains $(0,1), (-1,0), (2,1)$; $[(-1,4)]$ contains $(0,5), (-5,0), (2,7)$

8a. 100, 2100, 3149

8c. 10

9. $[(1,0)]$ is the set of all ordered pairs (x,y) satisfying the equation $x^2 + y^2 = 1$; that is, all points (x,y) lying on the unit circle; $[(2,1)]$ is the set of all ordered pairs (x,y) satisfying the equation $x^2 + y^2 = 5$; that is, all points (x,y) lying on the circle of radius $\sqrt{5}$ centered at the origin.

11a. R is reflexive, so aRa. Thus, $a \in [a]$.

12. There are 8 vertices corresponding to the 8 subsets of $\{1,2,3\}$. An edge exists between two vertices if and only if the subsets those vertices represent have the same number of elements.

14. Not an equivalence relation.

16b. $[x^3] = \{x^3 + c \mid c \in \mathbb{R}\}$; $[3x+2] = \{3x + c \mid c \in \mathbb{R}\}$; $[e^x] = \{e^x + c \mid c \in \mathbb{R}\}$

18. Interpretation of the symmetric property is incorrect.

Section 5.3 **2a.** All elements of the set

2c. b, g

2e. Yes

4a. $<$ is not reflexive on $\mathbb{R}$

4c. Not reflexive

5a. Divides is reflexive, antisymmetric and transitive.

5c. $\{1\}$, $\{x \mid x \in \mathbb{Z}^+, x \text{ is even }\}$

7. Consider all cases for $(a_1, b_1) \leq (a_2, b_2)$ and $(a_2, b_2) \leq (a_3, b_3)$.

9a. $\{(2, b) \mid b \geq 3\}$

9c. $\{(a, b) \mid a > 2, b \in \mathbb{Z}^+\}$

9e. $\{(a, 1) \mid a > 2\} \cup \{(a, 2) \mid a > 2\}$

11b. $(0, 1, 2) < (0, 2, 1)$

13. (a)

15a. -1

15c. Does not exist

15e. A

16. If (x_2, y_2) is straight above (x_1, y_1) or to the right of (x_1, y_1), then $(x_1, y_1) < (x_2, y_2)$.

17b. A set S is comparable to E if S contains only even integers, or, all even integers are elements of S (i.e., $E \subseteq S$).

20a. $R^{-1} \subseteq A \times A$

20b. (ii) $\{(S, T) \mid T \subseteq S$, where $S, T \subseteq \{a, b, c\}\}$

22. Suppose S has 2 greatest elements a and b. Each is a greatest element, so $a \leq b$ and $b \leq a$. By the antisymmetry property, $a = b$.

24a. $(\{2\}, \{a, b\})$, $(\{1, 2\}, \varnothing)$, $(\{1, 2\}, \{a\})$, $(\{1, 2\}, \{b\})$, $(\{1, 2\}, \{a, b\})$,

24c. $(\{1, 2\}, \{a, b\})$

Section 5.4 **1a.** $4 | (19 - 11)$; $19 = 4 \cdot 4 + 3$ and $11 = 2 \cdot 4 + 3$; $19 = 11 + 2 \cdot 4$

1c. $5 | (-15 - 0)$; $-15 = -3 \cdot 5$ and $0 = 0 \cdot 5$; $-15 = 0 + -3 \cdot 5$

2a. 7

2c. 0

2e. 5

2g. 0

3a. 5

3c. 1

4b. $0, 5, -5 \in [0]_5$; $1, 6, -4 \in [1]_5$; $2, 7, -3 \in [2]_5$; $3, 8, -2 \in [3]_5$; $4, 9, -1 \in [4]_5$

5a. $x = 4$

5c. $x = 1$

6. Notice that $168 = 2^3 \cdot 3 \cdot 7$ and $42 = 2 \cdot 3 \cdot 7$.

9a. $2 \equiv_3 8$

9c. $2 \equiv_3 8$ but $2 \not\equiv_9 8$

10b. Apply Theorem 5.38, as $\gcd(m, p) = 1$.

10d. $\gcd(a, m) = 1$ implies $ax + my = 1$ for some $x, y \in \mathbb{Z}$, so $ax \equiv_m 1$.

11b. 4

13a. Use the fact that $4 | 10^m$ if and only if $m \geq 2$.

13c. Consider the remainder when 10^m is dividing by 11; show it is either 1 or 10; $10 \equiv_{11} -1$.

14b. Divisible by 11.

16b. False

16d. True

17. Use the existence of multiplicative inverses modulo p for all positive integers less than p

19. Consider the Quotient-Remainder Theorem.

Chapter 6: Functions

Section 6.1 **1a.** Domain: $\mathbb{Z}$; range: $\mathbb{Z}$; possible co-domain: $\mathbb{Z}$, $\mathbb{R}$
1c. Domain: $\mathbb{Z}$; range: $\{0, 1, 2, \ldots, 16\}$; possible co-domain: $\{0, 1, 2, \ldots, 16\}$, $\mathbb{Z}$
1e. Domain: L; range: $\{x \in \mathbb{Z} \mid 0 \le x \le 26\}$; possible co-domain: $\{x \in \mathbb{Z} \mid 0 \le x \le 26\}$, $\mathbb{Z}$
2b. $g(-4)$ is undefined
2d. $0 \notin \mathbb{Z}^+$
4a. $g(x) = 1$ for all $x \in \mathbb{Z}$
4c. $g(x) = 1 + r$, where r is the remainder upon division by 3 of x
6a. $\{1, 2, 3, 4, 5\}$
6c. $\{2, 3, 4\}$
6e. $\{1, 2, 3, 4\}$
8a. Is a function
8c. Not a function
8e. Not a function
9b. $f(1) = 2$; $f(2) = 4$; $f(\{1, 2\}) = \{2, 4\}$; $f^{-1}(\{1, 2\}) = \{\pm\frac{1}{2}, \pm 1\}$
10a. $(4, 16]$
10c. $\{1, 2, 3, 4, 6\}$
11a. $\{5\}$
11c. $\{0, \pm 2, \pm 3, \pm 4\}$
12a. $g(-\pi) = 0$; $g(0) = \pi^2$; $g(\pi) = 4\pi^2$
12c. $g^{-1}(\{-\pi\}) = \varnothing$; $g^{-1}(\{0\}) = \{-\pi\}$; $g^{-1}(\{\pi\}) = \{-\pi \pm \sqrt{\pi}\}$
13a. e
13c. $\{\pi\}$
13e. $\mathbb{Z}^+ \times \mathbb{Z}^- \times \mathbb{R}$
13f. $\mathbb{Z}^+ \times \mathbb{Z}^- \times (\mathbb{R} - \mathbb{Q})$
15a. The set of all positive even integers
16a. In interval notation, $\mathrm{Dom}(f) = (-\infty, 0] \cup (1, 2) \cup (2, \infty)$.
16c. $\{5\}$
16e. $\{-3, -\frac{3}{2}, \sqrt{6}\}$
17. $\varnothing$
19. Because $x f_a y$ if and only if $x f y$, f_a is a function.
23. It is a true statement unless $X = \varnothing$.
26. Let $f : \mathbb{R} \to \mathbb{R}$ be given by $f(x) = x^2$ and $B = [-1, 1]$.
28a. If $x \in f^{-1}(C)$, then $f(x) \in C$. Then, $f(x) \in D$, meaning $x \in f^{-1}(D)$.

Section 6.2 **2a.** $f(a) = a + 4$
2c. $f(5) = 1$; $f(6) = 2$; $f(7) = 3$; $f(8) = 4 = f(9)$
2e. Not possible
4a. Neither
4c. Could be surjective but not injective
4f. Possibly both

4h. Both

5a. a

5c. $\frac{1}{8}$

6a. $\mathbb{R}^+$

6c. $\mathbb{Z}^+$

6e. $\mathcal{P}(\mathbb{Q}^- \times \mathbb{R})$

7a. Bijection

7c. Bijection

8a. Bijection

8c. Not injective

9b. Surjective

10a. Surjective $(0 \notin \mathbb{Z}^+)$

12. Injective (if $a1 = b1$, then $a = b$); not surjective $(0 \notin \mathrm{Ran}(S))$

14a. 4

14c. 4

15a. Not necessarily injective

16. Neither injective nor surjective

20a. If $f_A(x_1) = f_A(x_2)$, then $f(x_1) = f(x_2)$, meaning $x_1 = x_2$.

20c. For $f(a) \in f(A)$, $f_A(a) = f(a)$.

21. π_1 is injective if B has 0 or 1 elements; π_2 is injective if A has 0 or 1 elements

23. Injective because if $a \neq b$, then $f(a) \neq f(b)$. It need not be surjective.

25a. $f(a) = \{a\}$

25c. $f(S) = S$

25e. $f(S) = S$

Section 6.3 **1a.** 13

1c. 2

1e. 1

2a. 8

2c. $\frac{1}{32}$

2e. $x < 0$

3a. $(2, 1)$

3c. $(0, e^2)$

3e. $(1 + e, e)$

4a. $\{2, 4, 6, 8, \ldots\}$

5a. $(f \circ g)(x) = (2 - 3x)^3$; $(g \circ f)(x) = 2 - 3x^3$

5c. $(f \circ g)(x) = \frac{17x+6}{4x+2}$; $(g \circ f)(x) = \frac{14x+8}{15x+5}$

5e. $(f \circ g)(x) = \lfloor \lceil x \rceil \rfloor$; $(g \circ f)(x) = \lceil \lfloor x \rfloor \rceil$

6a. $2(\frac{1}{2}(x+5)) - 5 = x$; $\frac{1}{2}((2x-5) + 5) = x$

6c. $\frac{9}{5}(\frac{5}{9}(x - 32)) = x$; $\frac{5}{9}(\frac{9}{5}x + 32 - 32) = x$

6e. $\tan(\arctan(x)) = x$; $\arctan(\tan(x)) = x$

9a. f is not injective; $f(-2) = f(2)$

9c. f is not injective; $f(-1) = f(1)$

10a. If $a < b$, then $f(a) < f(b)$, and likewise, $g(f(a)) < g(f(b))$.

12. Restrict f to a set with a single element.
13a. The composition of bijections must be a bijection.
16. $g \circ f = \mathrm{Id}_X$ and $f \circ g = \mathrm{Id}_Y$
18. $(f^{-1} \circ g^{-1}) \circ (g \circ f) = f^{-1} \circ ((g^{-1} \circ g) \circ f) = f^{-1} \circ (\mathrm{Id}_Y \circ f) = f^{-1} \circ f = \mathrm{Id}_X$
20. $(\pi_1 \circ \pi)(a, b, c) = \pi_1(\pi(a, b, c)) = \pi_1(a, b) = a = \pi_1'(a, b, c)$
22a. Let $y \in Y$ with $x \in X$ so that $f(x) = y$. Then, $g(y) = (g \circ f)(x) = (h \circ f)(x) = h(y)$.
25. If $f(x) = x$, then $f^{-1}(x) = x$.

Chapter 7: Cardinality

Section 7.1 **1a.** 8; $f : \{\pm1, \pm2, \pm3, \pm4\} \to \{1, 2, \ldots, 8\}$ by $f(1) = 1$, $f(-1) = 2$, $f(2) = 3$, $f(-2) = 4$, etc.
1c. 50; if $x = 4t + 2$, map x to t $(1 \leq t \leq 50)$
1e. 9; $f((1, 1)) = 1$, $f((1, 2)) = 2$, $f((1, 3)) = 3$, etc.
3. Construct a bijection from the bijection $f : A \to \{1, 2, \ldots, n\}$ by enumerating the elements of A, and then taking those in $A \cap B$, construct a bijection from $A \cap B$ to $\{1, 2, \ldots, m\}$ $(m \leq n)$.
4b. Enumerate A and then use that enumeration to count the elements of $A - \{a_1, a_2, \ldots, a_n\}$.
6. Use induction on n and Theorem 7.4.
8. Use induction on n and Theorem 7.8.
12. There are $n!$ permutations on a set with n elements.
14a. Using Theorem 7.8, $|\{1\}| = 1$.
14b. Using Theorem 7.8, $|\{1, 2, \ldots, n\}| = n$.
15a. $|\mathcal{P}(A)| = 2^n$, so $|\mathcal{P}(A) \times \mathcal{P}(A)| = 2^n 2^n$.
15c. $|\mathcal{P}(A)| = 2^n$; cardinality of the power set of a set with 2^n elements is 2^{2^n}.
16b. Consider a constant function from an infinite set.
18a. What is the most that could be chosen so that all are different?
19b. Consider the 100 pairs of codes $\{4000, 4001\}, \{4002, 4003\}, \ldots, \{4198, 4199\}$ as the pigeonholes and the objects as the pigeons.
21a. Count the number of pairs of such points that are antipodal.
22. $2m$ consecutive integers contains m pairs of numbers differing by m.
24a. Count the number of minutes in a year.
25. 12 integers between 1 and 50 are divisible by 4.
27. Subdivide the square into 4 smaller squares of side length $\frac{1}{2}$. Two of the 5 points must lie in the same smaller square.

Section 7.2 **2a.** $f(2n) = 2n + 1$
2c. $f((1, n)) = 2n - 1$ and $f((2, n)) = 2n$
3a. $f(e^x) = x$
3c. $f(2^n) = n$ and $|\mathbb{Z}^+| = |\mathbb{Z}|$
4a. $\mathcal{P}(A) \times \mathcal{P}(B)$ is also finite.
4c. $\mathbb{Z} \times \mathbb{Z} \times \mathbb{Z}$ is countable by Theorem 7.28.

4e. Both P and $\mathbb{Z} - P$ are subsets of a countably infinite set.

5a. Let $A_i = \{m \in \mathbb{Z} \mid m = 4q + i \text{ where } q \in \mathbb{Z}\}$ $(i = 0, 1, 2, 3)$.

5c. It is not possible; subsets of $\mathbb{Z}$ that are not countably infinite are finite, and the union of four finite sets is finite.

6. First count the elements of A and then proceed to the elements of B.

9. (arbitrary intersection) Show A is a subset of a countable set.

11. (union) Use mathematical induction and results about the cardinality of the union of countable sets.

14a. If f is injective, then $|A| = |f(A)|$.

14c. Consider $f : \mathbb{Z}^+ \to \mathbb{Z}$ by $f(x) = x$; $g : \mathbb{Z}^+ \to \mathbb{Z}^+$ by $g(x) = |x|$.

16a. It is a finite union of countable sets.

16b. It is a countable union of countable sets.

Section 7.3 **1a.** $\mathbb{R}$, $\mathbb{R} \times \mathbb{R}$

1c. $\mathbb{Z}^+ \cup (0, 1)$; $\mathbb{Z}^+ \cup (1, 2)$

1e. $A = \mathbb{R}$; $B = \mathbb{R}^*$

1g. $A = \mathbb{R}$; $B = \mathbb{R}^+$

2a. False

2c. True; if it were countable, it can be shown that $\mathbb{R}$ is countable.

2e. True; by Theorem 7.20, $\mathbb{R}$ cannot be contained in a countably infinite set

2g. False; $\mathbb{R} \times \{0\}$ and $\mathbb{R} \times (\mathbb{R} - \{0\})$ partition $\mathbb{R}$.

4. Consider the contrapositive to the statement.

5b. $f(x) = 2x + 1$

6. Define a function on $(0, 1)$ mapping x to itself unless $x = \frac{1}{n}$ for $n \in \mathbb{Z}^+$, $n \geq 2$. In that case, map $\frac{1}{2}$ to 0, $\frac{1}{3}$ to 1, and $\frac{1}{n}$ to $\frac{1}{n-2}$ for $n \geq 4$.

9. $f(x) \neq (\frac{1}{3}, \frac{1}{3})$; $g((x, y)) \neq .101010\ldots$

11a. Use the Schröder-Bernstein Theorem

11b. Repeat the proof of Theorem 7.31, not allowing 5 for any of the digits (in addition to not allowing 0 or 9).

14. Use Cantor's diagonalization argument

15. Use the Schröder-Bernstein Theorem

17. Show $|\mathcal{P}(\mathbb{Z}^+)| = |[0, 1)|$.

Chapter 8: Introduction to Topology

Section 8.1 **1.** $\{\varnothing, \{1, 2\}\}, \{\varnothing, \{1\}, \{1, 2\}\}, \{\varnothing, \{2\}, \{1, 2\}\}, \{\varnothing, \{1\}, \{2\}, \{1, 2\}\}$

2. Topologies: (a), (c), (d), (e)

3. Topologies of size 0, 1, 4 and 9 do not exist.

4. $\{(-\frac{1}{n}, \frac{1}{n}) \mid n \in \mathbb{Z}^+\}$ **8.** If $\mathbb{R} - S$ is finite, then we can write $\mathbb{R} - S = \{x_1, x_2, \ldots x_n\}$.

9. The complement of every subset of a finite set is finite.

10. $\varnothing$ and $\mathbb{R}$ are elements of $\mathcal{T}$. If $x_1 < x_2$, then $(-\infty, x_1) \cap (-\infty, x_2) = (-\infty, x_1)$. Prove via induction that the intersection of n such intervals is an interval of this form as well.

12. Prove via induction on n, with the base case $n = 2$ holding by Definition

8.4.

12. $\{\{1\},\{2\},\{3\},\{4\},\{5\},\{6\}\}$ **14b.** $[0,1) \in \mathcal{B}$ while $(0,1) = \cup_{i=1}^{\infty}[\frac{1}{i},1)$

16. We know $\sqrt{2} \notin \mathbb{Q}$. Consider a union of elements of $\mathcal{T}$ such that their union equals $(\sqrt{2},\infty)$.

Section 8.2 **1.** Open: (a), (c), (d)

2. Open: (b), (d)

5a. $\{x\} \times (a,b)$, where $a,b,x \in \mathbb{R}$, $a < b$

5b. $[a,b) \times (c,d)$, where $a,b,c,d \in \mathbb{R}$, $a < b$, $c < d$. **7.** Use induction on n

9. Mimic the proof that $\mathcal{T}_A$ is closed under finite intersections, using Lemma 8.11.

10. Consider the possible intersections of a single interval (a,b) with $[0,2]$.

12. $\mathcal{T}_{\mathbb{Z}}$ is the discrete topology on $\mathbb{Z}$.

Section 8.3 **1a.** $\text{Int}(A) = \varnothing$; $\text{Cl}(A) = \{1,2,3,4\}$

1c. $\text{Int}(A) = \{1\}$; $\text{Cl}(A) = \{1,3,4\}$

1e. $\text{Int}(A) = \{1,3\}$; $\text{Cl}(A) = \{1,2,3\}$

2a. $\text{Int}(A) = (-1,1)$; $\text{Cl}(A) = [-1,1]$

2c. $\text{Int}(A) = (-4,0) \cup (2,4)$; $\text{Cl}(A) = [-4,0] \cup \{1\} \cup [2,4]$

2d. $\text{Int}(A) = [-1,1) = \text{Cl}(A)$

2f. $\text{Int}(A) = \varnothing$; $\text{Cl}(A) = \{1,2,3\}$

3. $\varnothing$, S

5. $\mathbb{R}$ and all finite subsets of $\mathbb{R}$

6. Consider $\varnothing$.

9. Use Theorem 8.20.

Section 8.4 **1.** Continuous: (a), (d), (e)

2a. Any definition will make f continuous.

2c. $f(3) = 2$; $f(4) = 4$

3a. $f(x) = x + 3$

3c. No such function exists

3e. $f(x) = x + 1$ (though any injective function will suffice)

4b. $g^{-1}(A)$ is open for any set $A \subseteq S$

5a. Consider multiple cases as we did for $f(x) = x^2$.

7. Use the Pasting Lemma, as the intersection of S_1 and S_2 is empty.

9. If $f : S_1 \to S_2$ and V is closed in S_2, then $S_2 - V$ is open. Thus, $f^{-1}(S_2 - V) = S_1 - f^{-1}(V)$ is open. Conversely, if U is open in S_2, $S_2 - U$ is closed and $f^{-1}(S_2 - U) = S_1 - f^{-1}(U)$ is closed.

11. If $\pi_1^{-1}(U) = U \times S_2$

13. Consider the inverse image of a general basis element of $S_1 \times S_2$.

14a. $f(2) = 2$; $f(3) = 1$.

Bibliography

[1] Colin Adams and Robert Franzosa. *Introduction to Topology: Pure and Applied*. Pearson Prentice Hall, 2008.

[2] Martin Aigner and Günter M. Ziegler. *Proofs from The Book*. Springer, sixth edition, 2018.

[3] Maria Rosa Antognazza. *Leibniz: An Intellectual Biography*. Cambridge University Press, 2011.

[4] Garrett Birkhoff. *Lattice Theory*. American Mathematical Society, 1940.

[5] George Boole. *The Mathematical Analysis of Logic: Being an Essay Towards a Calculus of Deductive Reasoning*. CreateSpace Independent Publishing Platform, 2016.

[6] Ronald S. Calinger. *Leonhard Euler: Mathematical Genius in the Enlightenment*. Princeton University Press, 2015.

[7] Georg Cantor. Ueber eine eigenschaft des inbegriffes aller reellen algebraischen zahlen. *Journal für die Reine und Angewandte Mathematik*, 77:258–262, 1874.

[8] Lewis Carroll. *Symbolic Logic: Complete & Illustrated*. CreateSpace Independent Publishing Platform, 2014.

[9] Lewis Carroll. *Through the Looking-Glass: Illustrated*. CreateSpace Independent Publishing Platform, 2016.

[10] Lewis Carroll. *Alice's Adventures in Wonderland*. CreateSpace Independent Publishing Platform, 2018.

[11] Mary Ann Caws. *Blaise Pascal: Miracles and Reason*. Reaktion Books, 2017.

[12] Arthur A. Clarke and Carl Gauss. *Disquisitiones Arithmeticae*. Yale University Press, 1965.

[13] Morton N. Cohen. *Lewis Carroll: A Biography*. Vintage, 1996.

[14] Paul Cohen. The independence of the continuum hypothesis. *Proceedings of the National Academy of Sciences of the United States of America*, 50:1143–1148, 1963.

[15] Paul Cohen. *Set Theory and the Continuum Hypothesis.* W. A. Benjamin, Inc., 1966.

[16] Joseph Dauben. *Georg Cantor: His Mathematics and Philosophy of the Infinite.* Princeton University Press, 1990.

[17] William Dunham. *Journey Through Genius: The Great Theorems of Mathematics.* Penguin Books, 1991.

[18] Howard Eves. *An Introduction to the History of Mathematics.* Cengage Learning, 1990.

[19] William Ewald and Wilried Sieg, editors. *David Hilbert's Lectures on the Foundations of Arithmetic and Logic, 1917-1933.* Springer, 2013.

[20] Stephen Gaukroger. *Descartes: An Intellectual Biography.* Oxford University Press, 1995.

[21] Kurt Gödel. *Consistency of the Axiom of Choice and of the Generalized Continuum Hypothesis with the Axioms of Set Theory.* Princeton University Press, 1940.

[22] David S. Gunderson. *Handbook of Mathematical Induction.* Discrete Mathematics and its Applications. CRC Press, 2011.

[23] Paul R. Halmos. *Naïve Set Theory.* Springer-Verlag, 1974. Reprint of the 1960 edition, Undergraduate Texts in Mathematics.

[24] Godfrey H. Hardy. *A Mathematician's Apology.* Cambridge University Press, 2012.

[25] Thomas L. Heath. *A Manual of Greek Mathematics.* Dover Publications, 2003.

[26] Thomas L. Heath and Euclid. *Euclid: The Thirteen Books of Elements, Volume 3, Books 10-13.* Dover Publications, 1956.

[27] Thomas L. Heath and Euclid. *The Thirteen Books of the Elements, Volume 1: Books 1-2.* Dover Publications, 1956.

[28] Thomas L. Heath and Euclid. *The Thirteen Books of the Elements, Volume 2: Books 3-9.* Dover Publications, 1956.

[29] Paul Hoffman. *The Man Who Loved Only Numbers: The Story of Paul Erdös and the Search for Mathematical Truth.* Hyperion Books, 1998.

[30] Karin Harman James. Sensori-motor experience leads to changes in visual processing in the developing brain. *Developmental Science,* 13(2):279–288, 2010.

[31] Des MacHale. *Comic Sections: Book of Mathematical Jokes, Humour, Wit and Wisdom.* Boole Press Ltd., 1993.

[32] Michael Sean Mahoney. *The Mathematical Career of Pierre de Fermat, 1601-1665.* Princeton University Press, 1994.

[33] Fyodor A. Medvedev. *Scenes from the History of Real Functions.* Birkhäuser, 2012.

[34] Leonard Mlodinow. *Euclid's Window : The Story of Geometry from Parallel Lines to Hyperspace.* Free Press, 2002.

[35] Augustus De Morgan. *Formal Logic: Or, The Calculus of Inference, Necessary and Probable.* Cambridge University Press, 2014.

[36] James Munkres. *Topology.* Prentice Hall, second edition, 2000.

[37] Jaroslav Nešetřil Ronald L. Graham and Steve Butler, editors. *The Mathematics of Paul Erdös. II.* Springer, New York, second edition, 2013.

[38] Simon Singh. *Fermat's Enigma: The Epic Quest to Solve the World's Greatest Mathematical Problem.* Anchor, 1998.

[39] Richard J. Trudeau. *Introduction to Graph Theory.* Dover Publications, 1994.

[40] Robin Wilson. *Four Colors Suffice: How the Map Problem Was Solved - Revised Color Edition.* Princeton University Press, 2013.

Index

Printed in the United States
by Baker & Taylor Publisher Services